THE

BIOPHILIA

HYPOTHESIS

ISLAND PRESS / Shearwater Books

Washington, D.C. · *Covelo, California*

A SHEARWATER BOOK

THE

BIOPHILIA

HYPOTHESIS

Edited by

Stephen R. Kellert

and

Edward O. Wilson

Grateful acknowledgment is expressed for permission to include the following previously published material. "Prayer," by Joseph Bruchac. Reprinted by permission of the author. Excerpts from *The Flight of the
Iguana*, by David Quammen. © David Quammen. Published by Delacorte Books, 1988. Excerpts from *Roman Vishniac* (Grossman); © 1974
Eugene Kinkead. Originally in *The New Yorker*.

LIBRARY OF CONGRESS
CATALOGING-IN-PUBLICATION DATA
The Biophilia hypothesis / edited by Stephen R. Kellert and
Edward O. Wilson
p. cm.
Includes bibliographical references and index.
ISBN 1-55963-148-1
1. Human ecology—Philosophy. 2. Philosophy of
nature. 3. Nature conservation—Philosophy. 4. Biological
diversity conservation—Philosophy. I. Kellert, Stephen R.
II. Wilson, Edward Osborne, 1929– .
GF21.B56 1993
179'.1—dc20 93-2021
CIP

Printed on recycled, acid-free paper.

Manufactured in the United States of America

10 9 8 7 6 5 4 3 2 1

Contents

THE

BIOPHILIA

HYPOTHESIS

Prelude: "A Siamese Connexion with a Plurality of Other Mortals"

Scott McVay

WE ARE POISED here at the beginning of a voyage of discovery reminiscent of the *Beagle*'s. We might begin with a poem called "Prayer" by Joseph Bruchac, an American of Abenaki and Czech origin:

> Let my words
> be bright with animals,
> images the flash of a gull's wing.
> If we pretend
> that we are at the center,
> that moles and kingfishers,
> eels and coyotes
> are at the edge of grace,
> then we circle, dead moons

about a cold sun.
This morning I ask only
the blessing of the crayfish,
the beatitude of the birds;
to wear the skin of the bear
in my songs;
to work like a man with my hands.[1]

Young Charles Darwin's voyage of nearly five years on H.M.S. *Beagle* provided the experience—and the journal a regular record of his observations—for *On the Origin of Species* (1859) nearly a quarter century later. Embedded in that fresh chronicle are less than two pages in the section on the Galápagos on "a most singular group of finches, related to each other in the structure of their beaks, short tails, form of body, and plumage: there are thirteen species . . ." He notes further: "The most curious fact is the perfect gradation in the size of the beaks in the different species of *Geospiza* . . ." Finally, and tellingly, he writes: "Seeing this gradation and diversity of structure in one small, intimately related group of birds, one might really fancy that from an original paucity of birds in this archipelago, one species had been taken and modified for different ends."[2]

It took a Darwin—nudged a bit by Wallace—to see the evolutionary message written in the different adaptations of the thirteen finches and then to propound the central thesis relating to our biological heritage. The question before us now is whether we as a society 133 years later are able to fully discern a central message of *our* time—one that has been conveyed by Edward Wilson with absolute clarity:

> The one process now going on that will take millions of years to correct is the loss of genetic and species diversity by the destruction of natural habitats. This is the folly our descendants are least likely to forgive us.[3]

Although oft cited and reported, the scale of the unfolding catastrophic loss of many and varied ecosystems through human activity is still only dimly perceived, for the link between the degradation of the biota and the diminishment of the human prospect is poorly understood.

Edward Wilson defines biophilia as "the innate tendency to focus on life and lifelike processes," noting that "to the degree that we come to understand other organisms, we will place greater value on them, and on ourselves." Yet until the biophilia hypothesis is more fully absorbed in the science and culture of our times—and becomes a tenet animating our everyday lives—the human prospect will wane as the rich biological exuberance of this water planet is quashed, impoverished, cut, polluted, and pillaged. The biological terrain must be better mapped, as the 1990 conference on the Amazon suggests, so that government and business leaders have better information on which to base decisions to shape sustainable development.

My conscious entry into the living tapestry of Earth was through the whale tribe, comprising some eighty-plus species who inhabit the watery parts. My mentor was Ishmael, who on one occasion in *Moby Dick* was linked by a monkey-rope to Queequeg, a seasoned harpooner, standing on the slippery dead whale's back where he was stripping off the beast's blubber. Ishmael reflected on this "humorously perilous business for both of us . . . so that for better or worse, we too, for the time, were wedded, and should poor Queequeg sink to rise no more, then both usage and honor demanded, that instead of cutting the cord, it should drag me down in his wake." Still further pondering led him to say:

> I saw that this situation of mine was the precise situation of every mortal that breathes; only in most cases, he, one way or other, has this *Siamese connexion with a plurality of other mortals.*[4]

That is my theme—our Siamese connexion, our interdependency with a plurality of other mortals—a thought which has quickened and intensified since Herman Melville penned those words over 140 years ago when he was half my age. Since then our numbers as a species have grown: from 1 billion to 5.5 billion. The exponential curve points toward 10 billion of us in less than half that time from now. And we are already straining the carrying capacity of the planet.

Our capacity for survival so far is impressive, but our perceptions about who we are and how we fit in and "whither we are tending" (Abraham Lincoln) are . . . spotty. In particular, our understanding of human/animal interactions is still woefully scanty. Even with 12 billion neurons upstairs—

an odd amazing 3 pounds of putty—we cannot easily imagine sensory systems that we lack ourselves. Until Donald Griffin was a senior at Harvard in 1938, for example, nobody had figured out that a bat could navigate by a cone or beam of sound emanating from its head. (An Italian, Spallanzani, had pieces of the puzzle 200 years earlier—he blinded bats and they returned to the church belfry—but the pieces did not yet come together.) Griffin's work enabled Arthur McBride to write in his notebook in 1948 that porpoises apparently navigate and food-find by sonar, too.

Some years ago I participated in studies of the behavior and communication of the bottlenose dolphin (or porpoise) for two years. As J. Allen Boone learned from his experience with an amazing German shepherd,[5] I learned from a precocious porpoise named Elvar about being in "right relation." He knew how to express *joy*—in a hundred modes—and *frustration* with his captive status . . . by not swallowing the last fish in an experiment (though he might fake it) and then, later, stuffing that fish down the drain to raise the level of water in the tank.

I'll describe briefly two encounters with porpoises during this time, heretofore unpublished, which reveal dimensions in these other mortals that inspire. And instruct.

It was April of 1964. I was working for a laboratory with research sites in Coconut Grove and St. Thomas. We were studying the brain, behavior, and communication of the bottlenose dolphin (*Tursiops truncatus*). What I am about to describe occurred outside the protocol of our regular work at the St. Thomas laboratory.

A bright, educated woman lived a few miles away. A connoisseur of art, she was an accomplished athlete who would sometimes spend hours swimming in the waters below her home. She puzzled at how a porpoise could purportedly save a struggling nonswimmer at sea (even though three cases were known from the literature and we had two further accounts in letters to the laboratory). First, one assumes that the drowning person would be thrashing and disoriented. Second, if he saw a fin, he might think it a shark. Third, even if this panicked person had the self-possession to grasp the dorsal fin, exhaustion would soon slacken his grip. She persistently queried the director of the institute, "How could a drowning person be rescued by a porpoise?"

The Biophilia Hypothesis

One Sunday she was invited to the lab with its tidal pool and a female porpoise perhaps three or four years old. What follows was recorded in air and underwater, and it was filmed.

The woman entered the water with this conundrum crowding out any other thought. She happened to lie face down in the water assuming "the dead man's float." From behind, the porpoise swam onto the woman's back and clasped its flippers firmly under her arms and began to propel her around the pool with powerful tail flukes. At first she resisted. She was unused to letting go or losing control. She noticed, however, that she could see and breathe. The weight and vertical stroking of the flukes lifted her head clear of the water as the two—joined by a belly-to-back Siamese connexion—made a circuit of the pool to the gasps of onlookers. She "let go." She told me she relaxed as deeply and as fully as she ever had. The porpoise made two complete circuits of the pool and then shot straight up in the air, releasing the woman gently and precisely on her knees on the cement lip of the pool. She said softly, "I understand."

In Coconut Grove our studies to map the sonic domain of these creatures took place on Monday through Saturday. On Sundays our family would sometimes travel to the Keys or the west coast of Florida or up the east coast. I located nine individuals (seven women and two men essentially unknown to one another) who "kept" porpoises under an array of circumstances. I urged each of them to keep a diary of observations, no detail being too trivial. One woman near Naples swam every day with a porpoise to whom she fed fresh pieces of fish. (Even though live fish swim through the incoming tide, a porpoise rarely reverts to eating them once it has been hand fed.)

After nearly one year of daily swims and feeding the porpoise from her hand, this woman had to travel to New York for four days. In her absence, the porpoise refused to take fish from anyone else and was in danger of becoming dehydrated since all water is absorbed from the fish consumed. When the woman returned, the porpoise was jubilant. He caught a live, wriggling fish in his teeth and offered it as a gift. The porpoise also kept her in the pool for hours, not wishing her to leave again. (You know how a devoted pet—or child—will eye you, sadly, when you pull out luggage for a trip.)

Independently, a couple of months later another woman on the east coast (unknown to the first) had a nearly identical experience. She went to New York for four days and the porpoise became despondent. It looked as though the porpoise would die of starvation—so *faithful* was he to the bonding which had occurred. But upon the woman's return a live fish was presented clasped in the teeth of the now animate porpoise.

In the case of these two porpoises, it appeared that they would not eat again but would rather endure death by dehydration than continue to live "unconnected." One hears of many instances among us where the death of a spouse seems to trigger the death of the other in a long-married couple. Through work funded by the Dodge Foundation at the Animal Medical Center by Susan Cohan, we know that the grief experience sensed by a person in the loss of a beloved pet can be as acute and lengthy as that felt after the death of a dear friend or mate. So, too, each of you knows how a dog or cat, upon the death or departure of the human to whom it is bonded, can grieve to death.

Keen observers have seen the world of animals as part of the larger sensibility of the planet. In *The Unexpected Universe*, Loren Eiseley put the theme succinctly: "One does not meet oneself until one catches the reflection in *an eye other than human*."[6] On a lighter note, but no less telling, Mark Twain reflected, "Heaven goes by favor. If it went by merit, you would stay out and your dog would go in."

Recent long-term research in the field has yielded a less fragmented picture of our fellow primates. Prominent among the women who possess the staying power that leads to a fuller perception of behavior and affiliations is Jane Goodall. Her work with chimpanzees in the Gombe Reserve of Tanzania has yielded singular information about their family structures, signaling systems, even some disturbing violent behavior, infanticides, beatings that she did not wish to publish until she was confident about their accuracy and did so only after twenty years of meticulous observations in the field.

Even though the chimpanzee is our closest cousin genetically with more than 98 percent convergence in the DNA, and we share a common ancestor, only one land mammal on earth has a larger brain than ours. (The largest ape brains are only half our 3-pounders.) That creature, with a brain three times as massive as ours, is the surpassing metaphor of both the continent of Africa and the subcontinent of India: the majestic elephant. Four men have done major studies of the elephant and gained international reputations: Iain Douglas-Hamilton, David Western, Richard Laws, and George Schaller. But none has spent more than a few years in the field.

One woman spent thirteen consecutive years studying one of the largest intact populations of elephants in the Amboseli National Park of southern Kenya. Cynthia Moss's observations of the matriarchy of elephants are contained in her book *Elephant Memories*. Moss explains: "The book is not about how I survived in the bush with 680 wild friends. It is about how *they* survived or succumbed to droughts, poachers, Masai warriors, disease, injuries, tourism, and even researchers. It is about their families, their relationships, their mates, and their offspring; their good times and bad times through the seasons and the years. I have merely gone along with them—a spectator at the banquet or a witness to less happy events."[7]

Remember the exuberant greeting that the porpoises gave each of the women upon their return from New York—a squirming live fish presented in their uplifted jaws. Cynthia Moss writes, "After 18 years of watching elephants I still feel a tremendous thrill at witnessing a greeting ceremony. Somehow it epitomizes what makes elephants so special and interesting. I have no doubt even in my most scientifically rigorous moments that the elephants are experiencing joy when they find each other again . . . elephantine joy." Long the stuff of legend and tribute, the actual field observations yield stories of such tenderness and tenacity that the "broken" elephants who carry logs as bidden are but a shadow of a shadow of their unbowed counterparts in the wild. Edward Wilson's reference to a tethered peccary comes to mind, its "repertory stunted by the impoverished constraints of human care . . . now a mute speaker trapped inside the unnatural clearing, like a messenger to me from an unexplored world."[8]

Here I should mention a few men whose Siamese connexion has been nurtured by a regard for creatures other than dolphins, apes, and elephants. Merlin Tuttle (of Bat Conservation International) has become the voice for our fellow mammal, the nocturnal bat, whose species number nearly a thousand, constituting almost a quarter of the world's mammal species. Not only do bats account for myriad forms of pollination among flowers, cacti, and fruit: "A single small insect-eating bat can eat a thousand insects each night. . . . Little brown bats, the most common of the forty-four North American species [and the subject of a major new work in 1994], they can catch hundreds of mosquitoes in an hour."[9] Our unfolding understanding of the bats' place in nature's scheme will be another measure of our desire to persist, for their role in pollination and seed dispersal is far greater than imagined. It may turn out that the sheer diversity of the bats' habits, faces, phonations, and feeding will offer a vivid litmus for the health of habitats and ecosystems.

Another "wing-ed," as the Native American would say, is the greylag goose celebrated by Konrad Lorenz in his classic *King Solomon's Ring*.[10] Who can forget his inimitable drawings of the imprinted baby geese trailing the inquisitive naturalist? Or the hilarious cockatoo-to-kaffee-klatsch communion of one bird alighting in the midst of a mound of powdered sugar (for strawberry dunking)—thoroughly dusting the assembled circle of matrons enjoying an afternoon gossip. Lorenz's lifelong study of these birds illuminates behavior previously unimagined.

At the Dodge Foundation we searched the land for an exceptional person to develop a national initiative to assist children with learning disabilities, kids who "fall through the cracks." Our scan led to a physician/clinician in Chapel Hill, North Carolina, who is a specialist in neurophysiological dysfunctions—as practical as he is charismatic. A Rhodes scholar from Brown, Melvin Levine acknowledges that his avocational interest in geese has transformed his professional interest in children:

> Over many years I have managed to accumulate and observe a rather substantial collection of many varieties of domestic and wild geese. The developmental variation within the gaggle has been as striking as it would be in any seventh grade classroom. . . . All too often it is the uniqueness of children that contributes to their learning problems. . . . These children are simply too complex to be characterized by simplistic labels, tidy systems of subtyping, or statistically generated syndromes.[11]

At his home, Sanctuary Farm, Levine has 150 geese of forty-three species from around the world. The two hours he spends with them early every morning inform his day.

The Biophilia Hypothesis

Another formidable gladiator on behalf of the great issues of our day—population, nuclear Armageddon, and extinction—is Paul Ehrlich, Bing Professor of Population Studies at Stanford. Where does he find his Siamese connexion? An invertebrate: the lepidoptera. For him and Vladimir Nabokov it was the butterfly. Every summer since 1961 Ehrlich and his wife, Anne, have observed the butterfly in a few meadows of western Colorado. His recent book, *New World, New Mind*, written with Robert Ornstein, is a tour of our plight in a world we are inadvertently disassembling with the same mental and sensory apparatus we had as hunter-gatherers. [12]

Karl von Frisch was regarded as an odd stick by his neighbors. He put out petri dishes with sugar water at various distances from the beehive. In a masterful piece of work that consumed most of his adult years, von Frisch described how the waggle dance of one foraging bee shows precisely the direction and distance of the sugar source to his fellows. When von Frisch was asked why he didn't study elephants, his answer suggests my theme—be in right relation with but one corner of the creation and the whole will become palpable and clear:

> The layman may wonder why a biologist is content to devote fifty years of his life to the study of bees and minnows without ever branching out into research on, say, elephants, or at any rate the lice of elephants or the fleas of moles. The answer to any such question must be that *every single species of the animal kingdom challenges us with all, or nearly all, the mysteries of life*. [13]

The recollections of the several bioaffiliations noted briefly here return us to the creator of the word *biophilia*, Edward Wilson, and a small book of the same name in 1984. That elegant explication had its roots in Wilson's lifelong study of ants which culminated in a monumental work, *The Ants*, with Bert Holldobler. Thus, their diligence in describing 8,800 species of ants is a revelatory example of how a sustained sharp focus on a single family can yield insights of value to all. What would happen if every elementary schoolchild chose a creature, whether an ant, a bee, cricket, dragonfly, spider, waterstrider, snake, frog, fly, beetle, or bat, to study and report on repeatedly during his or her first six years of school? The capacity for bioaffiliation in the rising generation would be boundless.

One of the great cognitive shocks to our consciousness as a species was delivered by a mild-mannered, often sickly fellow named Charles Darwin. He took his time in publishing *On the Origin of Species*, and it might not have appeared in 1859 without a prod from Alfred Russel Wallace. But what species was it that commanded Darwin's lifelong fascination? It wasn't the now famous finches of the Galápagos. What was it? It was the earthworm. David Quammen has written:

> Darwin spent forty-four years of his life, off and on, thinking about earthworms. This fact isn't something they bother to tell you in freshman biology. . . . The interest had begun back in 1837, when he was just home from his voyage on the *Beagle*, and it endured until very near the end of his life. He performed worm-related experiments that stretched across decades. Finally in 1881 he wrote a book about earthworms, a book in which the words "evolution" and "natural selection" are not (unless I blinked and missed them) even mentioned. That book is titled *The Formation of Vegetable Mould Through the Action of Worms, with Observations on Their Habits*. By "vegetable mould" he meant what today would be called humus, or simply topsoil. It was his last published work.[14]

Quammen continues:

> Darwin seems to have found something congenial about these animals.
> "As I was led to keep in my study during many months worms in pots filled with earth," he wrote, "I became interested in them, and wished to learn how far they acted consciously, and how much mental power they displayed. . . . Worms do not possess any sense of hearing. They took not the least notice of the shrill notes from a metal whistle, which was repeatedly sounded near them; nor did they of the deepest and loudest tones of a bassoon. They were indifferent to shouts, if care was taken that the breath did not strike them. When placed on a table close to the keys of a piano, which was played as loudly as possible, they remained perfectly quiet."
> But what mainly concerned Darwin was the collective and cumulative impact of worms in the wild. On this count, he made large claims for them. He knew they were numerous, powerful, and busy. A German scientist had recently come up with the figure 53,767 as the average earth-

worm population on each acre of the land he was studying, and to Darwin this sounded about right for his own turf too. Every one of those 53,767 worms, he realized, spent much of its time swallowing. It swallowed dead plant material for its sustenance, and it swallowed almost anything else in its path (including tiny rock particles) as it burrowed. . . . In many parts of England, he figured, the worm population swallowed and brought up ten tons of earth each year on each acre of land. Earthworms therefore were not only creating the planet's thin layer of fertile soil; they were also constantly turning it inside out. They were burying old Roman ruins. They were causing the monoliths of Stonehenge to subside and topple. . . . No wonder Darwin concluded: "Worms have played a more important part in the history of the world than most persons would at first suppose."[15]

The reverence for supposedly lower forms of life ripples through the senses of Roman Vishniac, microbiologist, physician, and microphotographer. He used to ladle, ever so gently, planktonic organisms from a pond in Central Park and take them back to his apartment/laboratory on West 81st Street. He took delight in watching their behavior under the microscope, for he "looks upon these minute, ubiquitous one-celled animals as friends and neighbors, deserving of civilized consideration and occupying a status equal to his own in the natural order of things. . . . 'Oh, what a variety of animals I can see in the contents of one Mason jar of pond water! One could take a trip around the world and not see as many kinds of animals or as many thrilling adventures as I see while I'm sitting in my chair before a microscope.'"[16]

Now these little fellows typically die within a day or two in a laboratory. Under Vishniac's care and grateful attentiveness they flourish. When he has completed his observations and taken photographs, with magnifications up to 2,000 times (some of which are in the permanent collection of the Museum of Modern Art), he returns the green water to the pond in Central Park making sure to return it to the exact spot from which he took it.

We have journeyed from whales and elephants to geese, to ants and bees, to earthworms and plankton. I cannot close this brief biogalactic tour, however, without saluting a woman whose regard for life bridged another chasm of our "mainstream" culture. Barbara McClintock, who finally received the Nobel Prize in her ninth decade, had such a regard for corn, or maize, such "a feeling for the organism," that her patient attuned attentiveness yielded an *astonishing* finding—jumping genes, a formidable contribution to mainstream science and to farming.[17] As Evelyn Keller wrote:

> In McClintock's working philosophy, the familiar virtues of *respect and humility* take on a new significance. To her, nature is characterized by a complexity that vastly exceeds the capacities of the human imagination. Organisms have a life and order of their own that scientists can only begin to fathom. "They do everything we [can think of], they do it better, more efficiently, more marvelously," McClintock says. It follows, therefore, that "trying to make everything fit into a set dogma won't work." McClintock believes that scientists must "listen to the material" and "let the experiment tell you what to do." This worldview implies a special attention to difference and idiosyncrasy. Each organism has an enduring uniqueness that must be respected. "No two plants are exactly alike. They're all different and as a consequence, you have to know that difference," she explains. "I don't feel I really know the story if I don't watch the plant all the way along. So I know every plant in the field. I know them intimately, and I find it a great pleasure to know them." From days, weeks, and years of patient observation comes what looks like privileged insight. The result, as one colleague described it, is an apparent ability to write the "autobiography" of every plant she works with.[18]
>
> Her vocabulary is consistently one of affection, kinship, and empathy. In speaking of her microscopic work with chromosomes, she says, "I actually felt as if I was right down there and these were my friends. . . . As you look at these things, they become part of you. And you forget yourself."[19]

The burning center of my own curiosity about nature for thirty years has been the whale. I wonder if the sperm whale—with a brain six times that of ours and an elegant click system that we have not yet begun to divine—is a philosopher king or the aquatic analog of the domestic cow or something in between? The data are far from conclusive. My research on the bottlenose dolphin and the humpback whale and the bowhead whale

The Biophilia Hypothesis

aimed at determining to what extent their signaling systems could reveal a mind of consequence. They live in, travel through, and food-find by sound—in the depths, at night, and through turbid waters. Yet only one species of the some eighty-plus members of the whale tribe seems to possess a signaling system possibly analogous to speech: *Orcinus orca*, the so-called killer whale. The late Michael Biggs and John Ford discovered that the signals of sixteen pods of these whales in Puget Sound fall into four linguistic clans. The research goes forward off Norway and in the Antarctic. But we do not need to find speech in our fellow creatures to affirm the wonder and sense of connection that can be discovered in almost any form of life—from the mighty elephant to the humble earthworm to the jumping genes of maize.

We can travel to the stars and what we find will not surpass Roman Vishniac's amazement in observing the living dance among the little creatures in a scoop of pond water. One evening in 1955 he was heard to say: "Some people think microscopic animals are all pretty much alike—but oh, no! They have individualities that make them different from one another, just like human beings."

At this point, Vishniac turned and bent over a microscope he had been working at earlier in the evening:

"Yes, yes!" he said eagerly. "Here comes a little animal who is full of curiosity. He wants to learn and see more, and is forever peering around his tiny landscape. Near him is another, who is interested only in searching for food. And now comes a third, who is more social. He hates to be alone, and is constantly running from one friend to another. Oh, and there's a tardigrade! What a cute little fellow! It makes me laugh to look at him. He is called the bear animalcule, because he looks so much like a teddy bear. And now a diatom asterionella comes along, twinkling like a star. What's this, a floscularia? Yes—a perfect specimen of an animal that is beautiful enough, in my opinion, to be called the Queen of the Microcosmos. She has long, fine hairs streaming from her head, now contracting and now expanding into rays of glory. But her beauty, sad to say, is only to attract her victims. As soon as some little animal approaches to explore her exquisite tresses, a sudden convulsion runs through her, as if she had touched an electric wire, and then her hairs form a menacing corona and push the prey into her body, in a way that is lovely and delicate and yet fearsome to watch. Then her grisly jaws open and start their brutal work. But now the Queen's place has been taken by a small *Encentrum lupus*, who attacks in the true wolf manner. His prey is the larger ciliates, like paramecium blepharisma, spirostomum, and even the mighty bursaria, one of the largest of the protozoa. Yes, just as I suspected, he is chasing a spirostomum, whose only thought is to run away and preserve the spark of life. And no wonder, for the encentrum's jaws can easily tear off large pieces of flesh, like whole beefsteaks, from the bodies of his victims. But one steak is not enough for him. Even while the first is still going through the grinding mill of his jaws, he continues to pursue his wounded prey. Greedy fellow! If he has torn off more than he can conveniently swallow at once, he draws in his head to create a pouch, and holds his surplus food in it until he is ready for it. And all the time he keeps up his pursuit. . . . There—he has caught the spirostomum and bitten him! Frantically, the poor little victim is attempting evasive action, using his cilia-like oars as he dodges this way and that. Now the encentrum has almost caught up with him again. Closer, closer. . . . Oh, how lucky! A heliozoan, *Actinosphaerium eichhorni*, has come between them just in time, and distracted the encentrum. The spirostomum is saved!"[20]

Many have worked in the field, and I expect each has stories to tell that reach beyond our everyday perceptions—and remind us of our newly discovered niche in the cosmos, stewards for a creation we did nothing to bring about but have an enormous responsibility to sustain. Like Ishmael, we are attached to a monkey-rope and have a Siamese connexion with a plurality of other mortals. We can see this connexion and feel it. As the California condor flaps out of existence on the western rim of our consciousness, the sun may be setting on us, too. Our persistence as a species will depend upon cognition of ourselves as part of nature and recognition of our new duty to see how much of creation can be sustained.

Part of the canon of peer-reviewed science is the assumption that the scientist is a "disinterested" observer who seeks to describe natural phenomena as objectively as possible. Papers in journals represent distillations of results of certain observations or experiments, but rarely do they describe the pathways that were fruitless or the zigzag route whereby a discovery was made. In contrast, Edison's daily notebooks are a treasure house of thousands of things that didn't work, leading to a few that did. *The Double Helix* by James Watson records in human terms the story of how a team at the Cavendish Labs discovered DNA. This appeared to be a refreshing new genre of first-person accounts of how science really occurs. Custom, however, seems stronger than our curiosity about how we learn new things.

As a consequence, even the notion of biophilia smacks of the "anthro-pomorphism" that custom has also sought to exclude from what is considered acceptable science. Yet even the examples offered here in this prelude suggest that any kind of deeper truth about the fellow organisms with whom we share space may require the shucking of this shibboleth if we are to understand the intricacies of other life-forms and how we may fit in. To advance our understanding, a greater intimacy in observation and reporting may be required.

One of the central aims of this book is to probe what an affinity for life— as experienced and articulated by a few forward observers through certain species and habitats—can mean for all of us. Are the perceptions grouped under the biophilia hypothesis *sufficiently strong* to have the slightest effect on the train of events that is dismantling the biological inheritance of this water planet, built over hundreds of millions of years, in a few decades? To what extent can an affinity for life urge moderation in our behavior?

One would like to think that, as Edward Wilson wrote: "The more we know of other forms of life, the more we enjoy and respect ourselves. . . . Humanity is exalted not because we are so far above other living creatures, but because knowing them well elevates the very concept of life."[21] One of the persisting questions, though, is whether these observations can be more broadly understood and felt—and whether they can influence our everyday actions.

It seems essential that the habits of mind and heart that evoked the notion of biophilia be assumed more broadly by universities, laboratories, think tanks, and government agencies that share responsibility for guiding our prospects. It is as though the gradations of beak and behavior of thir-

Prelude

teen Galápagos finches were brought to our attention, and now it is our fate to determine whether there will be a persistence of the species whose origin was so beautifully described by Charles Darwin.

NOTES

1. "Prayer" by Joseph Bruchac. Reprinted by permission of author.
2. *The Voyage of the Beagle* by Charles Darwin (London: Everyman's Library, n.d.), 104, pp. 364–365.
3. *Biophilia* by Edward O. Wilson (Cambridge: Harvard University Press, 1984), p. 121 and (definition) p. 1.
4. *Moby Dick; or the White Whale* by Herman Melville (New York: Dodd, Mead & Co., 1942), chap. 71, "The Monkey-Rope," p. 294. The emphasis is mine.
5. *Kinship with All Life* by J. Allen Boone (New York: Harper & Brothers, 1954).
6. *The Unexpected Universe* by Loren Eiseley (New York: Harcourt, Brace & World, 1964), p. 24. The emphasis is mine.
7. *Elephant Memories* by Cynthia Moss (New York: Morrow, 1988), pp. 124–125.
8. Wilson, *Biophilia*, p. 4.
9. Profile of Merlin Tuttle in *The New Yorker* by Diane Ackerman, 29 Feb. 1988, p. 42.
10. *King Solomon's Ring* by Konrad Lorenz (New York: Harper & Row, 1952).
11. From the preface to *Developmental Variation and Learning Disorders* by Melvin Levine, M.D. (Cambridge and Toronto: Educators Publishing Service, 1987), p. xi.
12. *New World, New Mind* by Robert Ornstein and Paul Ehrlich (New York: Doubleday, 1989).
13. From an article about Karl von Frisch in *Rockefeller Foundation Illustrated*, vol. 2, no. 1, Aug. 1974. The emphasis is mine.
14. *The Formation of Vegetable Mould Through the Action of Worms, with Observations on Their Habits* by Charles Darwin, vol. 16 of *The Works of Charles Darwin*, reprint of New York edition, 1893–1897. Quoted by David Quammen in "Thinking About Earthworms" in a collection of essays called *The Flight of the Iguana* by David Quammen (New York: Delacorte Press, 1988), pp. 11–12. Reprinted by permission of author.
15. Ibid., p. 13.
16. Profile of Roman Vishniac in *The New Yorker* by Eugene Kinkead, 2 July 1955, pp. 28–29.

17. *Working Woman* by Martin and Marian Goldman, Oct. 1983, p. 208.

18. "Women and Basic Research: Respecting the Unexpected" by Evelyn Fox Keller in *Technology Review*, Nov./Dec. 1984, p. 46. The emphasis is mine.

19. Ibid.

20. Kinkead, profile of Vishniac, pp. 29–30.

21. Wilson, *Biophilia*, p. 22.

Introduction

Stephen R. Kellert

PHILOSOPHERS, POETS, THE rarest of politicians, and even the occasional scientist have at times indulged in the effort to rationalize how human life is enriched by its broadest affiliation with the natural world—and, conversely, how the impoverishment of this relationship with nature could foster a less satisfactory existence.

In 1984, Edward O. Wilson published an extraordinary book, *Biophilia*, which sought to provide some understanding of how the human tendency to relate with life and natural process might be the expression of a biological need, one that is integral to the human species' developmental process and essential in physical and mental growth. Most simply put, Wilson (1984:1) defined biophilia as the "innate tendency to focus on life and life-like processes." The biophilia hypothesis proclaims a human dependence on nature that extends far beyond the simple issues of material and physical sustenance to encompass as well the human craving for aesthetic, intellectual, cognitive, and even spiritual meaning and satisfaction.

This daring assertion reaches beyond the poetic and philosophical articulation of nature's capacity to inspire and morally inform to a scientific claim of a human *need*, fired in the crucible of evolutionary development, for deep and intimate association with the natural environment, particularly its living biota. The biophilia notion compels us in Wilson's terms (1984:138–139) "to look to the very roots of motivation and understand why, in what circumstances and on which occasions, we cherish and protect life." The biophilia hypothesis necessarily involves a number of challenging, indeed daunting, assertions. Among these is the suggestion that the human inclination to affiliate with life and lifelike process is:

· Inherent (that is, biologically based)
· Part of our species' evolutionary heritage
· Associated with human competitive advantage and genetic fitness
· Likely to increase the possibility for achieving individual meaning and personal fulfillment
· The self-interested basis for a human ethic of care and conservation of nature, most especially the diversity of life

This book explores various elements of this compelling, eloquent, and provocative concept. We treat the biophilia notion as a hypothesis to underscore the need for systematic inquiry as the basis for putting some flesh on the bones of this bold proposition. The idea of a hypothesis, moreover, emphasizes the scientific convention that a proposition does not "exist" until proven otherwise. This cautious approach may help us avoid the inevitable suggestion that our exploration is but the disguised attempt to promote a romantic idealization of nature.

Despite this commitment to examine the theoretical and empirical evidence in support of the biophilia hypothesis, the richness and depth of the subject preclude the possibility of achieving any definitive "proof." We are forced to behave, instead, much like the blind men of the old allegory: convinced of the beast's existence but ready to confess to having little detailed understanding of its precise shape, form, content, structure, and function. Our labors will have been successful if we legitimize and stimulate future inquiry into this critical element of the human condition. Our grandest aspiration is to build the foundation and confidence for further systematic and deep examination of the biophilia hypothesis.

This effort has built upon several decades of important work regarding various aspects of the biophilia concept (even though this term was not specifically used): topics including the role of nature in human cognitive and mental development, the biological basis for diverse values of nature, the evolutionary significance of the human aesthetic response to varying landscapes and species, the sociobiological importance of human altruism and helping behavior, and the role of nature in human emotional bonding and physical healing, to mention but a sample.

The editors believe that the contributors are distinguished by the relevance of their prior work relating to the biophilia hypothesis, the outstanding quality of their scholarship, and the breadth of their disciplinary perspectives. We have proceeded with the conviction that the richness of the topic requires no less than a multidisciplinary consideration. This same diversity of talent and scholarship can, of course, represent an impediment to communication. The differing perspectives, drawing on varying epistemological traditions and vocabularies, can result in considerable challenges to the reader. Fortunately, we believe this group of very capable scholars has produced a volume distinguished by its overall coherence and a whole much greater than the sum of its parts.

The book's organization reflects this diversity of emphasis. Part One introduces the topic. In Chapter 1, Edward O. Wilson clarifies the biological basis of the biophilia concept by referring to it as a set of "learning rules," a type of prepared learning, rather than a simple instinct. He further elucidates the possible connection between biophilia and an ethic of nature conservation and protection. Chapter 2 by Stephen Kellert offers a taxonomy of presumably biologically based human values indicative of the biophilia tendency. This typology constitutes a heuristic device for describing the importance of nature in human evolution and development. Moreover, both Wilson and Kellert introduce the notion that antagonistic and even adversarial relationships to nature—what Roger Ulrich in this volume refers to as "biophobia"—can be regarded as an element of biophilia.

Part Two of the book, "Affect and Aesthetics," includes essays by Roger Ulrich, Judith Heerwagen and Gordon Orians, and Aaron Katcher and Gregory Wilkins. Each chapter addresses processes associated with the natural environment that condition human emotional, cognitive, and aesthetic development. These three chapters are further distinguished by the marshaling of empirical evidence and scientific proof in their investigations of the biophilia hypothesis. Roger Ulrich's chapter also offers important insight regarding the complementarity of negative and positive affiliations with nature as dialectical components of the biophilia phenomenon.

Part Three—"Culture"—provides an essential cross-cultural consideration of the biophilia hypothesis, particularly its expression among indigenous peoples in nonindustrial and non-Western societies. Richard Nelson's essay offers a moving and profound description of biophilia among northern indigenous peoples of North America whose cultures have retained their integrity and wholeness. His chapter also compels us to wonder discomfortingly if modern society's uncertainty regarding the biophilia hypothesis is but another expression of our contemporary estrangement from the natural world. Chapter 7 by Gary Nabhan and Sara St. Antoine offers a sobering reminder of the consequences of the erosion of biophilia tendencies among people in both tribal and industrial societies. Jared Diamond's chapter, based on extensive ethnographic study in New Guinea, presents uncertain evidence in support of the biophilia hypothesis in other cultures—although it is a powerful reminder of the extraordinary knowledge of natural process possessed by so-called primitive peoples.

Part Four of the book—"Symbolism"—consists of two essays that explore the role of nature, particularly animals, in human cognitive development and communication. Chapter 9 by Paul Shepard builds upon his seminal work in this area, focusing on the potentially negative impacts of the breakdown in the distinction between wild and domesticated nature in modern society. In Chapter 10, Elizabeth Lawrence provides an outstanding scholarly discussion of the symbolic uses of animals to facilitate communication and what she provocatively calls "cognitive biophilia." The bee, pig, and bat are chosen to elucidate how the human capacity for metaphorical expression and thought is enhanced by nature's rich tapestry of forms and kinds.

Part Five, "Evolution," explores connections between biophilia and human evolutionary development. Chapter 11 by Dorion Sagan and Lynn Margulis offers a provocative view of the relatively minor role, even in the modern context, of the human species in biological evolution. They further elucidate the possible connection between the biophilia concept and the notions of "Gaia" and "prototaxis" as generalized tendencies toward organismic symbiosis and the inherent inclination of species to behave in predictable ways toward one another. Chapter 12 by Madhav Gadgil discusses the possible relationship of biophilia and human cultural evolution, particularly the development of manufactured artifacts as reflections of the human fascination for complexity and diversity.

Part Six of the volume, "Ethics and Political Action," includes two chapters which examine the biophilia hypothesis in the contemporary context of moral relationships to nature and the imperatives of social change. In Chapter 13, Holmes Rolston explores the uncertain implications of the presumption of a biological basis for human values of nature, and the development of an ethic of care, respect, and concern for conserving the natural environment. Chapter 14 by David Orr offers a compelling argument for the political necessity of developing a new consciousness toward nature based on biophilia as a means of countering our current calamitous rush toward environmental destruction on a massive scale. Chapter 15 by Michael Soulé provides an important summary of needed research as an essential condition for the eventual scientific delineation and defense of the biophilia hypothesis.

Drafts of these chapters were initially presented in August 1992 at the Woods Hole Oceanographic Institute in Massachusetts. This meeting occurred because the editors believed that scientific inquiry of such a new and difficult subject required an initial opportunity for productive discussion and feedback. A highly attractive, retreat-like setting was chosen in the hope of stimulating deep and lively discussion. Our optimistic expectations were more than met by the reality of the institute's excellent facilities, enriched by the extraordinary beauty of Nantucket Sound, and the highly productive conversations eventually resulting in a much richer, deeper, and more compelling book.

We also gained much from the outstanding contributions of a small number of invited participants. We especially appreciated the insights of George Woodwell (executive director of the Woods Hole Research Center), who provided an inspiring perspective regarding our efforts and suggested that, "despite the crass and callous handling of our earthly trusteeship, a fundamental attraction between and among the organisms . . . is a reality." Carleton Ray, professor in the Department of Environmental Sciences at the University of Virginia, further offered the group stimulating reflections on the relationship of biophilia to human experience in the marine environment. We were ably assisted by the participation of three young scholars who served in the role of presentation responders: David Abrams of the State University of New York, Peter Kahn of Colby College, and Richard Wallace of the Marine Mammal Commission.

The discussions at Woods Hole were further stimulated by the insights of Barbara Dean of Island Press. More important, Barbara Dean's intellectual commitment and scholarly contributions to this project have been an integral aspect of the book from its inception to completion. She has served in very nearly the capacity of a third editor and only her modesty and humility prevent Ms. Dean from assuming this status.

A gathering of this scope and ambition is only possible because of the generosity, support, and inspiration of others. Particularly important, in all respects, has been Scott McVay, executive director of the Geraldine R. Dodge Foundation. Scott served as a critical participant at Woods Hole, assisted in providing the material support for this effort, and, of course, has offered inspiring guidance in the volume's Prelude.

Despite the wide divergence in perspectives and disciplinary backgrounds of the book's contributors, this undertaking has been bound by a common focus and a conviction regarding the importance and even urgency of the deliberation. The biophilia hypothesis represents for all of us a thesis of extraordinary intellectual elegance and challenge worthy of scientific daring and a spirit of courageous inquiry. The volume's contributors may be regarded as explorers of particularly uncharted territory and like all pioneers may expect a few arrows in their backs. Yet the intellectual risk is certainly justified by the worthiness of the task. As Wilson has suggested (1984:139), the object of this quest is no less than the possible truth that "we are human in good part because of the particular way we affiliate with other organisms" and, more broadly, nature. A central element of this effort has been the belief that the natural environment is critical to human meaning and fulfillment at both the individual and the societal level.

Our sense of urgency is prompted by the conviction that the modern onslaught upon the natural world is driven in part by a degree of alienation from nature. Our modern environmental crisis—the widespread toxification of various food chains, the multifaceted degradation of the atmosphere, the far-ranging depletion of diverse natural resources, and, above all, the massive loss of biological diversity and the scale of global species extinctions—is viewed as symptomatic of a fundamental rupture of human emotional and spiritual relationship with the natural world.

The mitigation of this environmental crisis may necessitate nothing less than a fundamental shift in human consciousness. David Orr provocatively refers to this change as the "biophilia revolution"—a love of life based on a knowledge and conviction that in our deepest affiliation with nature is the key to our species' most fundamental yearnings for a meaningful and fulfilling existence. As Aldo Leopold reminded us more than a generation ago (1966:239, 261): "All ethics so evolved rest upon a single premise: that the individual is a member of a community of interdependent parts. . . . The land ethic simply enlarges the boundaries of the community to include soils, waters, plants, and animals, or collectively: the land. . . . It is inconceivable . . . that an ethical relation to land can exist without love, respect, and admiration." An ethic of nature conservation and protection is no mere luxury or indulgence. It is the celebration of nature's capacity to enrich and enlarge our life's experience. Biological diversity and the ecological processes that make it possible are the crucibles in which our species' physical, mental, and spiritual being have been forged. If but for selfish reasons alone, the notion of biophilia prompts us to manifest an ethic of care, affection, and respect for nature. As Wilson himself has remarked (1984:115): "The more we know of other forms of life, the more we enjoy and respect ourselves. Humanity is exalted not because we are so far above other living creatures, but because knowing them well elevates the very concept of life." This volume represents but one fledgling attempt to lend scientific credence to this understanding of the human need to love life and engage it.

ACKNOWLEDGMENTS

I would like to thank a number of Yale University graduate students who provided invaluable assistance in preparing for the meeting at the Woods Hole

The Biophilia Hypothesis

Oceanographic Institute. Syma Ebbin, a doctoral student at the Yale School of Forestry and Environmental Studies, was extremely helpful and effective in handling the many logistical details and arrangements at Woods Hole. Susan Pufahl and Heather Merbs provided important assistance during the early planning stages of the project.

REFERENCES

Leopold, A. *Sand County Almanac*. New York: Oxford University Press, 1966.

Wilson, E. O. *Biophilia: The Human Bond with Other Species*. Cambridge: Harvard University Press, 1984.

Introduction

Part One

CLARIFYING

THE

CONCEPT

CHAPTER 1

Biophilia and the

Conservation Ethic

Edward O. Wilson

B IOPHILIA, IF IT exists, and I believe it exists, is the innately emotional affiliation of human beings to other living organisms. Innate means hereditary and hence part of ultimate human nature. Biophilia, like other patterns of complex behavior, is likely to be mediated by rules of prepared and counterprepared learning—the tendency to learn or to resist learning certain responses as opposed to others. From the scant evidence concerning its nature, biophilia is not a single instinct but a complex of learning rules that can be teased apart and analyzed individually. The feelings molded by the learning rules fall along several emotional spectra: from attraction to aversion, from awe to indifference, from peacefulness to fear-driven anxiety.

The biophilia hypothesis goes on to hold that the multiple strands of emotional response are woven into symbols composing a large part of culture. It suggests that when human beings remove themselves from the natural environment, the biophilic learning rules are not replaced by modern versions equally well adapted to artifacts. Instead, they persist from generation to generation, atrophied and fitfully manifested in the artificial new environments into which technology has catapulted humanity. For the indefinite future more children and adults will continue, as they do now, to visit zoos than attend all major professional sports combined (at least this is so in the United States and Canada), the wealthy will continue to seek dwellings on prominences above water amidst parkland, and urban dwellers will go on dreaming of snakes for reasons they cannot explain.

Were there no evidence of biophilia at all, the hypothesis of its existence would still be compelled by pure evolutionary logic. The reason is that human history did not begin eight or ten thousand years ago with the invention of agriculture and villages. It began hundreds of thousands or millions of years ago with the origin of the genus *Homo*. For more than 99 percent of human history people have lived in hunter-gatherer bands totally and intimately involved with other organisms. During this period of deep history, and still farther back, into paleohominid times, they depended on an exact learned knowledge of crucial aspects of natural history. That much is true even of chimpanzees today, who use primitive tools and have a practical knowledge of plants and animals. As language and culture expanded, humans also used living organisms of diverse kinds as a principal source of metaphor and myth. In short, the brain evolved in a biocentric world, not a machine-regulated world. It would be therefore quite extraordinary to find that all learning rules related to that world have been erased in a few thousand years, even in the tiny minority of peoples who have existed for more than one or two generations in wholly urban environments.

The significance of biophilia in human biology is potentially profound, even if it exists solely as weak learning rules. It is relevant to our thinking about nature, about the landscape, the arts, and mythopoeia, and it invites us to take a new look at environmental ethics.

How could biophilia have evolved? The likely answer is biocultural evolution, during which culture was elaborated under the influence of hereditary learning propensities while the genes prescribing the propensities were spread by natural selection in a cultural context. The learning rules can be inaugurated and fine-tuned variously by an adjustment of sensory thresholds, by a quickening or blockage of learning, and by modification of emotional responses. Charles Lumsden and I (1981, 1983, 1985) have envisioned biocultural evolution to be of a particular kind, gene-culture coevolution, which traces a spiral trajectory through time: a certain genotype makes a behavioral response more likely, the response enhances survival and reproductive fitness, the genotype consequently spreads through the population, and the behavioral response grows more frequent. Add to this the strong general tendency of human beings to translate emotional feelings into myriad dreams and narratives, and the necessary conditions are in place to cut the historical channels of art and religious belief.

Gene-culture coevolution is a plausible explanation for the origin of biophilia. The hypothesis can be made explicit by the human relation to snakes. The sequence I envision, drawn principally from elements established by the art historian and biologist Balaji Mundkur, is this:

1. Poisonous snakes cause sickness and death in primates and other mammals throughout the world.

2. Old World monkeys and apes generally combine a strong natural fear of snakes with fascination for these animals and the use of vocal communication, the latter including specialized sounds in a few species, all drawing attention of the group to the presence of snakes in the near vicinity. Thus alerted, the group follows the intruders until they leave.

3. Human beings are genetically averse to snakes. They are quick to develop fear and even full-blown phobias with very little negative reinforcement. (Other phobic elements in the natural environment include dogs, spiders, closed spaces, running water, and heights. Few modern artifacts are as effective—even those most dangerous, such as guns, knives, automobiles, and electric wires.)

4. In a manner true to their status as Old World primates, human beings too are fascinated by snakes. They pay admission to see captive specimens in zoos. They employ snakes profusely as metaphors and weave them into stories, myth, and religious symbolism. The serpent gods of cultures they have conceived all around the world are furthermore typically ambivalent. Often semihuman in form, they are poised to inflict vengeful death but also to bestow knowledge and power.

5. People in diverse cultures dream more about serpents than any other kind of animal, conjuring as they do so a rich medley of dread and magical power. When shamans and religious prophets report such images, they invest them with mystery and symbolic authority. In what seems to be a logical consequence, serpents are also prominent agents in mythology and religion in a majority of cultures.

Here then is the ophidian version of the biophilia hypothesis expressed in briefest form: constant exposure through evolutionary time to the malign influence of snakes, the repeated experience encoded by natural selection as a hereditary aversion and fascination, which in turn is manifested in the dreams and stories of evolving cultures. I would expect that other biophilic responses have originated more or less independently by the same means but under different selection pressures and with the involvement of different gene ensembles and brain circuitry.

This formulation is fair enough as a working hypothesis, of course, but we must also ask how such elements can be distinguished and how the general biophilia hypothesis might be tested. One mode of analysis, reported by Jared Diamond in this volume, is the correlative analysis of knowledge and attitude of peoples in diverse cultures, a research strategy designed to search for common denominators in the total human pattern of response. Another, advanced by Roger Ulrich and other psychologists, is also reported here: the precisely replicated measurement of human subjects to both attractive and aversive natural phenomena. This direct psychological approach can be made increasingly persuasive, whether for or against a biological bias, when two elements are added. The first is the measurement of heritability in the intensity of the responses to the psychological tests used. The second element is the tracing of cognitive development in children to identify key stimuli that evoke the responses, along with the ages of maximum sensitivity and learning propensity. The slithering motion of an elongate form appears to be the key stimulus producing snake aversion, for example, and preadolescence may be the most sensitive period for acquiring the aversion.

Given that humanity's relation to the natural environment is as much a part of deep history as social behavior itself, cognitive psychologists have been strangely slow to address its mental consequences. Our ignorance could be regarded as just one more blank space on the map of academic science, awaiting genius and initiative, except for one important circumstance: the natural environment is disappearing. Psychologists and other scholars are obligated to consider biophilia in more urgent terms. What, they should ask, will happen to the human psyche when such a defining part of the human evolutionary experience is diminished or erased?

There is no question in my mind that the most harmful part of ongoing environmental despoliation is the loss of biodiversity. The reason is that the variety of organisms, from alleles (differing gene forms) to species, once lost, cannot be regained. If diversity is sustained in wild ecosystems, the biosphere can be recovered and used by future generations to any degree desired and with benefits literally beyond measure. To the extent it is diminished, humanity will be poorer for all generations to come. How much poorer? The following estimates give a rough idea:

- Consider first the question of the *amount* of biodiversity. The number of species of organisms on earth is unknown to the nearest order of magnitude. About 1.4 million species have been given names to date, but the actual number is likely to lie somewhere between 10 and 100 million. Among the least-known groups are the fungi, with 69,000 known species but 1.6 million thought to exist. Also poorly explored are at least 8 million and possibly tens of millions of species of arthropods in the tropical rain forests, as well as millions of invertebrate species on the vast floor of the deep sea. The true black hole of systematics, however, may be bacteria. Although roughly 4,000 species have been formally recognized, recent studies in Norway indicate the presence of 4,000 to 5,000 species among the 10 billion individual organisms found on average in each gram of forest soil, almost all new to science, and another 4,000 to 5,000 species, different from the first set and also mostly new, in an average gram of nearby marine sediments. Fossil records of marine invertebrates, African ungulates, and flowering plants indicate that on average each clade—a species and its descendants—lasts half a million to 10 million years under natural conditions. The longevity is measured from the time the ancestral form

Biophilia and the Conservation Ethic

splits off from its sister species to the time of the extinction of the last descendant. It varies according to the group of organisms. Mammals, for example, are shorter-lived than invertebrates.

· Bacteria contain on the order of a million nucleotide pairs in their genetic code, and more complex (eukaryotic) organisms from algae to flowering plants and mammals contain 1 to 10 billion nucleotide pairs. None has yet been completely decoded.

· Because of their great age and genetic complexity, species are exquisitely adapted to the ecosystems in which they live.

· The number of species on earth is being reduced by a rate 1,000 to 10,000 times higher than existed in prehuman times. The current removal rate of tropical rain forest, about 1.8 percent of cover each year, translates to approximately 0.5 percent of the species extirpated immediately or at least doomed to much earlier extinction than would otherwise have been the case. Most systematists with global experience believe that more than half the species of organisms on earth live in the tropical rain forests. If there are 10 million species in these habitats, a conservative estimate, the rate of loss may exceed 50,000 a year, 137 a day, 6 an hour. This rate, while horrendous, is actually the minimal estimate, based on the species / area relation alone. It does not take into account extinction due to pollution, disturbance short of clear-cutting, and the introduction of exotic species.

Other species-rich habitats, including coral reefs, river systems, lakes, and Mediterranean-type heathland, are under similar assault. When the final remnants of such habitats are destroyed in a region—the last of the ridges on a mountainside cleared, for example, or the last riffles flooded by a downstream dam—species are wiped out en masse. The first 90 percent reduction in area of a habitat lowers the species number by one-half. The final 10 percent eliminates the second half.

It is a guess, subjective but very defensible, that if the current rate of habitat alteration continues unchecked, 20 percent or more of the earth's species will disappear or be consigned to early extinction during the next thirty years. From prehistory to the present time humanity has probably already eliminated 10 or even 20 percent of the species. The number of bird species, for example, is down by an estimated 25 percent, from 12,000 to 9,000, with a disproportionate share of the losses occurring on islands. Most of the megafaunas—the largest mammals and birds—appear to have been destroyed in more remote parts of the world by the first wave of hunter-gatherers and agriculturists centuries ago. The diminution of plants and invertebrates is likely to have been much less, but studies of archaeological and other subfossil deposits are too few to make even a crude estimate. The human impact, from prehistory to the present time and projected into the next several decades, threatens to be the greatest extinction spasm since the end of the Mesozoic era 65 million years ago.

Assume, for the sake of argument, that 10 percent of the world's species that existed just before the advent of humanity are already gone and that another 20 percent are destined to vanish quickly unless drastic action is taken. The fraction lost—and it will be a great deal no matter what action is taken—cannot be replaced by evolution in any period that has meaning for the human mind. The five previous major spasms of the past 550 million years, including the end-Mesozoic, each required about 10 million years of natural evolution to restore. What humanity is doing now in a single lifetime will impoverish our descendants for all time to come. Yet critics often respond, "So what? If only half the species survive, that is still a lot of biodiversity—is it not?"

The answer most frequently urged right now by conservationists, I among them, is that the vast material wealth offered by biodiversity is at risk. Wild species are an untapped source of new pharmaceuticals, crops, fibers, pulp, petroleum substitutes, and agents for the restoration of soil and water. This argument is demonstrably true—and it certainly tends to stop anticonservation libertarians in their tracks—but it contains a dangerous practical flaw when relied upon exclusively. If species are to be judged by their potential material value, they can be priced, traded off against other sources of wealth, and—when the price is right—discarded. Yet who can judge the *ultimate* value of any particular species to humanity? Whether the species offers immediate advantage or not, no means exist to measure what benefits it will offer during future centuries of study, what scientific knowledge, or what service to the human spirit.

At last I have come to the word so hard to express: spirit. With reference to the spirit we arrive at the connection between biophilia and the environ-

Biophilia and the Conservation Ethic

mental ethic. The great philosophical divide in moral reasoning about the remainder of life is whether or not other species have an innate right to exist. That decision rests in turn on the most fundamental question of all: whether moral values exist apart from humanity, in the same manner as mathematical laws, or whether they are idiosyncratic constructs that evolved in the human mind through natural selection. Had a species other than humans attained high intelligence and culture, it would likely have fashioned different moral values. Civilized termites, for example, would support cannibalism of the sick and injured, eschew personal reproduction, and make a sacrament of the exchange and consumption of feces. The termite spirit, in short, would have been immensely different from the human spirit—horrifying to us in fact. The constructs of moral reasoning, in this evolutionary view, are the learning rules, the propensities to acquire or to resist certain emotions and kinds of knowledge. They have evolved genetically because they confer survival and reproduction on human beings.

The first of the two alternative propositions—that species have universal and independent rights regardless of how else human beings feel about the matter—may be true. To the extent the proposition is accepted, it will certainly steel the determination of environmentalists to preserve the remainder of life. But the species-right argument alone, like the materialistic argument alone, is a dangerous play of the cards on which to risk biodiversity. The independent-rights argument, for all its directness and power, remains intuitive, aprioristic, and lacking in objective evidence. Who but humanity, it can be immediately asked, gives such rights? Where is the enabling canon written? And such rights, even if granted, are always subject to rank-ordering and relaxation. A simplistic adjuration for the right of a species to live can be answered by a simplistic call for the right of people to live. If a last section of forest needs to be cut to continue the survival of a local economy, the rights of the myriad species in the forest may be cheerfully recognized but given a lower and fatal priority.

Without attempting to resolve the issue of the innate rights of species, I will argue the necessity of a robust and richly textured anthropocentric ethic apart from the issue of rights—one based on the hereditary needs of our own species. In addition to the well-documented utilitarian potential of wild species, the diversity of life has immense aesthetic and spiritual value. The terms now to be listed will be familiar, yet the evolutionary logic is still relatively new and poorly explored. And therein lies the challenge to scientists and other scholars.

Biodiversity is the Creation. Ten million or more species are still alive, defined totally by some 10^{17} nucleotide pairs and an even more astronomical number of possible genetic recombinants, which creates the field on which evolution continues to play. Despite the fact that living organisms compose a mere ten-billionth part of the mass of earth, biodiversity is the most information-rich part of the known universe. More organization and complexity exist in a handful of soil than on the surfaces of all the other planets combined. If humanity is to have a satisfying creation myth consistent with scientific knowledge—a myth that itself seems to be an essential part of the human spirit—the narrative will draw to its conclusion in the origin of the diversity of life.

Other species are our kin. This perception is literally true in evolutionary time. All higher eukaryotic organisms, from flowering plants to insects and humanity itself, are thought to have descended from a single ancestral population that lived about 1.8 billion years ago. Single-celled eukaryotes and bacteria are linked by still more remote ancestors. All this distant kinship is stamped by a common genetic code and elementary features of cell structure. Humanity did not soft-land into the teeming biosphere like an alien from another planet. We arose from other organisms already here, whose great diversity, conducting experiment upon experiment in the production of new life-forms, eventually hit upon the human species.

The biodiversity of a country is part of its national heritage. Each country in turn possesses its own unique assemblages of plants and animals including, in almost all cases, species and races found nowhere else. These assemblages are the product of the deep history of the national territory, extending back long before the coming of man.

Biodiversity is the frontier of the future. Humanity needs a vision of an expanding and unending future. This spiritual craving cannot be satisfied by the colonization of space. The other planets are inhospitable and immensely expensive to reach. The nearest stars are so far away that voyagers would need thousands of years just to report back. The true frontier for humanity is life on earth—its exploration and the transport of knowledge about it into science, art, and practical affairs. Again, the qualities of life that validate the proposition are: 90 percent or more of the species of plants, animals, and microorganisms lack even so much as a scientific name; each of the species is immensely old by human standards and has been wonderfully molded to its environment; life around us exceeds in complexity and beauty anything else humanity is ever likely to encounter.

The manifold ways by which human beings are tied to the remainder of life are very poorly understood, crying for new scientific inquiry and a boldness of aesthetic interpretation. The portmanteau expressions "biophilia" and "biophilia hypothesis" will serve well if they do no more than call attention to psychological phenomena that rose from deep human history, that stemmed from interaction with the natural environment, and that are now quite likely resident in the genes themselves. The search is rendered more urgent by the rapid disappearance of the living part of that environment, creating a need not only for a better understanding of human nature but for a more powerful and intellectually convincing environmental ethic based upon it.

REFERENCES

I first used the expression "biophilia" in 1984 in a book entitled by the name (*Biophilia*, Harvard University Press). In that extended essay I attempted to apply ideas of sociobiology to the environmental ethic.

The mechanism of gene-culture coevolution was proposed by Charles J. Lumsden and myself in *Genes, Mind, and Culture* (Harvard University Press, 1981), *Promethean Fire* (Harvard University Press, 1983), and "The Relation Between Biological and Cultural Evolution," *Journal of Social and Biological Structure* 8(4) (October 1985):343–359. It represents an extension of theoretical population genetics in an effort to include the principles of cognition and social psychology.

Balaji Mundkur traced the role of snakes and mythic serpents in *The Cult of the Serpent: An Interdisciplinary Survey of Its Manifestations and Origins* (State University of New York Press, 1983).

Jared Diamond's study of Melanesian attitudes toward other forms of life and Roger S. Ulrich's review of psychological research on biophilia are presented elsewhere in this volume.

I have reviewed the measures of global biodiversity and extinction rates in greater detail in *The Diversity of Life* (Harvard University Press, 1992).

In evaluating the environmental ethic I have been aided greatly by the writings of several philosophers, including most notably Bryan Norton (*Why Preserve Natural Diversity?*, Princeton University Press, 1987), Max Oelschlaeger (*The Idea of Wilderness: From Prehistory to the Age of Ecology*, Yale University Press, 1991), Holmes Rolston III (*Environmental Ethics: Duties to and Values in the Natural World*, Temple University Press, 1988), and Peter Singer (*The Expanding Circle: Ethics and Sociobiology*, Farrar, Straus & Giroux, 1981).

CHAPTER 2

The Biological Basis for

Human Values of Nature

Stephen R. Kellert

THE BIOPHILIA HYPOTHESIS boldly asserts the existence of a biologically based, inherent human need to affiliate with life and lifelike processes (Wilson 1984). This proposition suggests that human identity and personal fulfillment somehow depend on our relationship to nature. The human need for nature is linked not just to the material exploitation of the environment but also to the influence of the natural world on our emotional, cognitive, aesthetic, and even spiritual development. Even the tendency to avoid, reject, and, at times, destroy elements of the natural world can be viewed as an extension of an innate need to relate deeply and intimately with the vast spectrum of life about us.

The hypothesis suggests that the widest valuational affiliation with life and lifelike processes (ecological functions and structures, for example) has conferred distinctive advantages in the human evolutionary struggle to adapt, persist, and thrive as individuals and as a species. Conversely, this notion intimates that the degradation of this human dependence on nature brings the increased likelihood of a deprived and diminished existence— again, not just materially, but also in a wide variety of affective, cognitive, and evaluative respects. The biophilia notion, therefore, powerfully asserts that much of the human search for a coherent and fulfilling existence is intimately dependent upon our relationship to nature. This hypothesized link between personal identity and nature is reminiscent of Aldo Leopold's alteration (1966:240) of Descartes's dictum of selfhood from "I think, therefore I am" (an anthropocentric conception of human identity) to "as a land-user thinketh, so is he" (a biocentric view of selfhood, recognizing Leopold's concept of land as a metaphor for ecological process).

This chapter explores the biophilia notion by examining nine fundamental aspects of our species' presumably biological basis for valuing and affiliating with the natural world. These hypothesized expressions of the biophilia tendency (regarded not as an instinct but as a cluster of learning rules) are referred to as the utilitarian, naturalistic, ecologistic-scientific, aesthetic, symbolic, humanistic, moralistic, dominionistic, and negativistic valuations of nature.

Before commencing the description of these basic values, it might be worth explaining briefly how these hypothesized categories of the basic human relationship to nature evolved in my work. This digression proceeds less from any personal indulgence than from a desire to indicate how the dimensions of the biophilia tendency became apparent as possibly universal expressions of the human dependence on nature.

A limited version of the typology of nine perspectives of nature was developed in the late 1970s as a way of describing basic perceptions of animals (Kellert 1976). This typology was employed in a study of nearly 4,000 randomly distributed Americans residing in the forty-eight contiguous states and Alaska (Kellert 1979, 1980, 1981). Expanded versions of the typology were subsequently used in researching human perceptions of varying taxa including wolves (Kellert 1986d, 1991a), marine mammals (Kellert 1986b, 1991b), diverse endangered species (Kellert 1986c), invertebrates (Kellert 1986a, 1992), and bears (Kellert 1993a); in analyzing the nature-related perspectives of diverse human groups such as hunters (Kellert 1978), birders (Kellert 1985b), farmers (Kellert 1984a) and the general public distinguished by age (Kellert 1985a), gender (Kellert 1987), socioeconomic status (Kellert 1983), and place of residence (Kellert 1981, 1984b); in exploring cross-cultural perspectives of nature and animals in Japan (Kellert 1991c), Germany (Schulz 1986; Kellert 1993b), and Botswana (Mordi 1991); and in examining historical shifts in perceptions of animals in Western society (Kellert 1985c).

The point of this digression is to note that in each study the value dimensions were revealed although they might vary, often greatly, in content and intensity. What began as merely the objective of describing variations in people's perceptions of animals gradually emerged as the possibility of universal expressions of basic human affinities for the natural world. The typology may be simply a convenient shorthand for describing varying perspectives of nature. Its occurrence, however, in a wide variety of taxonomic, behavioral, demographic, historic, and cultural contexts suggests the distinct possibility that these categories might very well be reflections of universal and functional expressions of our species' dependence on the natural world.

Classification of Values

The task of this chapter is to describe each of these categories as indicative of the human evolutionary dependence on nature as a basis for survival and personal fulfillment. As suggested, nine hypothesized dimensions of the biophilia tendency—the utilitarian, naturalistic, ecologistic-scientific, aesthetic, symbolic, humanistic, moralistic, dominionistic, and negativistic—are described here. This description is followed by a discussion of how this deep dependence on nature may constitute the basis for a meaningful and fulfilling human existence—that is, how the pursuit of self-interest may constitute the most compelling argument for a powerful conservation ethic.

Utilitarian

The utilitarian dependence on nature is both something of a misnomer and at the same time manifest. The possible inappropriateness of the term stems from the presumption that *all* the biophilia tendencies possess utilitarian value in the sense of conferring a measure of evolutionary advantage. The use of the utilitarian term here is restricted to the conventional notion of material value: the physical benefits derived from nature as a fundamental basis for human sustenance, protection, and security.

It has long been apparent that a biological advantage exists for humans in exploiting nature's vast cornucopia of food, medicines, clothing, tools, and other material benefits. What may constitute a major conservation development in recent years is the increasing recognition and detailed delineation of the potential and often unrealized material value of various genetic, biochemical, and physical properties of diverse plant and animal species (Myers 1978; Prescott-Allen 1986). Of particular significance has been the expanding realization of the "hidden" material value in nature represented by obscure species and unimpaired ecosystems, such as undiscovered organisms of the tropical rain forests, as potential repositories of material benefit as human knowledge expands to exploit the earth's vast genetic resource base (Eisner 1991).

Naturalistic

The naturalistic tendency may simplistically be regarded as the satisfaction derived from direct contact with nature. At a more complex and profound level, the naturalistic value encompasses a sense of fascination, wonder, and awe derived from an intimate experience of nature's diversity and complexity. The mental and physical appreciation associated with this heightened awareness and contact with nature may be among the most ancient motive forces in the human relationship to the natural world, although its recreational importance appears to have increased significantly in modern industrial society.

The naturalistic tendency involves an intense curiosity and urge for exploration of the natural world. This interest in direct experience of living diversity, and its possible evolutionary roots, is suggested by Wilson (1984:10, 76):

> Because species diversity was created prior to humanity, and because we evolved within it, we have never fathomed its limits. . . . The living world is the natural domain of the more restless and paradoxical part of the human spirit. Our sense of wonder grows exponentially; the greater the knowledge, the deeper the mystery and the more we seek knowledge

The Biological Basis for Human Values of Nature

to create new mystery. . . . Our intrinsic emotions drive us to search for new habitats, to cross unexplored terrain, but we still crave this sense of a mysterious world stretching infinitely beyond.

Discovery and exploration of living diversity undoubtedly facilitated the acquisition of increased knowledge and understanding of the natural world, and such information almost certainly conferred distinctive advantages in the course of human evolution. As Seielstad has remarked (1989:285): "The surest way to enrich the knowledge pool that will keep the flywheel of cultural evolution turning is to nourish the human spirit of curiosity." A genetic basis for this naturalistic tendency is suggested by Iltis (1980:3): "Involvement with nature . . . may be in part genetically determined; human needs for natural diversity . . . must be inherent. Man's love for natural colors, patterns and harmonies . . . must be the result . . . of . . . natural selection through eons of mammalian and anthropoid evolution."

The naturalistic tendency has been cited as providing an important basis for physical fitness and the acquisition of various "outdoor skills" such as climbing, hiking, tracking, and orienteering. The possession of these skills and associated states of mental and physical well-being have been empirically described for a variety of contemporary outdoor activities with a strong emphasis on the naturalistic experience (Driver and Brown 1983; Kaplan 1992). The mental benefits of these activities have been related to tension release, relaxation, peace of mind, and enhanced creativity derived from the observation of diversity in nature. The psychological value of the outdoor recreational experience is noted by Ulrich et al. (1991:203) in a review of the scientific literature: "A consistent finding in well over 100 studies of recreation experiences in wilderness and urban nature areas has been that stress mitigation is one of the most important verbally expressed perceived benefits." Kaplan (1983:155), drawing on extensive research of the naturalistic experience, concluded in a rather more subjective vein: "Nature matters to people. Big trees and small trees, glistening water, chirping birds, budding bushes, colorful flowers—these are important ingredients in a good life."

Ecologistic-Scientific

While important differences distinguish the scientific from the ecologistic relationship to nature, both perspectives similarly reflect the motivational urge for precise study and systematic inquiry of the natural world and the related belief that nature can be understood through empirical study. The ecologistic experience may be regarded as more integrative and less reductionist than the scientific, involving an emphasis on interconnection and interdependence in nature as well as a related stress on integral connections between biotic and abiotic elements manifest in the flow of energy and materials within a system.

The concept of ecology is, of course, a modern scientific formulation: Leopold (1966:176) proclaimed it "the outstanding scientific discovery of the twentieth century." Still, the notion of ecology encompasses far more than the conventional and narrow expression of scientific inquiry. Leopold, despite the previous assertion, recognized this possibility and remarked (1966:266): "Let no man jump to the conclusion that Babbitt must take his Ph.D. in ecology before he can 'see' his country. On the contrary, the Ph.D. may become as callous as an undertaker to the mysteries at which he officiates."

Still, the ecologistic experience of nature often involves a recognition of organizational structure and complexity barely discernible to the average person. This difficulty of perspective reflects the fact that most important ecological processes are prominently manifest at the bottom of biological food chains and energy pyramids often associated with the activities of invertebrate and microbial organisms. As invertebrates represent more than 90 percent of the planet's biological diversity, they perform most of the critical ecological functions of pollination, seed dispersal, parasitism, predation, decomposition, energy and nutrient transfer, the provision of edible materials for adjacent trophic levels, and the maintenance of biotic communities through mutualism, host-restricted food webs, and a variety of other functions and processes. Most people hardly recognize these ecological tendencies, let alone the species integral to their performance, preferring to direct their emotional and conscious awareness of nature to larger vertebrates and prominent natural features.

The human understanding of ecological function is thus at its initial stages of articulation and recognition through systematic inquiry and careful investigation. Nonetheless, the broad realization of ecological process has probably always been intuitively and empirically apparent to the astute human observer. An understanding of organismic and habitat interdependence has likely been the mark of certain figures throughout human history. Moreover, this ecological insight has probably conferred distinctive advantages in the meeting and mastering of life's physical and mental requirements—including increased knowledge, the honing of observational and recording skills, and the recognition of potential material uses of nature through direct exploitation and mimicry. The sense of nature's functional and structural interconnectedness may have further instilled in the prudent observer a cautious respect for nature likely to temper tendencies toward overexploitation and abuse of natural processes and species.

The scientific experience of nature, in contrast to the ecologistic, involves a greater emphasis on the physical and mechanical functioning of biophysical entities as well as a related stress on issues of morphology, taxonomy, and physiological process. The scientific perspective, as previously suggested, tends to be reductionistic: it focuses on constituent elements of nature often independent of the understanding of entire organisms or their relations to other species and natural habitats. Despite this restricted emphasis, often divorced from direct experiential contact with nature, the scientific outlook shares with the ecologistic an intense curiosity and fascination with the systematic study of life and lifelike processes. The depth and intensity of this pursuit of knowledge can often lead to a profound appreciation of nature's wonder and complexity. A sense of this wonder can be discerned in Scott McVay's description of such scientists as Wilson, Vishniac, and von Frisch (1987:5–6):

> I start with wonder, awe and amazement of the profusion of life. . . . E. O. Wilson . . . wrote that a genetic description of a mouse would fill every page of the Encyclopedia Britannica in every edition starting with the first printing in the 1750s to the present day. . . . Roman Vishniac [found] more wonder in a drop of pond water than in traveling to the most remote places on the planet. . . . Karl von Frisch . . . said that there was miracle enough in a single species to provide a life's work.

Such reflections suggest a derivative satisfaction from experiencing the complexity of natural process quite apart from its apparent utility or evolutionary advantage. Yet the actual and potential benefits of such awareness are also quite evident. One can imagine the value of vastly enhanced knowledge and understanding of nature conferred upon those who developed the capacities for precise observation, analysis, and detailed study of even a fraction of life's extraordinary diversity.

Aesthetic

The physical beauty of nature is certainly among its most powerful appeals to the human animal. The complexity of the aesthetic response is suggested by its wide-ranging expression from the contours of a mountain landscape to the ambient colors of a setting sun to the fleeting vitality of a breaching whale. Each exerts a powerful aesthetic impact on most people, often accompanied by feelings of awe at the extraordinary physical appeal and beauty of the natural world.

The human need for an aesthetic experience of nature has been suggested by the apparent inadequacy of artificial or human-made substitutes when people are exposed to them. This preference for natural design and pattern has been revealed in a variety of studies as Ulrich has noted (1983:109): "One of the most clear-cut findings in the . . . literature . . . is the consistent tendency to prefer natural scenes over built views, especially when the latter lack vegetation or water features. Several studies have [shown] that even unspectacular or subpar natural views elicit higher aesthetic preference . . . than do all but a very small percentage of urban views." Additional research suggests that this aesthetic preference for nature may be universally expressed across human cultures (Ulrich 1983:110): "Although far from conclusive, these findings . . . cast some doubt on the position that [aesthetic] preferences vary fundamentally as a function of culture."

Living organisms often function as the centrally valued element in people's aesthetic experience of nature. Unlike the previously described ecologistic-scientific emphasis on relatively obscure organisms, the aesthetic response is typically directed at larger, charismatic megavertebrate species. The basis for this aesthetic focus on relatively large animals is elusive yet, in all likelihood, critical to the understanding of the human attraction to and dependence on nature. Leopold (1966:137, 129–130) powerfully describes this aesthetic significance in alluding to the presence and absence of wildlife in the natural landscape:

The physics of beauty is one department of natural science still in the Dark Ages. . . . Everybody knows, for example, that the autumn landscape in the north woods is the land, plus a red maple, plus a ruffed grouse. In terms of conventional physics, the grouse represents only a millionth of either the mass or energy of an acre. Yet subtract the grouse and the whole thing is dead. An enormous amount of some kind of motive power has been lost. . . . My own conviction on this score dates from the day I saw a wolf die. . . . We reached the old wolf in time to watch a fierce green fire dying in her eyes. I realized then, and have known ever since, that there was something new to me in those eyes—something known only to her and to the mountain.

Leopold referred to this central aesthetic of animals in the landscape as its "numenon," its focus of meaning, in contrast to merely the "phenomenon" of a static and lifeless environment. This essential aesthetic is perhaps what George Schaller (1982) recognized in his reference to the Himalayas as "stones of silence" upon discovering the near extirpation of its endemic caprid fauna—in contrast to Leopold's revelation of the wolf's role in the landscape as requiring one to "think like a mountain." The animal in its contextual environment appears to confer upon its habitat vitality and animation, what Rolston (1986a) has called the essential wildlife aesthetic of "spontaneity in motion."

The biological advantage of the aesthetic experience of nature is difficult to discern, yet, as Wilson suggests (1984:104), "with aesthetics we return to the central issue of biophilia." The aesthetic response could reflect a human intuitive recognition or reaching for the ideal in nature: its harmony, symmetry, and order as a model of human experience and behavior. The adaptational value of the aesthetic experience of nature could further be associated with derivative feelings of tranquillity, peace of mind, and a related sense of psychological well-being and self-confidence. The aesthetic response to varying landscapes and species may also reflect an intuitive recognition of the greater likelihood of food, safety, and security associated with human evolutionary experience. Kaplan and Kaplan suggest, for example (1989:10): "Aesthetic reactions [to nature] . . . reflect neither a casual nor a trivial aspect of the human makeup. Rather, they appear to constitute a guide to human behavior that is both ancient and far-reaching. Underlying such reactions is an assessment of the environment in terms of its compatibility with human needs and purposes." Iltis has further argued for a genetic component in the human aesthetic response to nature (1973:5): "Human genetic needs for natural pattern, for natural beauty, for natural harmony, [are] all the results of natural selection over the illimitable vistas of evolutionary time." A more empirical delineation of this aesthetic preference for certain landscapes and species as a possible function of human evolutionary experience, associated with the likelihood of encountering food, safety, and security, is offered by Heerwagen and Orians. (See Chapter 4 in this volume and Orians 1980.)

Symbolic

The symbolic experience of nature reflects the human use of nature as a means of facilitating communication and thought (Lévi-Strauss 1970; Shepard 1978). The use of nature as symbol is perhaps most critically reflected in the development of human language and the complexity and communication of ideas fostered by this symbolic methodology. The acquisition of language appears to be enhanced by the engendering of refined distinctions and categorizations. Nature, as a rich taxonomy of species and forms, provides a vast metaphorical tapestry for the creation of diverse and complex differentiations. As Lawrence suggests (see Chapter 10) with reference to animals, though the notion can be more broadly extended to other categories of nature, "it is remarkable to contemplate the paucity of other categories for conceptual frames of reference, so preeminent, widespread, and enduring is the habit of symbolizing in terms of animals." Shepard further emphasizes the importance of animate nature as a facilitator of human language and thought (1978:249, 2):

> Human intelligence is bound to the presence of animals. They are the means by which cognition takes its first shape and they are the instruments for imagining abstract ideas and qualities. . . . They are the code images by which language retrieves ideas . . . and traits. . . . Animals are used in the growth and development of the human person, in those most priceless qualities we lump together as "mind." . . . Animals . . . are basic to the development of speech and thought.

The Biological Basis for Human Values of Nature

A limited indication of the symbolic function is reflected in the finding (Kellert 1983) that animals constitute more than 90 percent of the characters employed in language acquisition and counting in children's preschool books. Studies by Shepard (1978), Bettelheim (1977), Campbell (1973), Jung (1959), and others indicate the significance of natural symbols in myth, fairy tale, story, and legend as an important means for confronting the developmental problems of selfhood, identity, expressive thought, and abstraction.

An enduring question of modern life is the degree to which the human capacity for technological fabrication has provided an effective substitute for traditional natural symbols as the primary means of communication and thought. The unlikelihood of this possibility is suggested by the evolutionarily very short time period of modern industrial life relative to the long course of human evolution during which nature constituted the sole environment for our species' language development (Shepard 1978). More important, the dependence of the human psyche on highly varied and refined distinctions seems to be matched only by the extraordinary diversity, complexity, and vividness of the natural world as an extremely rich and textured system. Plastic trees, stuffed animals, and their fabricated kin seem but a meager substitute more likely to result in a stunted capacity for symbolic expression, metaphor, and communication.

Humanistic

The humanistic experience of nature reflects feelings of deep emotional attachment to individual elements of the natural environment. This focus, like the aesthetic, is usually directed at sentient matter, typically the larger vertebrates, although humanistic feelings can be extended to natural objects lacking the capacity for reciprocity such as trees and certain landscapes or geological forms.

The humanistic experience of strong affection for individual elements of nature can even be expressed as a feeling of "love" for nature, although this sentiment is usually directed at domesticated animals. Companion animals are especially given to the process of "humanization" of nature in the sense of achieving a relational status not unlike other humans might assume, even family members. The therapeutic mental and physical benefits of the companion animal have been documented in various studies, at times even resulting in significant healing benefits (Katcher and Beck 1983; Rowan 1989; Anderson et al. 1984; Chapters 3 and 5 in this volume).

The humanistic experience of nature can result in strong tendencies toward care and nurturance for individual elements of nature. From an adaptational viewpoint, the human animal as a social species, dependent on extensive cooperative and affiliational ties, may especially benefit from the interactive opportunities fostered by a humanistic experience of nature. An enhanced capacity for bonding, altruism, and sharing may be important character traits enhanced by this tendency. The use of companion animals for a variety of functional tasks, such as hunting and protection, may also contribute to evolutionary fitness through the acquisition of diverse skills and understandings of nature. This knowledge born of intimate human interaction with a nonhuman species is conveyed in Barry Lopez's description of semidomesticated wolves (1978:282):

> The wolves moved deftly and silently in the woods and in trying to imitate them I came to walk more quietly and to freeze at the sign of slight movement. At first this imitation gave me no advantage, but after several weeks I realized I was becoming far more attuned to the environment we moved through. I heard more . . . and my senses now constantly alert, I occasionally saw a deer mouse or a grouse before they did. . . . I took from them the confidence to believe I could attune myself better to the woods by behaving as they did—minutely inspecting things, seeking vantage points, always sniffing at the air. I did, and felt vigorous, charged with alertness.

Moralistic

The moralistic experience of nature encompasses strong feelings of affinity, ethical responsibility, and even reverence for the natural world. This perspective often reflects the conviction of a fundamental spiritual meaning, order, and harmony in nature. Such sentiments of ethical and spiritual connectedness have traditionally been articulated in poetry, religion, and philosophy, but today they can even be discerned in the modern discourse of scientific language, as suggested by Leopold's remarks (1966:222, 231):

Land is not merely soil; it is a fountain of energy flowing through a circuit of soils, plants, and animals. . . . A thing is right when it tends to preserve the integrity, stability, and beauty of the biotic community. It is wrong when it tends otherwise.

The moralistic perspective has often been associated with the views of indigenous peoples (see Chapter 6 in this volume). Booth and Jacobs (1990) describe important elements in the moralistic experience of nature among indigenous North Americans prior to European acculturation. They emphasize a fundamental belief in the natural world as a living and vital being, a conviction of the continuous reciprocity between humans and nature, and the certainty of an inextricable link between human identity and the natural landscape. This outlook is powerfully reflected in the words of Luther Standing Bear (1933:45):

We are of the soil and the soil is of us. We love the birds and beasts that grew with us on this soil. They drank the same water as we did and breathed the same air. We are all one in nature. Believing so, there was in our hearts a great peace and a willing kindness for all living, growing things.

A more Western articulation of this moralistic identification with nature, somewhat rationalized by the language of modern science, is offered by Loren Eiseley (1946:209–210):

It is said by men . . . that the smallest living cell probably contains over a quarter of a million protein molecules engaged in the multitudinous coordinated activities which make up the phenomenon of life. At the instant of death, whether of man or microbe, that ordered, incredible spinning passes away in an almost furious haste. . . . I do not think, if someone finally twists the key successfully in the tiniest and most humble house of life, that many of these questions will be answered, or that the dark forces which create lights in the deep sea and living batteries in the waters of tropical swamps, or the dread cycles of parasites, or the most noble workings of the human brain, will be much if at all revealed. Rather, I would say that if "dead" matter has reared up this curious landscape of fiddling crickets, song sparrows, and wondering men, it must be plain even to the most devoted materialist that the matter of which he speaks contains amazing, if not dreadful powers, and may not impossibly be, as Hardy has suggested, "but one mask of many worn by the Great Face behind."

From the perspective of this inquiry, the vexing question is the possible biological significance of a moralistic experience of nature. It might be supposed that a moralistic outlook articulated in a group context fostered feelings of kinship, affiliation, and loyalty leading to cooperative, altruistic, and helping behavior. Strong moralistic affinities for nature may also produce the desire to protect and conserve nature imbued with spiritual significance, as Gadgil (1990) has described for the nearly 6 percent of historic India regarded as sacred groves. It may be sufficient to suggest that a biological advantage is conferred on those who experience a profound sense of psychological well-being, identity, and self-confidence produced by the conviction of an ultimate order and meaning in life. The expression of this insight and its possibly pervasive significance is eloquently expressed by John Steinbeck (1941:93):

> It seems apparent that species are only commas in a sentence, that each species is at once the point and the base of a pyramid, that all life is related. . . . And then not only the meaning but the feeling about species grows misty. One merges into another, groups melt into ecological groups until the time when what we know as life meets and enters what we think of as non-life: barnacle and rock, rock and earth, earth and tree, tree and rain and air. And the units nestle into the whole and are inseparable from it. . . . And it is a strange thing that most of the feeling we call religious, most of the mystical outcrying which is one of the most prized and used and desired reactions of our species, is really the understanding and the attempt to say that man is related to the whole thing, related inextricably to all reality, known and unknowable. This is a simple thing to say, but a profound feeling of it made a Jesus, a St. Augustine, a Roger Bacon, a Charles Darwin, an Einstein. Each of them in his own tempo and with his own voice discovered and reaffirmed with astonishment the knowledge that all things are one thing and that one thing is all things—a plankton, a shimmering phosphorescence on the sea and the spinning planets and an expanding universe, all bound together by the elastic string of time.

The Biological Basis for Human Values of Nature

Dominionistic

The dominionistic experience of nature reflects the desire to master the natural world. This perspective may have been more frequently manifest during earlier periods of human evolution; its occurrence today is often associated with destructive tendencies, profligate waste, and despoliation of the natural world. Yet this view may be too narrow and associated with exaggerated dominionistic tendencies. Life, even in the modern era, may be regarded as a tenuous enterprise, with the struggle to survive necessitating some measure of the proficiency to subdue, the capacity to dominate, and the skills and physical prowess honed by an occasionally adversarial relationship to nature. Rolston's insight (1986b:88) is helpful:

> The pioneer, pilgrim, explorer, and settler loved the frontier for the challenge and discipline. . . . One reason we lament the passing of wilderness is that we do not want entirely to tame this aboriginal element. . . . Half the beauty of life comes out of it. . . . The cougar's fang sharpens the deer's sight, the deer's fleet-footedness shapes a more supple lioness. . . . None of life's heroic quality is possible without this dialectical stress.

Beyond an enhanced capacity to subjugate nature, the dominionistic experience may foster increased knowledge of the natural world. As Rolston's remarks intimate, the predator understands and even appreciates its prey to a degree no mere external observer can attain, and this perspective may be as true for the human hunter of deer or mushrooms as it is for the wolf stalking its moose or the deer its browse. While the survival value of the dominionistic experience may be less evident today than in the evolutionary past, one suspects a false arrogance in the denial of the human inclination to master nature in favor of strong emotional bonds of affection or kinship for life. The dominionistic experience of nature, like all expressions of the biophilia tendency, possesses both the capacity for functional advantage as well as exaggerated distortion and self-defeating manifestation.

Negativistic

The negativistic experience of nature is characterized by sentiments of fear, aversion, and antipathy toward various aspects of the natural world. Most advocates of conservation regard fear and alienation from the natural world as inappropriate and often leading to unwarranted harm and destruction. The potential biological advantage of avoiding, isolating, and even occasionally harming presumably threatening aspects of nature can, however, be recognized. (See Chapter 3 in this volume.) The disposition to fear and reject threatening aspects of nature has been cited as one of the most basic motive forces in the animal world. As Öhman suggests (1986:128): "Behaviors that can be associated with fear are pervasive in the animal kingdom. Indeed, one could argue that systems for active escape and avoidance must have been among the first functional behavior systems that evolved."

The human inclination to fear and avoid threatening aspects of nature has been particularly associated with reptiles such as snakes and arthropods such as spiders and various biting and stinging invertebrates. A predisposition to fear and avoid such creatures and other harmful elements of nature may have conferred some advantage during the course of human evolution resulting in its statistically greater prevalence. This potential has been described by Ulrich et al. in a review of the scientific literature (1991:206): "Conditioning studies have shown that nature settings containing snakes or spiders can elicit pronounced autonomic responses . . . even when presented subliminally." Schneirla (1965) further notes that the occurrence of "ugly, slimy, erratic" moving animals, such as certain snakes and invertebrates, provokes withdrawal responses among vertebrate neonates in the absence of overt or obvious threat.

Studies of human attitudes toward invertebrates (Kellert 1993c), as well as related research by Hardy (1988) and Hillman (1991), have discovered a variety of motivational factors in the human tendency to dislike and fear arthropods. First, many humans are alienated by the vastly different ecological survival strategies, spatially and temporally, of most invertebrates in comparison to humans. Second, the extraordinary "multiplicity" of the invertebrate world seems to threaten the human concern for individual identity and selfhood. Third, invertebrate shapes and forms appear "monstrous" to many people. Fourth, invertebrates are often associated with notions of mindlessness and an absence of feeling—the link between insects, spiders, and madness has been a common metaphor in human dis-

The Biological Basis for Human Values of Nature

course and imagination. Fifth, many people appear challenged by the radical "autonomy" of invertebrates from human will and control.

These sentiments of fear and alienation from nature can foster unreasonable human tendencies and the infliction of excessive harm and even cruel behavior on animals and other elements of nature. Singer (1977) has referred to this tendency as "specicide"—reflecting the willingness to pursue the destruction of an entire species, such as Lopez (1978) has described for the wolf in North America or might exist toward certain rodent, insect, and spider species. Hillman ruefully remarked in this regard (1991): "What we call the progress of Western Civilization from the ant's eye level is but the forward stride of the great exterminator."

Negativistic tendencies toward nature, given our modern technical prowess, have often resulted in the massive destruction of elements of the natural world. Yet the extent of today's onslaught on nature should not preclude one from recognizing its possible evolutionary origin or its continued biological advantage expressed at a more modest and even "rational" level. Fear of injury or even violent death in nature will continue to be an integral part of the human repertoire of responses to the natural world, and a realistic tension with threat and danger in nature is part of the challenge of survival. It might even be suggested that some measure of fear of the natural world is essential for the human capacity to experience a sense of nature's magnificence and sublimeness. The power of pristine nature to inspire and challenge human physical and mental development in all likelihood requires considerable elements of fear and danger.

Exploration

The presentation of nine, presumably biologically based, human valuations of nature represents an exploratory effort at supporting the biophilia hypothesis. While these descriptions certainly do not constitute "proof" of the biophilia complex, the typology may provide a heuristic approach for systematically examining the evolutionary basis of each of the suggested values. Each category of the typology is thought to represent a basic human relationship and dependence on nature indicating some measure of adaptational value in the struggle to survive and, perhaps more important,

TABLE 2.1. *A Typology of Biophilia Values*

Term	Definition	Function
Utilitarian	Practical and material exploitation of nature	Physical sustenance/security
Naturalistic	Satisfaction from direct experience/contact with nature	Curiosity, outdoor skills, mental/physical development
Ecologistic-Scientific	Systematic study of structure, function, and relationship in nature	Knowledge, understanding, observational skills
Aesthetic	Physical appeal and beauty of nature	Inspiration, harmony, peace, security
Symbolic	Use of nature for metaphorical expression, language, expressive thought	Communication, mental development
Humanistic	Strong affection, emotional attachment, "love" for nature	Group bonding, sharing, cooperation, companionship
Moralistic	Strong affinity, spiritual reverence, ethical concern for nature	Order and meaning in life, kinship and affiliational ties
Dominionistic	Mastery, physical control, dominance of nature	Mechanical skills, physical prowess, ability to subdue
Negativistic	Fear, aversion, alienation from nature	Security, protection, safety

to thrive and attain individual fulfillment. A summary of the biophilia values is presented in Table 2.1.

This chapter has relied on conceptual and descriptive analysis for delineating basic elements of the biophilia hypothesis. As suggested earlier, a limited empirical corroboration of the typology has been provided by the results of various studies, conducted by the author and others, of diverse cultures and demographic groups, human perceptions of varying taxa, and historical shifts in perspectives of nature. Although methodological problems preclude the assertion of this evidence as proof, these findings offer restricted support of the typology's occurrence. And although these results do not constitute a sufficient validation of the categories as biologically based expressions of human dependence on nature, their widespread empirical expression suggests the possibility that they may represent universal human characteristics. What appears to be relative is not the occurrence of the value types across cultures, taxa, and time but the content and intensity of this expression and its adaptational importance.

It has been argued in this chapter that each value type is indicative of our species' dependence on the natural world and represents a potential evolutionary advantage. It follows that their cumulative, interactive, and synergistic impact may contribute to the possibility of a more fulfilling personal existence. The effective expression of the biophilia need may constitute an important basis for a meaningful experience of self.

The conservation of nature is rationalized, from this perspective, not just in terms of its material and commodity benefits but, far more significantly, for the increased likelihood of fulfilling a variety of emotional, cognitive, and spiritual needs in the human animal. An ethical responsibility for conserving nature stems, therefore, from more than altruistic sympathy or compassionate concern: it is driven by a profound sense of self-interest and biological imperative. As Wilson suggests (1984:131): "We need to apply the first law of human altruism, ably put by Garrett Hardin: never ask people to do anything they consider contrary to their own best interests." Nature's diversity and healthy functioning are worthy of maintenance because they represent the best chance for people to experience a satisfying and meaningful existence. The pursuit of the "good life" is through our broadest valuational experience of nature. This deeper foundation for a conservation ethic is reflected in the words of René Dubos (1969:129):

> Conservation is based on human value systems; its deepest significance is the human situation and the human heart. . . . The cult of wilderness is not a luxury; it is a necessity for the preservation of mental health. . . . Above and beyond the economic . . . reasons for conservation, there are aesthetic and moral ones which are even more compelling. . . . We are shaped by the earth. The characteristics of the environment in which we develop condition our biological and mental being and the quality of our life. Were it only for selfish reasons, therefore, we must maintain variety and harmony in nature.

Clarifying the Concept

The converse of this perspective is the notion that a degraded relationship to nature increases the likelihood of a diminished material, social, and psychological existence. This chapter has intimated several possibilities in this regard, and it may be relevant to note the finding that significant abusers of nature, particularly those who inflict in childhood willful harm on animals, are far more likely in adulthood to reveal repeated patterns of violence and aggressive behavior toward other people (Kellert and Felthous 1985; Felthous and Kellert 1987). Indeed, presumably socially acceptable forms of destructive conduct toward nature may in retrospect come to be regarded as false and short-term benefits, as Leopold's lament of the last of the passenger pigeons suggests (1966:109):

> We grieve because no living man will see again the onrushing phalanx of victorious birds sweeping a path for spring across the March skies, chasing the defeated winter from all the woods and prairies. . . . Our grandfathers were less well-housed, well-fed, well-clothed than we are. The strivings by which they bettered their lot are also those which deprived us of pigeons. Perhaps we now grieve because we are not sure, in our hearts, that we have gained by the exchange. The gadgets of industry bring us more comforts than the pigeons did, but do they add as much to the glory of the spring?

A skeptical response to the assertion of the biophilia tendency as a biologically based human need to affiliate with nature is the view that this hypothesis is an expression of cultural and class bias. This view suggests that the assertions trumpeted here are but a romantic ideology of nature, paraded in the guise of biology, promoted for essentially elitist political and social reasons. Such a critique may claim that the biophilia hypothesis condemns, by implication, all those mired in poverty and trapped within urban walls to another stereotype of a less fulfilling human existence.

Abraham Maslow's (1954) notion of a hierarchy of needs may offer one response to this critique—implying the pursuit of self-realization through a broad valuational experience of nature as a higher order of human functioning. In other words, the biophilia tendency might become manifest once the basic human needs for survival, protection, and security have been realized. This argument, while superficially appealing, probably reflects a naive assumption of human functioning. People are typically inclined to pursue concurrently a wide range of simple to complex needs if they are not overwhelmed by the sheer necessity of confronting the material basis for survival (a relatively rare condition).

Any presumption of the relative unimportance of the biophilia tendency among persons of lower socioeconomic status or urban residence may, in itself, be an elitist and arrogant characterization. Nature's potential for providing a more satisfying existence may be less obvious and apparent among the poor and urban than the rich and rural, but this deprivation represents more a challenge of design and opportunity than any fundamental irrelevance of the natural world for a class of people. As Leopold noted (1966:266): "The weeds in a city lot convey the same lesson as the redwoods. . . . Perception . . . cannot be purchased with either learned degrees or dollars; it grows at home as well as abroad, and he who has a little may use it to as good advantage as he who has much." The capacity of nature to enrich and enlarge the human experience is a potential inherent in all but the most deprived and encapsulated within concrete walls. Society's obligation is not to bemoan the seeming "absence" of nature in the inner city or among the poor but to render its possibility more readily available. The presumption that only the materially advantaged and conveniently located can realize nature's value represents an arrogant characterization.

A more fundamental question is the recognition in modern society of the human need to affiliate deeply and positively with life's diversity. This is a complex issue too difficult to address here in detail. A partial response, however, may be provided by the results of the previously cited studies conducted in the United States and Japan. While these studies explore the biophilia hypothesis only indirectly, they offer circumstantial information regarding the modern relationship to the natural world among persons living in highly urban, technologically oriented, industrial societies. Insufficient space precludes all but a very brief summarization of these results, although more detailed information regarding the studies can be found elsewhere (Kellert 1979, 1981, 1983, 1991c, 1993b).

Both the United States and Japan have been described as nations with a pronounced appreciation for the natural world. Americans, for example, are known to be especially supportive of nature conservation: nearly 10 percent of the American public is formally affiliated with at least one environmental organization (Dunlap 1978), and American environmental legislation is recognized as among the most comprehensive and protective in the world (Bean 1983). Extensive outdoor recreational activity among Americans is reflected in nearly 300 million annual visits to national parks, and three-fourths of the public participates in some form of wildlife-related outdoor recreational activity (Foresta 1984; USFWS 1990).

Japanese culture too has been characterized as encouraging a strong appreciation for nature (Higuchi 1979; Minami 1970; Murota 1986; Watanabe 1974). Often cited expressions of this interest include the practices of Shintoism, flower arranging, plant cultivation (such as bonsai), the tea ceremony, certain poetry forms, rock gardening, and various celebrations of the seasons. Higuchi (1979:19) has described a Japanese view of nature "based on a feeling of awe and respect," while Watanabe (1974:280) has remarked on a Japanese "love of nature . . . resulting in a refined appreciation of the beauty of nature." Murota (1986:105) suggests: "The Japanese nature is an all-pervasive force. . . . Nature is at once a blessing and friend to the Japanese people."

Despite these assertions of an especially refined appreciation for nature in the United States and Japan, our research has revealed only limited concern for the natural world among the general public in both countries. Citizens of the United States and Japan typically expressed strong interest in nature only in relation to a small number of species and landscapes characterized by especially prominent aesthetic, cultural, and historic features. Furthermore, most Americans and Japanese expressed strong inclinations to exploit nature for various practical purposes despite the likelihood of inflicting considerable environmental damage. Most respondents revealed, especially in Japan, indifference toward elements of the natural world lacking any aesthetic or cultural value. Very limited knowledge and understanding of nature was found, particularly in Japan.

Japanese appreciation of nature was especially marked by a restricted focus on a small number of species and natural objects—often admired in a context emphasizing control, manipulation, and contrivance. This affinity for nature was typically an idealistic rendering of valued aspects of the natural environment, usually lacking an ecological or ethical orientation. This appreciation was described by one Japanese respondent as "a love of semi-nature," representing a largely emotional and aesthetic interest in using "the materials of seminature to express human feelings." Other respondents described it as a perspective of nature dominated by a preference for the artificial, abstract, and symbolic rather than any realistic experience of the natural world; a motivation to "touch" nature from a controlled and safe distance; an adherence to strict rules of seeing and experiencing nature intended to express only the centrally valued aspect; a desire to isolate favored aspects of nature in order to "freeze and put walls around it." Environmental features falling outside the valued aesthetic and symbolic boundaries tended to be ignored, dismissed, or judged unappealing (Saito 1983).

American respondents revealed a somewhat more generalized interest and concern for nature, especially among highly educated and younger Americans in comparison to similar demographic groups in Japan. On the other hand, nature appreciation among most Americans was largely restricted to particularly valued species and landscapes, while other aspects of the natural world were typically subordinated to strong utilitarian concerns. The great majority of Americans revealed little appreciation of "lower" life-forms, tending to restrict their appreciation to the large vertebrates.

In conclusion, most Americans and Japanese expressed a pronounced concern for only a limited number of species and natural objects. The biophilia tendency, as described here, was broadly evident only among a small segment of the population in both countries, most prominently the better educated and the young in the United States.

A New Basis for Conservation?

A largely conceptual argument has been offered here in support of the biophilia hypothesis. It appears that a variety of basic valuations of nature are consistent with the possibility of increased evolutionary fitness at both the individual and species levels. Each expression of the biophilia tendency—the aesthetic, dominionistic, ecologistic-scientific, humanistic, moralistic, naturalistic, symbolic, utilitarian, and even negativistic—has been depicted as potentially enhancing the basis for a profound development of self. A range of adaptational advantages has been cited as resulting from these basic experiences of nature—enhanced physical skills and material benefits, greater awareness, increased protection and security, opportunities for emotional gratification, expanded kinship and affiliational ties, improved knowledge and cognitive capacities, greater communication and expressive skills, and others.

A conservation ethic of care, respect, and concern for nature was regarded as more likely to emanate from the conviction that in our relationship to the natural world exists the likelihood of achieving a more personally rewarding existence. As Iltis has suggested (1980:3, 5), our mental and physical well-being may represent a far more compelling basis for nature conservation than the mere rationalization of enhanced material benefit:

> Here, finally, is an argument for nature preservation free of purely [material] utilitarian considerations; not just clean air because polluted air gives cancer; not just pure water because polluted water kills the fish we might like to catch; . . . but preservation of the natural ecosystem to give body and soul a chance to function in the way they were selected to function in their original phylogenetic home. . . . Could it be that the stimuli of non-human living diversity makes the difference between sanity and madness?

Iltis's question intimates the still tenuous state of our understanding of the biophilia phenomenon. The sophistication and depth of future inquiry may prove the measure of Iltis's response to his own question (1973:7):

> We may expect that science will [someday] furnish the objective proofs of suppositions about man's needs for a living environment which we, at present, can only guess at through timid intuition; that one of these days we shall find the intricate neurological bases of why a leaf or a lovely flower affects us so very differently than a broken beer bottle.

The importance of this recognition of our basic human dependence on nature is suggested by the meager appreciation of the natural world evinced among the general public in modern Japan and the United States. The great majority of people in these two leading economic nations recognized to only a limited extent the value of nature in fostering human physical, cognitive, emotional, and spiritual development. Most Americans and Japanese expressed an aloofness from the biological matrix of life, restricting their interest to a narrow segment of the biotic and natural community. This narrow emphasis on certain species and landscapes is clearly an insufficient basis for a fundamental shift in global consciousness—one capable of countering the contemporary drift toward massive biological impoverishment and environmental destruction.

REFERENCES

Anderson, R., B. Hart, and L. Hart. 1984. *The Pet Connection*. Minneapolis: University of Minnesota Press.

Bean, M. 1983. *The Evolution of National Wildlife Law*. New York: Praeger.

Bettelheim, B. 1977. *The Uses of Enchantment*. New York: Vintage Books.

Booth, A., H. Booth, and H. M. Jacobs. 1990. Ties that bind: Native American beliefs as a foundation for environmental consciousness. *Env. Ethics* 12: 27–43.

Campbell, J. 1973. *Myths to Live By*. New York: Viking Press.

Driver, B., and P. Brown. 1983. Contributions of behavioral scientists to recreation resource management. In I. Altman and J. Wohlwill (eds.), *Behavior and the Natural Environment*. New York: Plenum Press.

Dubos, R. 1969. *Ecology and Religion in History*. New York: Oxford University Press.

Dunlap, R. 1978. *Environmental Concern*. Monticello, Ill.: Vance Bibliographies.

Eiseley, L. 1946. *The Immense Journey*. New York: Random House.

Eisner, T. 1991. Chemical prospecting: a proposal for action. In H. Bormann and S. Kellert (eds.), *Ecology, Economics, Ethics: The Broken Circle*. New Haven: Yale University Press.

Felthous, A., and S. Kellert. 1987. Childhood cruelty to animals and later aggression against people. *Amer. J. Psychiat.* 144:710–717.

Foresta, R. 1984. *America's National Parks and Their Keepers*. Baltimore: Johns Hopkins University Press.

Gadgil, M. 1990. India's deforestation: patterns and processes. *Soc. Nat. Res.* 3:131–143.

Hardy, T. 1988. Entomophobia: the case for Miss Muffet. *Entom. Soc. Amer. Bull.* 34:64–69.

Higuchi, K. 1979. *Nature and the Japanese*. Tokyo: Kodansha International.

Hillman, J. 1991. *Going Bugs*. Gracie Station, N.Y.: Spring Audio.

Iltis, H. 1973. Can one love a plastic tree? *Bull. Ecol. Soc. Amer.* 54:5–7, 19.

————. 1980. Keynote address. *Trans. Symp.: The Urban Setting: Man's Need for Open Space*. New London: Connecticut College.

Jung, C. 1959. *The Archetype and the Collective Unconscious*. New York: Pantheon Books.

Kaplan, R. 1983. The role of nature in the urban context. In I. Altman and J. Wohlwill (eds.), *Behavior and the Natural Environment*. New York: Plenum Press.

Kaplan, S. 1992. The restorative environment: nature and human experience. In D. Relf (ed.), *The Role of Horticulture in Human Well-Being and Social Development*. Portland, Ore.: Timber Press.

Kaplan, R., and S. Kaplan. 1989. *The Experience of Nature: A Psychological Perspective*. Cambridge: Cambridge University Press.

Katcher, A., and A. Beck. 1983. *New Perspectives on Our Lives with Companion Animals*. Philadelphia: University of Pennsylvania Press.

Kellert, S. 1976. Perceptions of animals in American society. *Trans. N.A. Wild. & Nat. Res. Conf.* 41:533–546.

————. 1978. Characteristics and attitudes of hunters and anti-hunters. *Trans. N.A. Wild. & Nat. Res. Conf.* 43:412–423.

————. 1979. *Public Attitudes Toward Critical Wildlife and Natural Habitat Issues*. Washington: U.S. Government Printing Office.

————. 1980. *Activities of the American Public Relating to Animals*. Washington: U.S. Government Printing Office.

————. 1981. *Knowledge, Affection and Basic Attitudes Toward Animals in American Society*. Washington: U.S. Government Printing Office.

————. 1983. Affective, evaluative and cognitive perceptions of animals. In I. Altman and J. Wohlwill (eds.), *Behavior and the Natural Environment*. New York: Plenum Press.

————. 1984a. Public attitudes toward mitigating energy development impacts on western mineral lands. *Proc. Issues & Tech. Mang. Impacted Western Wildlife*. Boulder: Thorne Ecological Institute.

————. 1984b. Urban American perceptions and uses of animals and the natural environment. *Urb. Ecol.* 8:209–228.

————. 1985a. Attitudes toward animals: age-related development among children. *J. Env. Educ.* 16: 29–39.

————. 1985b. Birdwatching in American society. *Leis. Sci.* 7:343–360.

————. 1985c. Historical trends in perceptions and uses of animals in 20th century America. *Env. Rev.* 9:34–53.

————. 1986a. The contributions of wildlife to human quality of life. In D. Decker and G. Goff (eds.), *Economic and Social Values of Wildlife*. Boulder: Westview Press.

————. 1986b. Marine mammals, endangered species, and intergovernmental relations. In M. Silva (ed.), *Intergovernmental Relations and Ocean Resources*. Boulder: Westview Press.

————. 1986c. Social and perceptual factors in the preservation of animal species. In B. Norton (ed.), *The Preservation of Species*. Princeton: Princeton University Press.

————. 1986d. The public and the timber wolf in Minnesota. *Trans. N.A. Wild. & Nat. Res. Conf.* 51:193–200.

————. 1987. Attitudes, knowledge, and behaviors toward wildlife as affected by gender. *Bull. Wild. Soc.* 15:363–371.

————. 1991a. Public views of wolf restoration in Michigan. *Trans. N.A. Wild. & Nat. Res. Conf.* 56:152–161.

————. 1991b. Public views of marine mammal conservation and management in the northwest Atlantic. *Int. Mar. Mamm. Assn. Tech. Rpt.* 91-04. Guelph, Ontario.

————. 1991c. Japanese perceptions of wildlife. *Cons. Biol.* 5:297–308.

————. 1993a. Public attitudes toward bears and their conservation. In C. Servheen (ed.), *Proc. 9th Int. Bear Conf.* Missoula: U.S. Fish and Wildlife and Forest Services.

————. 1993b. Attitudes toward wildlife among the industrial superpowers: United States, Japan, and Germany. *J. Soc. Iss.* 49:53–69.

————. 1993c. Values and perceptions of invertebrates. Submitted to *Cons. Biol.*

Kellert, S., and A. Felthous. 1985. Childhood cruelty toward animals among criminals and noncriminals. *Hum. Rel.* 38:1113–1129.

Leopold, A. 1966. *A Sand County Almanac*. New York: Oxford University Press.

Lévi-Strauss, C. 1970. *The Raw and the Cooked*. New York: Harper & Row.

Lopez, B. 1978. *Of Wolves and Men*. New York: Scribner's.

Luther Standing Bear. 1933. *Land of the Spotted Eagle*. Lincoln: University of Nebraska Press.

Maslow, A. 1954. *Motivation and Personality*. New York: Harper & Row.

McVay, S. 1987. A regard for life: getting through to the casual visitor. *Phil. Zoo Rev.* 3:4–6.

Minami, H. 1970. *Psychology of the Japanese People*. Honolulu: East-West Center.

Mordi, R. 1991. *Attitudes Toward Wildlife in Botswana*. New York: Garland Publishing.

Murota, Y. 1986. Culture and the environment in Japan. *Env. Mgt.* 9:105–112.

Myers, N. 1978. *The Sinking Ark*. New York: Pergamon Press.

Öhman, A. 1986. Face the beast and fear the face: animal and social fears as prototypes for evolutionary analyses of emotion. *Psychophysiol.* 23:123–145.

Orians, G. 1980. Habitat selection: general theory and applications to human behavior. In J. Lockard (ed.), *The Evolution of Human Social Behavior*. New York: Elsevier.

Prescott-Allen, C., and R. Prescott-Allen. 1986. *The First Resource*. New Haven: Yale University Press.

Rolston, H. 1986a. Beauty and the beast: aesthetic experience of wildlife. In D. Decker and G. Goff (eds.), *Economic and Social Values of Wildlife*. Boulder: Westview Press.

————. 1986b. *Philosophy Gone Wild*. Buffalo: Prometheus Books.

Rowan, A. 1989. *Animals and People Sharing the World*. Hanover, N.H.: University Press of New England.

Saito, Y. 1983. The aesthetic appreciation of nature: Western and Japanese perspectives and their ethical implications. Doctoral thesis, University of Michigan. University Microfilms, Ann Arbor.

Schaller, G. 1982. *Stones of Silence*. New York: Viking Press.

Schneirla, T. 1965. *Principles of Animal Psychology*. Englewood Cliffs, N.J.: Prentice-Hall.

Schulz, W. 1986. Attitudes toward wildlife in West Germany. In D. Decker and G. Goff (eds.), *Economic and Social Values of Wildlife*. Boulder: Westview Press.

Seielstad, G. 1989. *At the Heart of the Web*. Orlando: Harcourt Brace Jovanovich.

Shepard, P. 1978. *Thinking Animals: Animals and the Development of Human Intelligence*. New York: Viking Press.

Singer, P. 1977. *Animal Liberation*. New York: Avon Books.

Steinbeck, J. 1941. *Log from the Sea of Cortez*. Mamaroneck, N.Y.: P. P. Appel.

Ulrich, R. 1983. Aesthetic and affective response to natural environment. In I. Altman and J. Wohlwill (eds.), *Behavior and the Natural Environment*. New York: Plenum Press.

Ulrich, R., et al. 1991. Stress recovery during exposure to natural and urban environments. *J. Env. Psych.* 11:201–230.

U.S. Fish and Wildlife Service (USFWS). 1990. *1990 National Survey of Hunting, Fishing and Wildlife-Associated Recreation*. Washington: Department of the Interior.

Watanabe, H. 1974. The conception of nature in Japanese culture. *Science* 183:279–282.

Wilson, E. O. 1984. *Biophilia: The Human Bond with Other Species*. Cambridge: Harvard University Press.

Part Two

AFFECT

AND

AESTHETICS

CHAPTER 3

Biophilia, Biophobia,

and Natural Landscapes

Roger S. Ulrich

THE BELIEF THAT contact with nature is somehow good or beneficial for people is an old and widespread notion. The gardens of the ancient Egyptian nobility, the walled gardens of Persian settlements in Mesopotamia, and the gardens of merchants in medieval Chinese cities indicate that early urban peoples went to considerable lengths to maintain contact with nature (Shepard 1967; Hongxun 1982). During the last two centuries, in several countries, the idea that exposure to nature fosters psychological well-being, reduces the stresses of urban living, and promotes physical health has formed part of the justification for providing parks and other nature in cities and preserving wilderness for public use (Parsons 1991; Ulrich et al. 1991). These notions might be considered early forms of the biophilia hypothesis. But E. O. Wilson's interpretation of biophilia (1984) is not limited to the proposition that humans are characterized by a tendency to pay attention to, affiliate with, or otherwise respond positively to nature. His definition of biophilia also includes the proposition that there is a partly genetic basis for humans' positive responsiveness to nature.

This chapter examines the biophilia hypothesis in the context of theory and empirical findings relating to psychological, physiological, and certain health-related responses associated with *viewing* natural landscapes. Much of the discussion focuses on the effects of people's visual experiences with such "nonanimal" nature as landscapes dominated by vegetation, water, or other nature content. The first section outlines an evolutionary perspective that serves in subsequent sections as a framework for examining behavioral science research findings relevant to the biophilia hypothesis. One prominent feature of the conceptual perspective is the position that theoretical arguments for a genetic component to biophilia gain plausibility if a genetic predisposition in humans for *biophobic* responsiveness to certain dangerous nature phenomena is likewise postulated. In line with this position, some of the early sections survey research findings which suggest there is a partly innate basis for negative or biophobic responses to certain nature stimuli such as snakes.

The survey of empirical evidence for a genetic role in biophobia provides a springboard for next advancing theoretical notions concerning biophilia. This exploration leads to some of the major sections of the chapter, which discuss conceptual arguments and research findings relating to three general types of biophilic responses to unthreatening natural landscapes: liking/approach responses; restoration or stress recovery responses; and enhanced high-order cognitive functioning. The discussion of biophilic responses then leads to a consideration of human benefits that might be lost when natural landscapes are eliminated. The final section discusses research that would expand our understanding of nature's beneficial effects on people and suggests research approaches that might shed empirical light on a possible genetic role in biophilia.

An Evolutionary Perspective

The speculation that positive responses to natural landscapes might have a partly genetic basis implies that such responses had adaptive significance during evolution. In other words, if biophilia is represented in the gene pool it is because a predisposition in early humans for biophilic responses to certain natural elements and settings contributed to fitness or chances for survival. A basic conceptual argument in this chapter is that both the *rewards* and the *dangers* associated with natural settings during human evolution have been sufficiently critical to favor individuals who readily learned, and then over time remembered, various adaptive responses— both *positive/approach* (biophilic) responses and *negative/avoidance* (biophobic) responses—to certain natural stimuli and configurations. This perspective explicitly recognizes that the natural habitats of early humans contained dangers as well as advantages. A general argument influencing the content and organization of the chapter is that theoretical propositions for an innate predisposition for biophilia gain plausibility and consistency if they also postulate a corresponding genetic predisposition for adaptive biophobic responses to certain natural stimuli that presumably have constituted survival-related threats throughout human evolution.

The notion that fears and even phobic responses to certain natural stimuli have an evolutionary basis is not new. Perhaps not surprisingly, Charles Darwin (1877) may have been the first to advance this hypothesis. The following sections on biophobia focus mainly on responses to fear-relevant animals and give comparatively little coverage to the physical environment. This approach is in keeping with the point that many risks for early humans were related to predators. By contrast, the later sections on biophilia focus mainly on responses to natural landscapes, an approach which is consonant with the proposition that many critical survival-related advantages for early humans (food, water, security) were tied to characteristics of the physical environment. Important survival advantages were related to animals as well; other chapters in this volume address biophilic responses to animals.

It is suggested that a partly genetic basis for biophilia and biophobia should be reflected in *biologically prepared learning*—and possibly in particular characteristics of responses to certain natural stimuli (such as very short reaction times) that may not be evident for learning and response characteristics with respect to modern and urban stimuli. As initially proposed by Seligman (1970, 1971), prepared learning theory holds that evolution has predisposed humans and many animal species to easily and quickly learn, and persistently retain, those associations or responses that foster survival when certain objects or situations are encountered. For example, it might be hypothesized that humans should readily acquire, and then not forget, adaptive fear/avoidance responses to such risk-relevant stimuli or situations as snakes, spiders, and heights. Although the recent large-scale transformation of environments in industrialized countries has largely eliminated the real danger of the objects of fears and phobias, fear/avoidance responses might nonetheless persist because they are represented in the gene pool. Seligman's theory suggests that prepared learning should not be evident for stimuli that were comparatively neutral during evolution—that is, did not have major importance for survival as either threats or advantages. It is worth emphasizing that prepared learning theory does not postulate that adaptive responses to prepared natural stimuli should appear spontaneously or in the absence of learning. Rather, some conditioning is necessary to elicit a response which is then characterized by resistance to "extinction," or forgetting.

Although Seligman (1970) suggests that positive responses might be biologically prepared for some objects, nearly all empirical tests of preparedness theory have focused on aversive reactions, especially fears and phobias, with respect to fear-relevant stimuli. The following sections survey the findings from preparedness research and from behavior-genetic studies that are relevant to biophobia. The term biophobia is defined here as a partly genetic predisposition to readily associate, on the basis of negative information or exposure, and then persistently retain fear or strong negative/avoidance responses to certain natural stimuli that presumably have constituted risks during evolution. We begin with a survey of biophobia findings because research on biophobia is comparatively advanced, has shed light on a genetic role in responses to nature, and has yielded theory and ingenious research methods that could prove important in future research on biophilia.

Biophobia

There is considerable evidence from clinical psychology and psychiatry (Costello 1982; and McNally 1987) that the majority of phobic occurrences involve strong fears with respect to certain objects and situations that have threatened humans throughout evolution (snakes, spiders, heights, closed spaces, blood). This finding appears to hold across industrialized societies for which data are available. The most common phobic fears in Western societies may be fears of snakes and spiders. For developing nations, sound data are lacking on the prevalence of phobias and the objects of phobias. Here it is relevant to note that in the absence of sound data for a society, one cannot infer from observed cultural practices such as the creation of cobra shrines in India (Chapter 12 in this volume) that snake fears and phobias are not common in that country. To the contrary, there are grounds for arguing on the basis of psychological theory that such shrines and associated rituals may represent complex active coping efforts to reduce the anxiety associated with a dangerous and fascinating creature.

Research Findings

The notion that biologically prepared learning plays a role in biophobia has received support from scores of experiments performed in different countries by several investigators, most notably Arne Öhman and his associates in Sweden and Norway (Öhman, Erixon, and Löfberg 1975; Öhman 1979; Öhman, Dimberg, and Öst 1985). Nearly all studies have used variants of an ingenious Pavlovian conditioning approach pioneered by Öhman. These experiments typically involve comparisons between defense or aversive responses conditioned (that is, learned through repeated exposure or experience) to slides of fear-relevant and fear-irrelevant or neutral stimuli. Responses are usually assessed by recording autonomic nervous system indicators such as skin conductance and heart rate. In the first part of a typical experiment, defense responses are conditioned by showing either fear-relevant stimuli (such as snakes, spiders) or neutral stimuli (geometric figures) and pairing each slide presentation with an aversive stimulus ("unconditioned" stimulus) that usually is an electric shock having some resemblance to a bite. This phase of the experiment makes it possible to compare the fear-relevant and neutral stimuli with respect to the speed and magnitude of acquisition of a defense/aversive response. Following the initial acquisition phase, the same stimuli are presented ten to forty additional times but without the reinforcement of electric shock. This "extinction" phase allows comparison of the fear-relevant and neutral stimuli in terms of resistance to forgetting the defense/aversive response acquired earlier.

In a review of such conditioning experiments, McNally (1987) concluded that, on balance, findings indicate that conditioned responses are sometimes though not reliably acquired more quickly—but are consistently more resistant to forgetting (extinction)—for certain premodern risk stimuli such as snakes and spiders than for neutral or fear-irrelevant stimuli. Put differently, Seligman's prepared learning theory receives at best equivocal support regarding the ease or speed of response acquisition. Biologically prepared learning receives considerable support, however, from the consistently greater resistance to forgetting that is evident for responses acquired to natural stimuli such as snakes and spiders that probably have constituted threats to humans throughout evolution.

At this point it might be argued that the greater persistence of responses conditioned to snakes and spiders stems not from biologically prepared learning but rather from "regular" learning of fear associations, including ingrained cultural conditioning that these stimuli are strongly negative and dangerous. This possibility has been tested in conditioning experiments that exposed individuals to fear-relevant premodern natural stimuli, such as snakes and spiders, and to far more dangerous (and arguably more culturally conditioned) modern stimuli such as handguns and frayed electrical wires (Cook, Hodes, and Lang 1986; Hugdahl and Karker 1981). Findings have provided additional support for the notion of biological preparedness because defense responses conditioned to the dangerous modern stimuli extinguished or were forgotten more quickly than responses to snakes and spiders.

Vicarious Acquisition of Adaptive Responses

A noteworthy aspect of biophobia is that responses can be acquired through various types of "vicarious" conditioning or learning experiences. Several studies have shown that merely telling the subject that shock will be administered is alone sufficient for effective acquisition of responses to fear-relevant but not to fear-irrelevant natural stimuli (see Hugdahl 1978). In a vicarious conditioning study Hygge and Öhman (1978) exposed individuals to an allegedly phobic experimenter/actor who reacted fearfully to slides of either fear-relevant natural stimuli (snakes, spiders, rats) or fear-irrelevant natural stimuli (such as berries). Findings indicated that people acquired much more persistent defense reactions when watching the experimenter's reactions to fear-relevant in contrast to fear-irrelevant slides. Similarly, Mineka and her associates have performed several studies with rhesus monkeys that have yielded strong evidence of vicarious conditioning of fear/aversive responses to fear-relevant stimuli (such as toy snakes, toy crocodiles) but not to fear-irrelevant stimuli such as toy rabbits (Mineka et al. 1984; Cook and Mineka 1989, 1990).

The research on humans suggests that simply observing another person's fearful or strongly aversive reaction to a presumably biologically prepared natural object—or even receiving information regarding a possibly aversive consequence (such as a shock) of exposure to the object—can be sufficient to condition adaptive defense/aversive responses. An important implication of these findings is that vicarious acquisition may greatly enhance the adaptive or survival-related significance of biologically prepared responses for a *group* of humans or primates. To illustrate this argument, consider the example of an early human in a hunting and gathering group who is bitten by a poisonous snake. Although the bite experience would presumably condition in the person a persistent disposition to respond with fear/avoidance to snakes, this response would have no adaptive value for the person if the venom proved fatal. But other members of the group might acquire unforgettable fear/avoidance responses by having witnessed the bite episode, by having observed the effects on the person, or by receiving vivid information from others about the episode and its painful, fatal consequences. To the extent that such vicarious conditioning occurred throughout the members of the band, the fatal episode would conceivably advance the group's survival chances.

In light of the findings indicating the efficacy of threatening information in conditioning defense/aversive responses, it is not too great a speculation to suggest that one critical adaptive function of certain vivid oral folklore (Chapter 7 in this volume), mythology, or other culturally transmitted information focusing on certain dangerous creatures or objects (Chapter 6 in this volume) might be to vicariously condition adaptive fear/avoidance responses. In this way certain cultural traditions, in combination with biologically prepared learning, might serve as highly efficient means for achieving acquisition of adaptive responses throughout a society. Losses in the form of injuries or deaths would be reduced because effective conditioning would occur without the need for people to have direct—dangerous—encounters with biophobic stimuli. Repeated exposures achieved by the telling and retelling of vivid stories or myths in certain cultures about prepared dangerous natural phenomena might be considered to have rough parallels to the repeated acquisition trials in conditioning studies that are so effective in conditioning persistently retained aversive responses to certain natural stimuli.

Nonconscious Processing of Biophobic Stimuli

Another noteworthy aspect of biophobia has emerged from a few studies that raise the possibility of nonconscious automatic processing for prepared fear-relevant stimuli but not fear-irrelevant stimuli. In a series of publications Öhman has advanced detailed theoretical arguments proposing that as a remnant of evolution, humans have a biologically controlled predatory defense system with a capacity for very quick, automatic, or "unconscious" processing of certain cues—such as paired forward-looking eyes—signaling the presence of approaching snakes and other predators (Öhman 1986). To evaluate empirically the nonconscious processing hypothesis, Öhman and his associates have used a variant of the aversive Pavlovian conditioning approach that incorporates "backward-masking" methods. Most of these experiments begin with a phase in which aversive / defense responses are conditioned in the usual manner to slides of premodern risk stimuli (such as snakes) and slides of neutral or risk-irrelevant stimuli. In a later stage of the experiment, however, the same conditioned slides are displayed *subliminally* (15–30 milliseconds) before being "masked" by a slide of another stimulus or setting that can be recognized and otherwise consciously processed.

Findings from these backward-masking experiments have consistently suggested that, following initial conditioning, *subliminal* presentations of natural settings containing snakes or spiders can elicit strong defense / aversive reactions in normal or nonphobic persons (Öhman 1986; Öhman and Soares 1993a). In the case of people with animal fears or phobias, it can be assumed that conditioning has occurred long before the experiment. In this regard, a recent study of snake-fearful, spider-fearful, and nonfearful control individuals found that a masked subliminal presentation was alone sufficient to elicit defense responding in the fearful groups to their particular feared stimulus (Öhman and Soares 1993b). These studies suggest the following conclusion: coherent fear/defense responding, evident in both physiological and affective indicators, can occur without recognition or even conscious awareness of quite specific natural threat stimuli. There is no such response to neutral or fear-irrelevant natural stimuli.

Regarding the issue of very quick responses to biologically prepared stimuli, Dimberg (1990) has used the technique of facial electromyography (Cacioppo, Tassinary, and Fridlund 1990) to show that specific emotional reactions to presumably prepared stimuli (snakes, spiders, angry faces, happy faces) are readily detectable within 400 milliseconds or less following presentation of the stimuli. This extremely fast emotional/physiological responding is very difficult to reconcile plausibly with a purely "controlled" conscious cognitive perspective on human/nature interactions (Ulrich et al. 1991).

Preparedness to Depth/Spaciousness in Landscapes

The discussion of biophobia up to this point has focused on biologically prepared responses to animals. A functional-evolutionary perspective, however, implies that certain properties of the physical environments of early humans probably had major influences on risk probabilities and survival chances. In this regard it is reasonable to propose that throughout evolution visual *depth/spaciousness* characteristics of natural environments have affected such important risk-related factors as surveillance, proximity to hidden threats, and escape opportunities. From this argument it follows that humans may be biologically prepared to respond with moderate dislike/avoidance or wariness to spatially restricted settings that might contain hidden dangers and constrain opportunities to escape (Ulrich 1983). A partly genetic predisposition for acquiring *strong* persistent fear/avoidance responses would be maladaptive, however, because it would strongly inhibit exploiting the refuge's advantages (Appleton 1975; Chapter 4 in this volume) and the food opportunities associated with many enclosed settings. These arguments are broadly consistent with findings from many studies of liking or aesthetic preferences for natural landscapes that indicate people in Western and Eastern societies rather consistently dislike spatially restricted environments but respond with greater liking to settings having moderate to high visual depth or openness (Ulrich 1983, 1986a). Arguably, most people at one time or another have experienced a scare or other aversive unconditioned stimulus in an enclosed setting that might condition a prepared disposition to respond subsequently with moderate dislike and wariness to spatially restricted situations.

In the discussion of biophilia in later sections, several arguments will be advanced to explain why people should respond positively to spatially open, savanna-like landscapes. (See also Chapter 4 in this volume.) One critical advantage of open savannas during evolution presumably was the comparatively low risk associated, for instance, with lower probabilities of encountering close hidden predatory threats. Whereas threat-related episodes may have been relatively infrequent on the savanna, an evolutionary perspective implies the hypothesis that in the event an early human did experience a close call, injury, or other traumatic experience (a child lost or separated from its parent, for example) on the open savanna, that person should acquire an "unforgettable" adaptive response disposition of cautiousness and reduced liking. A variation of this argument suggested by Klein (1981) contends that it would be adaptive for infants or small children to associate fear or alarm with being in the open separated from their mothers, where they would be vulnerable to predators and other dangers.

To evaluate the hypothesis that biologically prepared learning may play a role in responses to gross depth/spaciousness properties of natural settings, Ulrich, Dimberg, and Öhman (1993) performed an aversive classical conditioning experiment using a laboratory environmental depth simulator. Through a hooded viewing port, people observed slides of either low-depth (1 meter) or high-depth (more than 100 meters) natural scenes displayed on a large back-projection screen positioned only 1 meter in front of their eyes. All scenes were dominated by green vegetation, and the low-depth and high-depth exemplars were equivalent in complexity and brightness. The subjects' arms rested comfortably on a flat surface that extended forward to the screen, so that their fingertips were positioned near the lower part of the screen but were not visible. Electrodes affixed to the fingertips administered a bite-like shock as the unconditioned stimulus.

The findings suggest that autonomic defense/aversive responses conditioned to the high-depth settings were significantly more resistant to forgetting than for the low-depth settings. For unreinforced (no shock) presentations, affective self-ratings were significantly more positive for high-depth than for low-depth scenes. Following administration of the shock, there were pronounced negative shifts in affective responses to both spatial categories, although ratings remained more positive for high-depth than low-depth environments. The results provide some support for the preliminary conclusion that humans may be biologically prepared to acquire a persistent posture of defense/cautiousness but not strong fear/ avoidance following an aversive experience in a high-depth, spatially open, natural environment. The findings appear relevant to our understanding of agoraphobia (interpreted narrowly here as fear of open spaces), which is among the most debilitating and costly of phobias.

Twin Studies and Behavior-Genetic Approaches

In addition to conditioning studies, research on human twins, including studies using behavior-genetic approaches, has yielded convincing evidence that biophobia has a partly genetic basis. As a general context for these findings, it is relevant to mention that during the last decade twin studies have suggested that genetic factors play a major role in a wide range of human characteristics and traits, including obesity, personality, and physiological (autonomic) reactivity. As one example, a review of several twin studies (Loehlin et al. 1988) concluded that genetics may account for about half the variation in neuroticism.

Regarding biophobia, an initial wave of family history and twin studies produced a pattern of evidence suggesting strongly that some fears or phobias are familial and partly genetic in origin (Rose et al. 1981; Moran and Andrews 1985; Fyer et al. 1990). But most of these studies did not use sufficiently advanced statistical methods, or large enough samples, to permit either elucidation of the relative contributions of genetic and environmental risk factors for different types of phobias or an evaluation of preparedness theory. Recently these limitations have been overcome in a few studies using large twin samples and advanced multivariate statistical methods. Before discussing these findings and their implications for biophobia, a brief description of behavior-genetic methods is in order.

Behavior-genetic research typically focuses on a response, trait, or variable of interest (such as snake phobia, agoraphobia, emotionality, personality trait) that is characterized by considerable variability among persons (Gabbay 1992). And if the study is to enable one to make detailed inferences, including insights into causal relationships, a second requirement is a large sample of persons (several hundred to a few thousand) for whom genetic similarity/dissimilarity between individuals or pairs can be determined. The sample typically consists of monozygotic and dizygotic twins, but it is also possible to use a very large sample that includes family members or relatives. If these requirements are met, a data collection phase is undertaken. This process usually involves personal interviews to gain information regarding the presence/absence and magnitude of the variable in question, as well as collecting other types of information such as the person's age at onset of a phobia. Finally, the data are analyzed using multivariate genetic statistical models (such as LISREL) that identify and may allow causal insights concerning the contribution of genetic effects, familial (common) environmental factors, and individual-specific environmental effects (Heath et al. 1989; Neale et al. 1989).

Using the approach outlined here, Kendler and his associates (1992) studied the genetic epidemiology of different types of phobias in a sample of 2,163 American female twins. One-third (33.4 percent) of the individuals gave a lifetime personal history that indicated some type of phobia. Findings from different multivariate models converged generally to indicate that genetic factors play a major role in animal phobias (fears of snakes, spiders, bugs, or bats) and in agoraphobia: estimates of heritability ranged from 30 percent for animal phobias to 40 percent for agoraphobia. Kendler and his associates concluded that the results strongly support an interpretation of inherited biophobia "proneness," because the familial clustering of any type of phobia stemmed in large part from genetic factors but not from familial or common environmental factors (that is, not from environmental factors such as a common home situation experienced by both twins in a pair).

Importantly, the results were consistent with biologically prepared learning theory because "individual-specific" aversive or traumatic experiences appear to play a critical role in the triggering of phobias (Kendler et al. 1992). The traumatic events were highly specific for animal phobias—for example, an aversive experience with a snake would be linked with snake phobia. By comparison, the pathogenic experiences for agoraphobia were less specific. The latter finding is perhaps not surprising from an evolutionary perspective, because aversive or traumatic experiences in the open savanna might stem from widely different events or threats, ranging from predator attacks to being lost as a child.

Finally, the findings confirmed results from previous studies (such as Marks 1969) reporting that the onset ages for different phobias are quite different. For snake, spider, and other animal phobias, onset typically occurred during childhood (about 70 percent of phobias began by ten years of age). Age at onset was latest for agoraphobia (about 60 percent of onsets occurred between fifteen and thirty years of age). It is pertinent to mention that plausible functional-evolutionary explanations have been advanced to account for differences in the onset ages of different phobias. For example, Öhman, Dimberg, and Öst (1985) have argued that a biologically prepared readiness for early childhood onset of animal fears was adaptive for premodern humans because young children are especially vulnerable to snakes and other predators.

Summary of Biophobia Findings

To summarize briefly, findings from many laboratory conditioning experiments support the notion that humans are biologically prepared to acquire and especially to not "forget" adaptive biophobic (fear/avoidance) responses to certain natural stimuli and situations that presumably have presented survival-related risks throughout evolution. It appears that an important adaptive feature of biophobia is that comparatively unforgettable responses to certain fear-relevant but not fear-irrelevant natural stimuli can be acquired through vicarious conditioning or learning experiences. Moreover, recent findings suggest that processing of biologically prepared fear-relevant natural stimuli can be very fast and may often occur automatically or "unconsciously." Although the vast majority of conditioning studies have focused on fear-relevant predatory stimuli, there is limited evidence that humans might also be biologically prepared to acquire a persistent posture of defense/cautiousness following an aversive experience in a high-depth, spatially open environment such as a savanna.

Behavior-genetic studies and other research on human twins have produced convincing evidence that genetic factors play a major role in biophobia. As well, recent behavior-genetic findings support an inherited biophobia "proneness" interpretation and are clearly consistent with biologically prepared learning theory. Accordingly, the notion that biophobia is partly genetic and manifested in prepared learning has received support both from conditioning studies as well as behavior-genetic research. This convergence lends considerable credibility to a biologically prepared learning interpretation of biophobia. It is noteworthy that several investigators working in different fields have used entirely different research approaches (aversive conditioning and behavior-genetic methods) yet have reached broadly similar conclusions.

If humans have a partly genetic predisposition to biophobia—that is, to respond fearfully or aversively to certain living things and natural situations—should one begin to doubt the plausibility of the *biophilia* hypothesis? In fact, the implications of the biophobia research are quite the opposite. If the many years of sophisticated scientific inquiry into biophobia had instead produced no evidence of a genetic role in negative responses to stimuli that were critical survival-related dangers throughout evolution, it would be implausible now to postulate a partly genetic predisposition for *positive* responsiveness to advantageous natural stimuli. The findings suggesting a robust genetic role in biophobia imply tenability and even optimism for the biophilia hypothesis, which now becomes our focus.

Biophilia and Natural Landscapes

As described earlier, this discussion of biophilia concentrates on natural physical environments rather than animals. At the outset it should be reiterated that theory and research on biophilia are less developed than for biophobia. Approaches such as behavior-genetic methods have not yet been tried that might yield direct evidence regarding the possible role of genetic factors. Moreover, in sharp contrast to the large body of conditioning studies relating to biological preparedness and biophobia, there is a lack of research that has directly tested prepared learning theory with respect to positive or biophilic responses to nature. The latter deficiency is partly due to the fact that *positive* Pavlovian conditioning studies are usually more difficult to perform than aversive conditioning experiments, because producing an immediate, strongly positive, unconditioned stimulus in the laboratory is more problematic than providing an immediate negative unconditioned stimulus.

But there is a more fundamental reason for the lack of scientific research on biophilia: psychology, with its impressive repertoire of theories and scientific research methods, has shown little interest in studying human transactions with natural environments. In this regard, environmental psychology is a small and peripheral subfield within psychology. Disciplines that focus on natural settings, such as landscape architecture, are small and lack expertise in behavioral science methods and other research approaches that are necessary for rigorously investigating many aspects of the biophilia hypothesis. Additionally, no funding agency in the United States (with the limited exception of the USDA Forest Service) has assigned importance to funding or stimulating research on the possibly beneficial human effects associated with experiencing natural environments. These factors help to account for the current embryonic state of much biophilia-related research. Only limited beginnings have been made in investigating certain issues that might have considerable scientific and social significance, such as identifying the human benefits or values that could be lost when natural areas are eliminated.

These obstacles notwithstanding, scientific research on certain aspects of positive responding to natural environments has gradually expanded and improved over the last two decades, especially in the area of aesthetic preferences for landscapes. As will be evident from the empirical findings surveyed in later sections, certain consistent cross-cultural patterns in aesthetic preferences provide circumstantial support for the hypothesis that biophilia has a partly genetic basis. Without studies directly testing the genetic hypothesis and the preparedness notion (such as conditioning experiments), however, the conceptual propositions advanced here regarding biophilia are necessarily more speculative and general than those proposed for biophobia.

Broadly speaking, the conceptual arguments advanced here represent the second half of a "symmetric" biophobia/biophilia framework that assigns major importance to biologically prepared learning. Regarding biophilia, the basic proposition is that certain rewards or advantages associated with natural settings during evolution were so critical for survival as to favor the selection of individuals with a disposition to acquire, and then retain, various adaptive positive/approach responses to unthreatening natural configurations and elements. From this it follows that as a remnant of evolution, modern humans might have a biologically prepared readiness to learn and persistently retain certain positive responses to nature but reveal no such preparedness for urban or modern elements and configurations. This implies that one general approach for scientifically evaluating aspects of the biophilia hypothesis is to investigate possible differential characteristics of positive responsiveness to natural settings and stimuli in contrast to urban or modern environments and stimuli.

It is proposed that biologically prepared learning may play a role in at least three general adaptive positive (biophilic) responses to unthreatening natural landscapes: liking/approach responses; restoration or stress recovery responses; and enhanced high-order cognitive functioning when a person is engaged in a nonurgent task. Although there is considerable research relating to the first type of positive responsiveness—liking/approach—and a limited but growing amount relating to restoration, the presumed role of genetic factors in these two types of biophilic responding has not yet received direct empirical support. The third type of positive response—enhanced high-order cognitive performance—is proposed only tentatively because research to empirically evaluate this response is still in progress. There is no suggestion here that these three types of positive responding represent a comprehensive list of biophilic responses to natural environments. Research on biophilia is at an early stage, and future studies may well reveal other important types of positive responding.

Affect and Aesthetics

Liking/Approach Responses

It is suggested here that humans have a partly genetic predisposition to readily acquire and then persistently retain liking/attention/approach responses to natural elements and configurations that favored survival because they were associated with primary necessities such as food, water, and security. Orians (1980, 1986) has provided a convincing explanation why *savanna* environments, compared to other habitats such as rain forests and deserts, had major advantages for early humans from the standpoint of providing much more favorable combinations of these primary necessities. (See also Chapter 4 in this volume.) These points are consistent with considerable evidence indicating that much of human evolution took place in savannas. Open savannas were better suited than other habitats to early humans having upright posture, bipedal locomotion, and free-swinging arms. (For an overview of anthropological research on these issues see Lumsden and Wilson 1983.) Compared to the rain forest, savannas offered more abundant plant and animal food for ground-dwelling humans as well as lower risk because of visual openness, escape opportunities (Appleton 1975; Chapter 4), surveillance (Appleton 1975), and lower probabilities of encountering close hidden predatory threats. Put in terms of the earlier discussion of biophobia, most savannas in comparison to rain forests were characterized by lower levels of *biophobic* properties, including less spatial enclosure and fewer close encounters with snakes, spiders, and other fear-relevant stimuli. In view of these important survival-related advantages, Orians (1980), Appleton (1975), and others have proposed that modern humans retain a partly genetic predisposition to like or visually prefer natural settings having savanna-like or parklike properties such as spatial openness, scattered trees or small groupings of trees, and relatively uniform grassy ground surfaces. As suggested in the earlier discussion of biophobia, humans might also be biologically prepared to acquire a persistent posture of cautiousness (but not strong fear/avoidance or dislike) toward savannas in association with a presumably infrequent threat-related episode in a spatially open setting. Although such an episode might "unforgettably" temper a person's liking responses to open scenes, the person should still prefer open savanna-like environments to spatially restricted settings.

A functional-evolutionary perspective further implies that people should respond positively to natural settings having water and spatial openness. There is considerable evidence from excavations in East Africa that even early hominids often located their camps at the edge of water (Leakey 1980; Brown et al. 1985). The survival-related advantages would have included immediate availability of drinking water, security and defense advantages, attraction of animals that could be hunted, and in some locations (seacoast, estuary, salmon river) extremely high food productivity associated with fish, shellfish, and crustaceans. Coss and Moore (1990) have argued that the capacity to find drinking water has acted as a major source of selection during evolution. Accordingly both modern children and adults evidence strong preferences for scenes with water and are sensitive to certain optical properties of water in landscapes, especially glossiness.

A more general functional-evolutionary prediction is that certain broad classes of natural elements—especially water, green vegetation, and flowers—should be visually preferred over most modern synthetic elements such as glass and concrete (Ulrich 1983; Kaplan and Kaplan 1989). These natural elements should tend to elicit liking and attention because throughout evolution they have directly and indirectly signaled either the certain presence or the likelihood of finding two survival necessities: water and food. Regarding vegetation, preferences should tend to be higher for settings having green or somewhat verdant vegetation in contrast to vegetation having colors and forms characteristic of arid or desert environments where food and water would be harder to find.

Research Findings

During the last twenty years a large research literature, running to hundreds of studies internationally, has focused on affective responses to natural and urban landscapes. (For reviews or collections of articles see Zube, Brush, and Fabos 1975; Daniel and Vining 1983; Ulrich 1983, 1986a; Smardon 1988; Kaplan and Kaplan 1989; Nasar 1988; Ribe 1989.) Nearly all these studies have obtained data using affective or emotion-laden rating scales;

among the most common have been preference (liking), pleasantness, and scenic beauty. Findings obtained from these and other verbal scales usually are highly correlated (see Zube, Pitt, and Anderson 1975).

The vast majority of studies have exposed people to *simulations* of landscape scenes, usually color slides or photographs, rather than to real environments. Several studies have assessed the validity of using slides and other simulations by comparing on-site ratings of real scenes with responses to simulations of the same settings. Whereas most studies have found that on-site ratings of settings that are static correlate highly with ratings of color slides, the validity issue has not been fully resolved. (See, for example, Taylor, Zube, and Sell 1987; Hull and Stewart 1992.)

In accord with the conceptual arguments outlined above, several studies of liking/preference for diverse samples of *unspectacular* natural scenes have found that European, North American, and Asian adult groups consistently respond with high liking to environments that are savanna-like or parklike in appearance. (See Rabinowitz and Coughlin 1970; Ulrich 1977, 1983; Ruiz and Bernáldez 1982; Hultman 1983; Yi 1992.) Such savanna-like views, which can be located in rural areas, urban fringe locations, or even cities (as in parts of New York City's Central Park), are typically characterized by moderate to high depth or openness, relatively smooth or uniform-length grassy vegetation or ground surfaces, and scattered trees or small groupings of trees. These findings are paralleled by results from a significant research literature that has focused specifically on liking/preference responses to forest landscapes. (For a review see Ribe 1989.) Many studies carried out in different countries have clearly indicated that observers prefer forest settings having some similarities to savanna-like or parklike settings, including visual openness and uniform ground cover associated with large-diameter mature trees and relatively small amounts of slash and downed wood. (See, for example, Daniel and Boster 1976; Arthur 1977; Patey and Evans 1979; Savolainen and Kellomäki 1984.)

Another reliable finding consonant with the earlier conceptual arguments is that natural settings with water features elicit especially high levels of liking or preference. (See Shafer, Hamilton, and Schmidt 1969; Brush and Shafer 1975; Civco 1979; Penning-Rowsell 1979; Bernáldez, Abelló, and Gallardo 1989; Chokor and Mene 1992.) It should be noted that young children show strong positive responses to water (Zube, Pitt, and Evans 1983). There is considerable empirical evidence to support the conclusion of Zube and his colleagues that water is a dominant element of the visual landscape that nearly always enhances positive responding (Zube, Pitt, and Anderson 1975)—unless the water configuration involves risk (a stormy sea) or contains clearly visible pollution (Ulrich 1983; Lang and Greenwald 1987).

By contrast, some of the properties consistently associated with *low* preference for natural settings include sharply restricted depth as well as disordered high complexity and rough ground textures that obstruct movement (such as a forest setting with small, closely spaced trees, large amounts of slash or downed wood, and a visually impenetrable dense understory). In this regard, Chokor and Mene (1992) found that diverse groups of urban and rural dwellers in Nigeria accorded lower preference to a spatially restricted view of dense tropical rain forest than to more spatially open rain forest scenes. In the case of temperate or northern forest environments, clear-cut areas are very much less preferred by North American and European groups than large-diameter tree stands affording some visual openness. (See, for example, Rutherford and Shafer 1969; Daniel and Boster 1976; Echelberger 1979; Hultman 1983; Savolainen and Kellomäki 1984; Ribe 1989.) Likewise, Chokor and Mene's Nigerian research (1992) found that people responded with low preference to a natural landscape with large areas of vegetation destroyed by oil exploration activities. Another characteristic that can sharply reduce liking in natural physical environments is the presence of a judged threat or risk (Ulrich 1983). But certain people such as sensation-seekers (Zuckerman, Ulrich, and McLaughlin 1993), including many young males (Bernáldez, Abelló, and Gallardo 1989), may respond somewhat positively or with less dislike to risk-evoking properties or to characteristics appraised as challenging (abrupt terrain, turbulent water).

Extent of Agreement on Liking / Preference

Scientific studies on landscape preferences began to appear about 1970 in various social science and design disciplines that have traditionally emphasized learning and culture as preeminent determinants of human preference, thought, and behavior. Because learning was assumed to be the key mechanism shaping responses to landscapes, it was widely anticipated that studies would reveal great differences between groups and individuals in preferences for natural landscapes as a function of such variables as rural versus urban background and especially culture. Although some studies have reported statistically significant variations as a function of variables such as age (Zube, Pitt, and Evans 1983), ethnicity, and the sensation-seeking personality trait (Zuckerman, Ulrich, and McLaughlin 1993), these differences usually are small compared to the percentage of variance associated with group similarities, which in turn can be related to physical properties of the landscapes. Accordingly, on balance, the pattern of findings that has emerged over the last two decades runs directly counter to the initial expectation of wide differences as a function of learning or experience-related variables. The overarching conclusion supported by this large body of research is that similarities in responses to natural scenes usually far outweigh the differences across individuals, groups, and diverse European, North American, and Asian cultures. (See, for example, Shafer and Tooby 1973; Daniel and Boster 1976; Ulrich 1977; Wellman and Buhyoff 1980; Hull and Revell 1989.)

As an example, a recent study by Yi (1992) investigated the roles of cultural and occupational differences in influencing the natural landscape preferences of diverse groups of South Koreans and Texans, including farmers, ranchers, and nonfarmer urban groups. Individuals were shown a collection of color photographs depicting diverse natural settings in Korea and Texas. The collection included several scenes from Korea and Texas that contained features having strongly positive associations for one of the cultures but not the other—for example, a Korean landscape with a distinctive mountain known to Koreans but not Texans as the site of a famous Buddhist temple. Despite stacking the deck in this manner in favor of cultural influences, Yi's results reveal high agreement among all groups in their aesthetic preferences. Differences attributable to culture and occupation were statistically significant but comparatively minor, accounting for little of the variance. It should be mentioned that the groups were similar in according especially high preference to landscapes having water features or savanna-like characteristics.

A few studies have compared diverse groups or cultures with respect to

Biophilia, Biophobia, and Natural Landscapes

preferences for *living* in diverse natural landscapes depicted by slides or photographs. Compared to affect-saturated aesthetic preferences, ratings of living preferences probably involve more deliberate cognition or evaluation and, accordingly, should be more strongly influenced by personal experience and other learning. Not surprisingly, some of this research has found comparatively wide variations between different cultures in preferred environments for living (Sonnenfeld 1967). In a study of living preferences that is particularly relevant to the biophilia hypothesis, Balling and Falk (1982) displayed color photographs of diverse natural biomes to American groups that ranged widely in age from young children to adults. Their findings indicate that the most preferred environment for the youngest children in the study (ages seven and eight) was the savanna.

Aesthetic Preferences for Natural vs. Urban Scenes

Findings from scores of studies on preferences for natural versus urban scenes have provided a pattern of circumstantial yet persuasive support for the genetic aspect of the biophilia hypothesis. A clear-cut finding in this research is a strong tendency for diverse European, North American, and Asian groups to prefer natural landscape scenes over urban or built views, especially when the latter lack natural content such as vegetation and water. (See, for example, Kaplan, Kaplan, and Wendt 1972; Zube, Pitt, and Anderson 1975; Wohlwill 1976, 1983; Bernáldez and Parra 1979; Ulrich and Addoms 1981; Hull and Revell 1989.) Even mediocre natural scenes consistently receive higher ratings than do all but a very small percentage of unblighted built settings lacking nature. Levels of preference for natural settings are usually so much higher than for urban views that the distributions of scores for the two domains hardly overlap (Kaplan, Kaplan, and Wendt 1972; Ulrich 1983). This pattern is evident even in a study that compared aesthetic liking for undistinguished natural scenes and comparatively attractive Scandinavian townscapes (Ulrich 1981). Similarly, a study by Chokor and Mene (1992) of landscape preferences in a developing nation—Nigeria—found that diverse groups of urban and rural dwellers responded with higher preference to natural scenes than to various urban scenes lacking nature. In the same study, however, a suburban scene largely dominated by natural content (large trees, flowerbeds, verdant plants, landscaping) but containing upper-income residences outscored certain comparatively wild landscapes such as a view of a dense rain forest (Chokor and Mene 1992).

Several studies have found that the introduction of certain artificial elements into natural landscapes has strongly detrimental effects on visual preference—for example, electrical transmission towers and power lines, large advertising signs or billboards, and prominent concrete or asphalt road surfaces (see Clamp 1976; Hull and Bishop 1988). Further, a person's aesthetic preferences for natural landscapes can be strongly and negatively affected by urban air pollution that discolors the atmosphere in the natural area, degrades visual detail, reduces visual range, and in some instances kills or stunts vegetation, thereby altering the landscape's appearance and perhaps reducing biodiversity (Latimer, Hogo, and Daniel 1981).

This general area of research has also provided strong evidence that people respond in fundamentally different ways to natural versus built contents and settings, irrespective of other visual properties such as depth and complexity. Many studies employing multivariate procedures such as multidimensional scaling or factor analysis have suggested that natural versus built features have a central role in influencing perception and categorization of outdoor environments. (See Kaplan, Kaplan, and Wendt 1972; Ward 1977; Wohlwill 1983.) These methods have shown that for groups studied in different countries, natural versus built groupings of landscape scenes typically emerge as prominent dimensions when affective ratings are obtained for diverse samples of views. This research has also shed light on the visual configurations and elements that people respond to as "natural." In this regard, the "natural" domain appears to be broad for people in industrialized societies, extending considerably beyond wilderness to include many obviously human-made settings such as pastures, fields planted in cereal crops, wooded parks, and even golf courses. In very general terms, European, North American, and Japanese adult groups tend to respond to scenes as natural if the landscape is predominantly vegetation, water, and mountains, if artificial features such as buildings, automobiles, and advertising signs are absent or inconspicuous, and if the dominant visual contours or edges are curvilinear or irregular rather than starkly rectilinear or regular. (See Ródenas, Sancho-Royo, and Bernáldez 1975; Ulrich 1983, 1986a; Wohlwill 1983.)

Biophilia, Biophobia, and Natural Landscapes

Several studies performed in the United States, Europe, Japan, and to a limited extent Africa have compared aesthetic preferences for urban scenes with and without natural elements. Most of this work has focused on the preference effects of trees and other vegetation in urban or built environments. In general, findings have shown that people usually accord higher liking to urban scenes containing nature than to similar urban settings lacking nature. (For surveys of studies see Ulrich 1986a; Smardon 1988; Schroeder 1989.) For instance, the presence of trees and associated vegetation substantially increases liking for such urban settings as residential areas and streets, commercial streets and strips, and parking lots. (See, for example, Nasar 1983; Schroeder and Cannon 1983; Anderson and Schroeder 1983; Asakawa 1984; Lambe and Smardon 1986; Sheets and Manzer 1991; Chokor and Mene 1992.) Several investigators have found that urban parks having savanna-like characteristics are especially preferred visual amenities in urban areas. (See Ulrich and Addoms 1981; Herzog, Kaplan, and Kaplan 1982.) Likewise, Kennedy (1989) reports that even long-term residents of the desert city of Tucson accorded high preference to scenes with separated trees, smooth grassy ground textures, and other savanna-like qualities. In the same study, Kennedy found that Tucson residents accorded higher preference to settings containing green, comparatively verdant vegetation than to scenes with cacti and brownish desert vegetation. In certain urban areas where crime is a major problem, learned fear/risk associations intensify negative responding to settings having dense foreground vegetation that blocks surveillance (Schroeder and Anderson 1984; Hull and Harvey 1989).

Implications for the Biophilia Hypothesis

The large research literature on aesthetic preferences for landscapes offers strong direct support for part of the biophilia hypothesis in the sense that diverse groups and cultures have been found to respond positively to unthreatening natural settings. Moreover, the main findings in this area are broadly compatible with the proposition that biophilia may have a partly genetic basis, although the support is indirect or circumstantial. Consistent with an adaptive-evolutionary perspective, the findings reveal a convincing pattern of positive responsiveness to certain natural settings (savanna-like environments, settings with water) that presumably offered major survival advantages for early humans from the standpoint of providing favorable combinations of such primary necessities as food, water, and security. A few studies have converged in reporting that high preferences for savanna-like scenes and water are evident in young children. Moreover, diverse groups and cultures reflect agreement in responding with lower liking to natural environments that would have been less favorable during evolution because of lower availability of food and water as well as greater exposure to *biophobic* or risk-relevant properties. There are other salient findings consistent with aspects of the biophilia hypothesis: the strong tendency across different groups and cultures to prefer unspectacular or even mediocre natural scenes over urban settings lacking nature; the pattern for certain broad classes of natural content (vegetation, water) to elicit more positive responses than artificial contents; and the central role that natural versus human-made content appears to play in perceptual categorization of physical environments.

Despite the convincing pattern of evidence indicating positive responsiveness to certain natural settings that would have favored survival during evolution, this research offers at best indirect support for the notion that biophilia may have a partly genetic basis. Some caution in interpreting the findings is warranted in part because of limited data for non-Western cultures. Moreover, the methods used in landscape preference research are inadequate for clarifying the possible role of genetic factors. In this regard, alternative explanations for the findings cannot be convincingly ruled out—such as the argument that learning experiences might be much more similar across different groups and cultures than the anthropological literature suggests. Nonetheless, in the face of steadily mounting evidence that there may be considerable correspondence across Western and some non-Western cultures in terms of positive aesthetic responsiveness to natural landscapes, cultural and other learning-based perspectives increasingly show clear weaknesses. Arguably, the overall record of findings on landscape preferences is more plausibly reconcilable with a conceptual perspective that encompasses both learned and genetic influences than with an exclusively learning-based explanation.

Biophilia, Biophobia, and Natural Landscapes

Restorative Responses

Daily living for early humans was no extended picnic in a serene pastoral environment. It involved fatiguing and often demanding activities to obtain the necessities for survival and sometimes involved stressful encounters with threats. As the earlier discussion of biophobia emphasized, predators were a critical threat. Other risks and stressful episodes were linked directly or indirectly to the physical environment (encountering a violent storm, getting lost, not finding water). Additional stressful situations may have arisen frequently because of aggressive encounters with other humans related to establishing dominance hierarchies and social order (Öhman 1986) and in some instances because of hostile conflicts with other human groups (Chagnon 1977).

Functional-Evolutionary Perspectives

These comments imply that acquiring a capacity for restorative responding to certain natural settings had major advantages for early humans including, for instance, fostering the recharge of physical energy, rapid attenuation of stress responses following an encounter with a dangerous threat, and perhaps rapid reduction of aggression following antagonistic contacts with other humans (Ulrich et al. 1991). In this perspective, one key function of the restorative category of biophilic responses can be characterized as compensatory. That is, a capacity for restorative responding would enhance survival chances in part because of its role in promoting recovery from fatigue and other deleterious effects stemming from behaving adaptively in a previous demanding situation. In this view, restorative responding would have allowed early humans to regain the capacity to respond effectively in a subsequent situation.

To illustrate this line of reasoning, consider the somewhat extreme example of a savanna-dwelling early human who encounters a dangerous carnivore, is pursued by the animal, and escapes by climbing a tree. Following the theoretical arguments of a framework set out in detail elsewhere (Ulrich 1983; Ulrich et al. 1991), it is proposed that a critical part of the individual's initial level of responding would be a quick-onset emotional reaction comprising fear and interest/attention. This rapid-onset emotional reaction would have major influences on subsequent conscious processing and play a central role in initiating adaptive mobilization in physiological systems (such as the autonomic, neuroendocrine, skeletomuscular) and in very quickly motivating appropriate avoidance or flight behavior (Ulrich et al. 1991). The immediate benefit to the individual would be great (survival), but there would also be costs evident in, among other response modes, a strongly negative emotional state and energy-draining physiological mobilization or arousal. If the predator left the area and the threat dissipated, the theory suggests that the adaptive compensatory need would be for restoration—a "breather" from stress (Ulrich 1983; Ulrich et al. 1991). In this example, the several benefits of restoration would include, for instance, a shift toward a more positively toned emotional state, mitigation of deleterious effects of physiological mobilization (reduced blood pressure, lower levels of circulating stress hormones), and the recharging of energy expended in the physiological arousal and behavior. This recharging of energy could in turn be important, for example, in sustaining activities to exploit food, water, or other advantages of the area. Further, because responding to fatiguing challenges or stressors is sometimes accompanied by persistent declines in cognitive functioning or performance (Glass and Singer 1972; Hockey 1983), restoration could be evident in gains in cognitive performance (Kaplan and Kaplan 1989; Hartig, Mang, and Evans 1991).

Among the components of an adaptive constellation of restorative responses in many contexts should be attention/interest to the natural setting accompanied by liking or increased levels of positive affects, reduced levels of negatively toned feelings such as fear and anger, and reductions in physiological arousal (such as sympathetic nervous system activity) from high levels to more moderate ranges (Ulrich 1979, 1981, 1983). Following a functional-evolutionary perspective, it can also be predicted that such restoration should occur fairly rapidly—often within minutes rather than several hours, depending on the intensity and duration of the demanding situation and associated stress response (Ulrich et al. 1991). Because of the major advantages for early humans of restorative responding, it is proposed that modern humans might have a biologically prepared readiness to acquire, and then retain, restorative responses to many unthreatening nat-

Biophilia, Biophobia, and Natural Landscapes

ural settings but reveal no such prepared responsiveness for most urban or built elements and configurations.

Although most natural scenes should tend to be more effective than most built settings or modern stimuli in fostering restoration, certain natural settings should not be relaxing. A setting that contains a biophobic stimulus such as a snake, for instance, should elicit fear/avoidance responding that is stressful rather than restorative. On the other hand, properties linked with security or low risk should be characteristic of natural visual settings that are effective in producing restoration. Some of these security-related properties may include spatial openness, calm or slowly moving water, and conceivably a small contained fire. (It is perhaps justified to speculate that the safety and other advantages of a campfire for early humans were sufficiently critical to favor individuals with a predisposition to respond to such stimuli with attention, liking/approach, and restoration.) Along with security, perhaps restoration would be fostered by settings associated with comparatively high availability of food and water (savanna-like settings, for example). It seems warranted to speculate that for early humans this property might have contributed to restoration in part because there would be less anxiety about what would be eaten tomorrow.

Stress-Reducing Effects of Outdoor Recreation

Although stress can be defined in a number of ways, here it is construed as the process by which a person responds psychologically, physiologically, and often with behavior, to a situation that is demanding or threatens well-being (Evans and Cohen 1987). Whereas certain short-term, mildly stressful situations can sometimes improve human performance (Hockey 1983), stress is interpreted here mainly as a negative condition that should be mitigated over time to prevent detrimental effects on psychological well-being, performance, and health. Although the terms *stress recovery* and *restoration* are used interchangeably, restoration is a broader concept that is not limited to stress recovery situations or to recovery from excessive physiological arousal and negatively toned emotional "excitement" (anxiety), but could also refer to recuperation from understimulation or prolonged boredom (Ulrich 1981, 1983; Ulrich et al. 1991).

A large body of research on recreational experiences has shown convincingly that leisure activities in natural settings are important for helping people cope with stress as well as in meeting other needs unrelated to stress. Findings from more than 100 studies of recreationists in wilderness areas have shown that stress reduction consistently emerges as one of the key perceived benefits. (For surveys see Knopf 1987; Ulrich, Dimberg, and Driver 1991.) Similarly, most studies on urban parks and other urban natural settings have found that restoration from stress is a key perceived benefit (Ulrich and Addoms 1981; R. Kaplan 1983; Schroeder 1989). Although this research offers strong evidence that outdoor recreation fosters restoration, the role of the natural environment per se has not been clarified. Recreation experiences are usually complex, and it is difficult to disentangle the stress-mitigating contribution of the natural environment from the effects of other mechanisms such as physical exercise. Moreover, the extent to which restorative effects may hold across diverse groups of people is clouded in many studies because of the use of self-selected samples of recreationists.

Nonetheless, there are indications in some recreation studies that part of the restoration benefit stems from exposure to natural surroundings (Ulrich and Parsons 1992). These findings tend to confirm the earlier conceptual argument that certain natural configurations and elements are more effective than others in eliciting restoration. Specifically, recreationists tend to report such states as "relaxation" and "peacefulness" in association with exposure to settings having savanna-like properties or a water feature. In this regard, a few park studies have found significant associations between stress mitigation ratings and questionnaire items relating to an area's savanna-like appearance and natural content—scattered trees, grass, open space (Ulrich and Addoms 1981). Schroeder (1986) has found that the most common feelings reported by visitors to the Morton Arboretum near Chicago were tranquillity or serenity, feelings most often linked to experiences with areas having openness, lush vegetation, and large trees. A Swedish study of a wide variety of park types found that diverse groups of users responded to savanna-like settings, including those with water, as "peaceful" (Grahn 1991).

Hartig and his associates have reported the restorative effects of experiencing a parklike nature area while controlling for certain stress-reducing variables such as physical exercise (Hartig, Mang, and Evans 1991). They

Biophilia, Biophobia, and Natural Landscapes

first produced stress in individuals with a demanding cognitive task and then measured recovery effects of either (1) a forty-minute walk in an urban fringe nature area dominated by trees and other vegetation, (2) walking for an equivalent period in a comparatively attractive, safe urban area, or (3) reading magazines or listening to music for forty minutes. Their findings suggest that people randomly assigned to the nature walk reported more positively toned emotional states than those assigned to the other two conditions—and performed better on a cognitive task (proofreading).

Using quite a different approach, Francis and Cooper-Marcus (1991) studied a sample of university students living in the San Francisco area to ascertain the settings they sought out when they felt stressed or depressed. A considerable majority of them (75 percent) cited outdoor places that were either natural environments or urban settings dominated by natural elements or configurations (such as wooded urban parks, places offering scenic views of a natural landscape, locations at the edge of water such as lakes or the ocean).

Restorative Effects of Viewing Natural Settings

Direct evidence of the restorative influence of natural settings has come from a few studies that have analyzed the effects on stressed individuals of viewing different outdoor scenes. Consistent with the functional-evolutionary perspective outlined earlier, these findings suggest that viewing unthreatening natural landscapes tends to promote faster and more complete restoration from stress than does viewing unblighted urban or built environments lacking nature. Some of this research has controlled for the possible effects of relatively content-independent visual properties of outdoor settings (complexity, information rate, depth), and accordingly it appears that differences in natural versus human-made contents play a role in the differential restoration influences reported. In general, this small but growing area of research suggests that biophilic responding to natural landscapes extends far beyond aesthetic preference or liking to include broadly positive shifts in emotional states and positive changes in activity levels in physiological systems.

One early restoration study focused on groups of American university students who were experiencing mild stress because of a challenging final exam (Ulrich 1979). A self-rating questionnaire was used to assess stress recovery associated with either viewing a diverse sample of color slides of rather ordinary rural natural settings dominated by green vegetation (no savanna-like settings were included) or exposure to unblighted urban views lacking natural elements such as trees and water. The findings suggest that the undistinguished natural views held the subjects' attention more effectively and fostered greater psychological restoration—indicated by greater reductions in negative feelings such as fear and anger/aggression and much greater increases in positive affects. Honeyman (1992) has replicated this study with the addition of a third recovery condition consisting of urban scenes containing prominent vegetation. Her results suggest that these urban settings produced more recovery than the urban scenes lacking nature. In a study performed in Sweden using *unstressed* university students (Ulrich 1981), self-rating data similarly suggest that everyday natural scenes held the subjects' attention more effectively through a lengthy viewing session and produced more positively toned emotional states than did Scandinavian townscapes lacking nature. Importantly, these self-rating findings were broadly convergent with results obtained in the same study by recording brain electrical activity (EEG) in the alpha frequency range. These electrocortical data suggest that people were more wakefully relaxed during exposure to the natural landscapes. Sheets and Manzer (1991) also used unstressed subjects in an investigation of emotional responses and cognitive appraisals of American urban street scenes with and without prominent trees and other landscaping. Their results suggest that the presence of vegetation subtantially and positively changed responses to street views and that higher levels of positively toned feelings were reported during exposure to scenes with vegetation. Herzog and Bosley (1992) found that a group of unstressed American students' affective appraisals of landscapes for "tranquillity" were highest for scenes depicting large bodies of water with relatively calm surfaces. (Savanna-like scenes were not evaluated in the study.)

In a study that used a number of measurement techniques for assessing the stress-reducing effects of experiencing natural versus urban environments, 120 persons were first shown a stressful movie and then randomly assigned to a recovery condition that consisted of viewing one of six different color/sound videotapes of natural settings or urban environments lacking nature (Ulrich et al. 1991). Data concerning stress recovery during the environmental presentations were obtained from self-ratings of affective states and four physiological measures: heart rate, skin conductance, muscle tension (frontalis), and pulse transit time (a noninvasive measure that correlates highly with systolic blood pressure). Findings from all measures, verbal and physiological, converged in indicating that recuperation from stress was much faster and more thorough when people were exposed to the natural settings (a grassy, parklike landscape and a setting with a prominent water feature). Regarding the self-rating findings, people exposed to the natural settings, in contrast to the urban environments, had much higher levels of positive feelings and lower levels of anger/aggression and fear. In the case of the physiological indicators, greater recovery influences of the natural settings were suggested by lower levels of skin conductance fluctuations, lower blood pressure (longer pulse transit times), and greater reductions in muscle tension. It is noteworthy that there were directionally different cardiac responses to the natural versus urban settings, suggesting that perceptual intake/attention was higher during the exposures to nature. The overall pattern of physiological findings raises the possibility that responses to the natural settings may have had a salient parasympathetic nervous system component. Parasympathetically dominated responding is associated with sustained yet nontaxing perceptual sensitivity with respect to the external environment, as well as restoration or maintenance of bodily resources (Lacey and Lacey 1970). There was no evidence of pronounced parasympathetic involvement during the recovery phase presentations of urban environments. Another finding warranting attention is the rapidity with which restoration occurred during the exposures to nature. After only four to six minutes of exposure to natural versus urban settings, significantly greater recovery was evident in all physiological measures.

The finding that even short-term visual contacts with unthreatening natural landscapes can promote stress recovery has also emerged from a few studies in which acutely stressed patients in health care settings were exposed for short periods, such as ten minutes, to views of nature. In a pilot study by Heerwagen and Orians on patient anxiety in a dental fears clinic (Heerwagen 1990), data that included heart rate measurements as well as affective self-ratings suggest that patients felt less stressed on days when a large mural depicting a spatially open natural landscape was hung on a wall of the waiting room in contrast to days when the wall was blank. Coss (1990) has studied the effects of displaying different ceiling-mounted pictures to acutely stressed patients who were lying on gurneys in a presurgical holding room. His findings suggest that after only three to six minutes, patients exposed to "serene" pictures (primarily displaying water or other aspects of nature) had systolic blood pressure levels that were 10 to 15 points lower than patients exposed to either a control condition of no picture or an aesthetically pleasing "exciting" outdoor scene (such as a sailboard rider leaning into the wind). In a study of patients who were about to undergo dental surgery, Katcher and his associates found that a short period of visual contemplation of a different configuration of nature content—an aquarium with fish—significantly reduced anxiety and discomfort and increased scores for patient compliance during surgery (Katcher, Segal, and Beck 1984).

A preliminary study of the effects of different types of wall art on psychiatric patients in a Swedish hospital has yielded additional insights concerning the positive influence of natural scenes (Ulrich 1986b). Short-term patients, some of whom were clinically anxious, were studied in a ward extensively decorated with paintings and prints reflecting a wide variety of styles and subject matter. Interview data suggest that patients responded positively to wall art dominated by natural content (a rural landscape, a vase of flowers) but tended to react negatively to abstract paintings and prints in which the content was either ambiguous or unintelligible. An analysis of records kept during a fifteen-year period yielded information regarding strongly negative patient responses and actions directed to paintings and prints. These actions included strong complaints to the staff and even physical attacks (such as tearing the picture from the wall and smashing the frame)—dramatic actions given that these patients were considered to be unaggressive and not at all prone to violence. (The ward was not locked.) Seven paintings and prints had been the targets of such attacks, and all of them showed a consistent pattern of abstract content. During the fifteen-year history of the ward, apparently no attack had been directed at a picture depicting nature.

Health-Related Effects of Viewing Natural Scenes

Some of the studies surveyed in the preceding section suggest that short-term exposure to unthreatening natural scenes can promote recovery from mild and even acute stress. Conceivably the restorative effects of natural views are often greatest when people experience high levels of stress or anxiety and are obliged to spend long periods in confined situations such as hospitals, prisons, and certain high-stress work environments (Ulrich 1979). In these and other settings, long-term or frequent views of unthreatening nature may have persistent positive effects on psychological, physiological, and even behavioral components of stress. In time, these effects may be manifested in higher levels of wellness.

In this regard, findings from a few studies of hospitals and prisons suggest that prolonged exposure to window views of nature can have important health-related influences. Prisons and especially hospitals provide some of the best opportunities for scientific research in real environments on the effects of viewing nature, because it is sometimes possible in these settings to control for other variables that influence wellness (such as exercise) and to use a variety of health-related data that are collected as a matter of routine. A study examined patients recovering from gall bladder surgery in a Pennsylvania hospital to evaluate whether assignment to a room with a window view of a natural setting might have therapeutic influences (Ulrich 1984). Recovery data were compared for pairs of patients who were closely matched for variables that could influence recovery such as age, sex, weight, tobacco use, and previous hospitalization. The patients were assigned essentially randomly to rooms that were identical except for window view: one member of each pair overlooked a small stand of deciduous trees; the other had a view of a brown brick wall. Patients with the natural window view had shorter postoperative hospital stays, had far fewer negative com-

Affect and Aesthetics

ments in nurses' notes ("patient is upset," "needs much encouragement"), and tended to have lower scores for minor postsurgical complications such as persistent headache or nausea requiring medication. Moreover, the wall-view patients required many more injections of potent painkillers, whereas the tree-view patients more frequently received weak oral analgesics such as acetaminophen. Somewhat similarly, findings from a questionnaire study of patients who were severely disabled by accidents or illness (and hence were presumably stressed) suggest that an especially highly preferred category of hospital window views included scenes dominated by natural content (Verderber 1986).

In a prison study, Moore (1982) examined the need for healthcare services by inmates whose cells looked out onto the prison yard versus those who had a view of nearby farmlands and forests. He reported that the inmates with natural views were less likely to report for sick call. Likewise, West (1985) found that cell window views of nature—compared to views of prison walls, buildings, or other prisoners in cells—were associated with lower frequencies of health-related stress symptoms such as headaches and digestive upsets.

In an extension of this line of research, Outi Lundén and I (1990) investigated whether exposure to visual stimulation in hospital intensive care units, including simulated natural views, promotes wellness with respect to the postoperative courses of open-heart surgery patients. At Uppsala University Hospital in Sweden, 166 patients who had undergone open-heart surgery involving a heart pump were randomly assigned to a visual stimulation condition consisting of a nature picture (either an open view with water or a moderately enclosed forest scene), an abstract picture dominated by either curvilinear or rectilinear forms, or a control condition consisting of either a white panel or no picture at all. Our findings suggest that the patients exposed to the open view of water experienced much less postoperative anxiety than the control groups and the groups exposed to the other types of pictures. The comparatively enclosed forest setting with shadowed areas did not reduce anxiety significantly compared to the control conditions. The rectilinear abstract picture was associated with *higher* anxiety than the control conditions. Future reports stemming from this research will present findings based on a wide variety of indicators of well-ness, both physiological (such as blood pressure) and behavioral (such as use of painkillers and postsurgical length of stay).

Implications for the Biophilia Hypothesis

Research on recreation experiences in wilderness and urban parks has yielded evidence consistent with the biophilia hypothesis by suggesting that restoration benefits stem at least partly from exposure to natural sur-roundings. These findings are compatible with earlier conceptual argu-ments in the sense that restoration tends to be linked especially to natural settings having savanna-like properties or nonturbulent water features. Convincing evidence of restoration has emerged from a smaller body of re-search that has analyzed the emotional and in some cases physiological ef-fects on stressed and unstressed persons of viewing outdoor environments. This area of research has shown that restorative responding to natural land-scapes involves a shift toward a more positively toned emotional state and may include positive changes in physiological activity. These changes tend to be accompanied by sustained attention or perceptual intake that may block or reduce stressful or worrisome thoughts. Findings from a few stud-ies suggest that in certain prolonged high-stress situations, restorative bio-philic responding may include important health-related influences.

Restorative or stress-reducing responses to natural scenes have a num-ber of characteristics consistent with the functional-evolutionary perspec-tive outlined earlier. These features, which presumably have been exceed-ingly adaptive during human evolution, include the quickness of recovery influences, effective reduction of negatively toned affects such as fear and aggression, reduction of taxing and deleterious sympathetic nervous sys-tem mobilization (such as reduced blood pressure), and the possibility of pronounced parasympathetic nervous system involvement that would be associated with the maintenance or recharging of energy. Another finding consistent with the functional-evolutionary perspective is that unthreaten-ing natural settings foster more complete and faster stress recovery than built or modern settings lacking nature. Moreover, the conceptual argu-ments are reconcilable with indications from a few studies that spatially open natural settings, including those with savanna-like properties or relatively calm water, are more effective in eliciting restorative responding than natural settings having properties associated with lower security, such as low depth.

These findings are also broadly consonant with the proposition that restorative responding to unthreatening natural landscapes might have a partly genetic basis. It should be mentioned that one influential theoretical perspective in psychology—arousal or stimulation theory (Berlyne 1971; Mehrabian and Russell 1974)—has been directly tested as an alternative explanation for the restorative effects of nature and found to have major shortcomings (Ulrich 1981; Ulrich et al. 1991). And on the basis of limited evidence from analysis of cardiac responses, it does not appear that restorative influences stem from an elaborated, active process of cognition entailing positive associations or memories with respect to natural environments (Ulrich et al. 1991). Additionally, certain cultural and other learning-based arguments—such as those which take into account presumed positive conditioning effects of media advertising for automobiles, retail stores, and other modern elements—are difficult to reconcile with the findings suggesting greater restorative effects of natural compared to urban settings. This area of biophilia research is still at an early stage of development, however, and findings are too sparse to cast serious doubt on certain other learning-based explanations. In this embryonic area there is a conspicuous lack of research on different cultures, diverse demographic groups, and young children. More fundamentally, no study has yet used a conditioning approach to test the prepared learning hypothesis, nor has any research employed behavior-genetic methods that would enable a comparatively direct evaluation of whether genetic factors play a role in restorative responding to nature.

Effects on High-Order Cognitive Functioning

It has been known for decades that the negative manifestations of stress can include reduced performance on cognitive tasks (Glass and Singer 1972; Hockey 1983). Most studies have investigated performance declines for "low-order" cognitive tasks such as proofreading that require narrowly focused attention to a restricted set of information and do not involve the integration of diverse information. As might be expected, this research has also shown that recovery from stress can be accompanied by gains in performance on low-order tasks. Concerning nature and cognitive performance, the influential nineteenth-century landscape architect and planner Frederick Law Olmsted wrote presciently about his intuitive conviction that viewing nature could produce stress recovery and lead to restored mental performance or recovery from mental fatigue (1865). Somewhat similarly, R. and S. Kaplan (1989; R. Kaplan and Talbot 1983) have speculated that exposure to nature can promote recovery from mental fatigue stemming from work situations involving prolonged, directed, effortful attention. A study by Hartig and his associates suggests that greater restoration from stress fostered by natural versus urban settings is manifested in performance gains for a low-order task, proofreading (Hartig, Mang, and Evans 1991).

Quite apart from performance gains related to recovery from stress and fatigue, it is proposed tentatively that because of critical survival-related advantages during evolution, humans might have a partly genetic predisposition for enhanced *high-order* cognitive functioning when engaged in nonurgent tasks (such as toolmaking) in certain natural settings. "Higher-order" cognitive functioning involves integrating diverse material or associating in a flexible way previously unrelated information or concepts. Higher-order functioning is required for forming remote associations and for creative problem solving. Effective associational functioning is widely considered to have a central role in creativity, as reflected in the associational theory of creativity (Mednick 1962). Put simply, this influential theory holds that creativity involves the novel combination of typically unrelated elements.

The notion that exposure to natural settings may enhance high-order cognitive functioning is proposed tentatively as a "candidate" category of biophilic responding. The previous discussions of restorative and liking / approach responding were able to draw on considerable empirical research. As research to evaluate the cognitive integration or creativity hypothesis is still in progress, however, this section will be largely speculative. Despite the lack of relevant empirical evidence, it seems warranted to dis-

Affect and Aesthetics

cuss the conceptual arguments because of their far-reaching implications in terms of possibly justifying substantially higher estimates of the value, both nonmonetary and monetary, of natural environments.

A Functional-Evolutionary Perspective

Many have emphasized that increases in cognitive resources, such as the capacity for long-term memory and language, have been critically advantageous during evolution and have played a central role in the rapid progress of humans and culture. (See, for example, Lumsden and Wilson 1983.) Arguably, along with these and other advances in cognitive capacity, an increased capacity for high-order cognitive functioning or creativity must have been critically important in the progress of humans and culture. An increased ability for flexibly integrating diverse information, for applying previously learned information in effective ways to new situations, and for creative problem solving must have played an absolutely central role in the crescendo of innovation that has driven much of human progress, especially during the last 50,000 years or so. Put differently, in the absence of a capacity for high-order cognitive functioning or creativity, other important advances such as long-term memory, capacity for language, brain specialization, and knowledge transmission through culture together seem inadequate for explaining how so much impressive innovation has occurred through human history: in toolmaking, weapons, hunting strategies, food storage and transport, plant domestication, and so on. Other advances account for the remembering, transmission, and often the application of innovations, but they neither explain creativity nor adequately account for the origin of innovations. Postulating an increased human capacity for high-order cognitive functioning quite plausibly accounts for much technological and other innovation.

Arguably, the advantages of high-order cognitive functioning during evolution have been so critical as to favor the selection of creative individuals. In a hunting and gathering band of early humans, one person capable of advanced creative problem solving and integrative thinking could increase the survival chances for the entire group. Groups with creative members would presumably advance more rapidly and have better survival chances than groups comprised of individuals whose cognitive functions were restricted to low to moderate-order tasks.

If natural selection has favored the development of high-order thinking capabilities in humans generally, what role might natural physical surroundings play both in fostering and perhaps hindering a person's performance on high-order tasks? To answer this question, it is necessary to digress briefly from these functional-evolutionary speculations and discuss cognitive science research showing that one's emotional state has a profound effect on virtually all aspects of thinking, including performance on higher and lower-order tasks. The basic underlying argument, proposed recently by a colleague of mine (L. G. Tassinary, pers. comm., March 1991), is that because certain natural settings have been found to elicit positive emotional states, exposure to such environments may facilitate creative problem solving or high-order cognitive functioning via their ability to alter one's emotional state.

Effects of Emotional States on Creative Problem Solving

A growing series of studies, notably by Isen and her colleagues, have demonstrated convincingly that positive versus negative emotional states have reliably different effects on the recall of information from memory and on creative problem solving in general. (For a survey see Isen 1990.) Positive emotional states readily cue a diverse range of information with positive associations as well as considerable "neutral" information or associations. Negatively toned feelings are much less effective in cuing the recall of information. Negative emotions such as sadness cue a comparatively small amount of negative—but not neutral—information that has little connection with other information (Isen 1985).

A second important finding regarding positive emotional states is that in addition to cuing the retrieval of much larger amounts of better-interconnected information, positive feelings in contrast to negative feelings facilitate remote associations, integration, perception of relatedness among different material, and creativity (Isen et al. 1985). Different studies have found that positive feelings significantly increase people's scores on tests of creativity and high-order functioning, whereas negative feelings *lower* performance (Isen 1990). Negative emotions restrict the focus of attention, impede the integration of information, and accordingly hinder creativity. Positive feelings do not, however, appear to increase performance on lower-order tasks that require a narrow focus of attention and rejection of associations. In some cases, positive emotions may actually reduce performance on lower-order tasks.

Nature, Positive Emotional States, and Creativity

As discussed in earlier sections, a growing number of studies have found that unthreatening natural environments are effective in eliciting broadly positive shifts in emotional states among unstressed as well as stressed individuals (Ulrich 1979, 1981; Hartig, Mang, and Evans 1991). These findings, in combination with the cognitive science research just surveyed, suggest the plausible hypothesis that exposure to unthreatening natural environments should facilitate creativity and high-order cognitive functioning in general.

Returning to one of the main functional-evolutionary arguments in the chapter, recall that during human evolution the environments that were preferred and restorative were those that offered abundant primary necessities such as food and water in combination with security. Arguably, such environments would probably also be those in which early humans experienced positive emotional states more frequently and those which allowed more opportunity for focused but nonurgent tasks where creativity was extremely advantageous (as in toolmaking). Perhaps it is not too great a speculation to propose that individuals would have been favored who experienced positive feelings and accordingly were significantly "smarter," in terms of creativity and high-order functioning in general, when they engaged in nonurgent tasks or activities in savanna-like environments, in open settings with nonturbulent water, and conceivably in proximity to a small fire. Perhaps modern humans, as a partly genetic remnant of evolution, tend to have more positive emotional states and accordingly are "smarter" in creative thinking when exposed to most unthreatening natural settings compared to most built environments lacking nature—especially when exposed to natural environments with the features just cited.

If these notions were to receive empirical support in future research, it would then be reasonable to investigate, for instance, whether researchers in high-tech firms might tend to be more creative and hence often more productive in workplaces having, say, extensive window views of a parklike (savanna-like) landscape or a view with a water feature. Likewise, it might be conjectured that these employees would tend to be smarter in creative problem solving on a Monday following a weekend trip in a natural environment that produced lingering positive feelings. Do molecular biologists working at the Salk Institute in La Jolla have better ideas if their windows overlook the Pacific, for instance, or if they take a morning walk on the beach before going to the laboratory? There is anecdotal information suggesting that several Nobel Prize–winning ideas may have occurred to researchers during walks or other contacts with nature. Perhaps a parklike or savanna-like college campus yields benefits that go beyond aesthetic preference and in some cases restoration from stress and fatigue to include the advantage of supporting positive emotional states that facilitate creativity. These speculations aside, I am currently involved in a research project directed by L. G. Tassinary that is investigating the basic hypothesis that exposure to unthreatening natural settings, including savanna-like environments, induces positive shifts in emotional states and accordingly increases scores on an associational test of creativity. Research is needed to determine whether nature is as effective as, or possibly more effective than, other types of positive stimulation or reinforcement (eating a tasty dessert, listening to certain works by Mozart) that have been shown in previous studies to elicit positive feelings and produce *short-term* increases in creativity (Isen 1990). If nature is shown to enhance creativity during short-term exposures, the central issue will then become whether nature scenes are more effective in sustaining higher creativity during experiences of long duration. It seems likely that the creativity enhancement effect fairly soon wears off, for instance, if a worker listens to the same work by Mozart all day. In the event that humans have a partly genetic predisposition for higher creativity when exposed to nature, however, the enhancement effect might not attenuate completely if one's work environment provides visual access to a savanna-like setting or a spatially open nature area with a water feature.

Affect and Aesthetics

Implications for Nature Valuation

What do the foregoing sections on biophilic responses to natural land-scapes imply for the nonmonetary and monetary valuation of nature? A starting point for this discussion is provided by the widely accepted position in valuation and economic theory that to make a sound comprehensive estimate of the social or monetary values of any "good"—whether a public or private good—it is necessary to have information regarding the good's various attributes or dimensions of value (Ulrich 1988). In the case of many natural environments, whether the settings are public or private goods, one can fairly easily define a limited number of economic value attributes and on this basis generate sound estimates concerning certain *short-term* economic benefits that can be realized by eliminating, developing, or exploiting natural resources. Examples of such economic values include revenues derived from clear-cutting old-growth forest in the American Northwest, lease revenues associated with coal strip-mining on wilderness lands, or clear-cutting an area of tropical rain forest, selling the timber, and using the land for pasture.

In sharp contrast to these well-defined short-term economic values, we know little about most of the long-term monetary and nonmonetary values that are lost when natural areas are eliminated or seriously damaged. Many natural environments are goods that almost certainly have numerous important nonmonetary and monetary value attributes about which little information is available and which in many cases have not even been identified, much less measured with any precision (Ulrich 1988). Our present state of knowledge means that neither decision makers nor the public can make well-informed assessments regarding what may likely be large non-monetary and often monetary value losses that accompany the elimination of nature—losses that may often considerably outweigh the short-term economic gains. If natural environments have values that are not identified—or if they yield important benefits that are not considered in value estimates because they are poorly understood—then it follows that natural environments may be grossly undervalued. In this perspective, research on several important types of biophilic responding should contribute to a deeper understanding of some of the nonmonetary and monetary values of nature and accordingly provide the basis for sounder value estimates that will often be far higher than previous estimates.

In assessing the values of any physical environment, whether natural or urban, it is relevant to evaluate the extent to which substitutes are available that can generate benefits or values similar to those of the environment. If the environment is distinctive, or if substitutes are not readily available, the environment typically will be assigned greater value. The research surveyed in earlier sections on various biophilic responses provides the basis for insights concerning the availability of substitutes for some of the many value attributes of natural environments. In the case of visual or aesthetic preferences, recall that there is a consistent tendency across diverse groups and cultures to prefer even mediocre natural scenes over the vast majority of urban or built views lacking nature. Certain natural views—such as savanna-like scenes and settings with water features—elicit much higher aesthetic liking than nearly all urban landscapes; indeed, only a relative handful of urban views (such as New York City's skyline at night) achieve comparable scores. Hence the overwhelming majority of urban landscapes lacking prominent nature are clearly not substitutes for natural environments from the standpoint of the liking / preference category of biophilic responding. To the extent that natural areas in different countries are eliminated by most forms of urban development, there is a largely nonsubstitutable loss of landscape visual quality or aesthetic preference benefits.

Regarding the substitution issue and restorative biophilic responding, recall that even unspectacular natural settings can promote stress recovery faster and more completely than urban environments lacking nature. During exposure to unthreatening nature, the constellation of restorative responses involves, among other components, a broad shift in feelings toward a more positively toned emotional state, sustained attention or perceptual intake, and positive changes in activity levels in different bodily systems. In terms of the general magnitude of recovery suggested by physiological measures and affective self-ratings (Ulrich et al. 1991), certain non-environmental substitutes are available for natural settings. Oral tranquilizers can produce greater relaxation, for instance, although the onset is not as rapid as for exposure to natural views. Moreover, after receiving training in self-relaxation techniques, some people are able to achieve restoration

Affect and Aesthetics

from stress similar to or greater than levels elicited by unspectacular natural environments.

The question of finding substitutes for nature becomes more problematic, however, if consideration is restricted to physical environments (Ulrich 1988). Although research currently is limited to a few studies, the findings intimate that it may prove quite difficult to identify urban settings lacking nature that have stress recovery effects matching those of even everyday natural settings. Irrespective of whether the built or modern content is, for instance, a lively city street or a well-designed architectural facade, it seems unlikely that such visual stimuli will prove to be adequate substitutes for, say, a natural setting with slowly moving water from the standpoint of eliciting a parasympathetically dominated response. Few built or urban settings lacking nature may match the capabilities of most natural landscapes from the standpoint of eliciting a combination of responses that includes positive feelings, sustained yet nontaxing perceptual intake, and recharge or maintenance of energy. If future behavior-genetic studies or other approaches were to show that restorative biophilic responses have a partly genetic basis, then one could expect to find that certain response characteristics are largely unique to natural elements and configurations. Savanna-like views and natural scenes with water, for example, might prove to be consistently more effective than urban or built content in eliciting attention / interest and positive emotional responses that are comparatively sustained or resistant to habituation. Again, if urban scenes are found to be unsatisfactory substitutes with respect to producing restorative influences, then it would often be warranted to assign considerably higher values to natural environments when the issue is the proposed destruction of nature and its replacement with, for instance, urbanization.

Biophilia, Biophobia, and Preservation

There is a growing recognition that the earth is experiencing an unprecedented crisis with respect to the destruction of biodiversity in innumerable ecosystems and geographical areas (Wilson 1992). Among the many issues relevant to preserving biodiversity, the public's emotion-laden attitudes toward different natural environments play a role in motivating political and other support for reducing the destruction of nature and the extinction of living things. The earlier sections on biophilic and biophobic responses imply that it might be comparatively easy to educate or foster positive emotion-saturated attitudes with respect to certain environments. It may be more difficult, however, in the case of the terrestrial environment that is most critical for preserving biodiversity: the tropical rain forest.

Early forms of humans left the rain forest and moved out into the savanna for a number of sound adaptive reasons (Orians 1980, 1986). For a bipedal, ground-dwelling creature with upright posture, savanna environments were much more advantageous than either rain forests or deserts from the standpoint of offering more abundant food and water. Although savannas have much less biodiversity than rain forests, their biomass productivity can be rather high, and this is related to the high productivity of food that is readily accessible to ground-dwelling humans. By contrast, much of the great biodiversity and biomass of tropical rain forests are concentrated in the forest canopy far above the ground, a location that represented a disadvantage for ground-dwelling early humans. In other words, the great biodiversity of rain forests that is so valued today for a number of very important reasons (Wilson 1992) did not represent a correspondingly high survival-related advantage for early humans.

The high biodiversity of rain forests undoubtedly had another drawback for early humans: its association with greater risk because of its higher levels of *biophobic* properties, including spatial enclosure and higher probabilities of encountering close hidden threats, including snakes, spiders, and other fear-relevant stimuli. By comparison, most savannas probably were characterized by lower levels of biophobic properties and risk, including much less spatial enclosure, more surveillance and escape opportunities, and fewer close encounters with snakes and other predators. As well, certain diseases may have presented a greater risk in hot moist rain forests than in the drier savannas. On the basis of these arguments, it was proposed early in the chapter that modern humans may retain as a remnant of evolution a partly genetic or biophilic disposition to respond positively to environments with savanna-like properties and to spatially open natural settings with water features. This implies the possibility that it might prove comparatively easy to foster public support for preserving such environments by providing information and visual images that elicit feelings and perhaps emotion-laden attitudes that are distinctly positive and tend to be persistently retained because they are in part genetically primed or biologically prepared.

But this functional-evolutionary perspective also implies that biologically prepared learning may hinder (or at best play a mixed role in) attempts to foster public appreciation for tropical rain forests. One reason is that rain forests lack certain combinations of visual properties, such as spatial openness and uniform grassy ground cover, for which humans might be biologically prepared to acquire and persistently retain positive responses. Another important reason is that visual images and ecological information about rain forests are likely to elicit certain genetically primed *biophobic* responses and possibly affect-saturated attitudes that are negatively toned. Recall that there is convincing evidence that fear / avoidance responses to risk-relevant stimuli such as snakes and spiders are partly genetic and manifested in prepared learning. Accordingly, it seems likely that exposing the public to images of spatial enclosure, creepy-crawly creatures, snakes, or other fear-relevant stimuli may elicit strong attention with respect to tropical rain forests, but the emotional tenor of people's responses, attitudes, and knowledge will often be partly negative. In light of research suggesting a robust genetic role in biophobia, even well-conceived education programs may achieve only limited success in fostering public appreciation of certain risk-relevant properties and living things in the rain forest because of the difficulty in overcoming a biologically prepared disposition to respond negatively.

Attempts to promote public appreciation for tropical rain forests probably will be more successful if visual images and other information about the *biophilic* properties of rain forests are also given prominent attention. These elements include, for instance, verdant vegetation, flowers or blooms, and benign attractive animals such as birds. It would be appropriate to include visual images and other information that clearly depict the environmental harm that accompanies the destruction of rain forests and living things. In this regard, recall that scenes of clear-cut settings or barren deforested areas—including rain forests—are disliked. Views containing dead animals elicit strongly negative emotional responses, even disgust

(Lang and Greenwald 1987). Hence it seems likely that portraying in a vivid but accurate manner the consequences of destroying tropical rain forests could produce strong emotion-saturated public attitudes against such destructive activities.

Research Needs and Promising Directions

E. O. Wilson's biophilia hypothesis can be interpreted as consisting of two broad propositions: first, that humans are characterized by a tendency to respond positively to nature; second, that this disposition has a partly genetic basis. At a very general level, this definition implies two corresponding categories of research needs regarding natural landscapes: studies to increase our understanding of the benefits associated with exposure to natural landscapes and research that tests the proposition that biophilic responses are partly genetic. These two basic research needs represent largely different agendas with respect to the questions addressed and especially the most appropriate research methods. Certain research methods have excellent potential for shedding light on possible health-related benefits of experiencing natural settings, for instance, but are ineffective for investigating the genetic component of the biophilia hypothesis.

The first category of research would involve a program of studies to extend and deepen our understanding of the positive responses cited in this chapter, especially restorative and health-related effects. This direction warrants high priority because of its great promise for identifying key human benefits derived from contact with nature, as well as strengthening the rationale for preserving nature by establishing credible links between natural environments and public health. As an example, one potentially important line of research is suggested by previous findings showing that the stress-reducing effects of viewing natural settings are evident in central nervous system indicators such as blood pressure. These findings make it very likely that future studies will find that restorative or stress-reducing effects are also expressed in the endocrine system as salutary reductions in levels of stress hormones such as cortisol, epinephrine, and perhaps norepinephrine (Parsons 1991; Ulrich et al. 1991; Frankenhaeuser 1980). High levels of circulating stress hormones have a variety of deleterious health-related effects, including suppression of immune system functioning (Parsons 1991; Kennedy, Glaser, and Kiecolt-Glaser 1990). Accordingly, if restorative responses were found to involve the neuroendocrine system, then it could be reasonably anticipated, for instance, that people experiencing stress for a lengthy period would benefit from frequent or prolonged exposure to natural environments with their stress-reducing influences that produce enhanced immune system functioning and hence over time foster higher levels of wellness or health. If a program of future studies were to demonstrate that these stress-reducing influences involve positive physiological changes such as reduced blood pressure, lower levels of stress hormones, and even enhanced immune system function, then in many countries it would be relatively straightforward to reconcile the human benefits of nature contacts with legal interpretations of public health and welfare. In this way the legal system could become a more effective vehicle for preserving natural environments. As an important accompaniment to this general research direction, there is a conspicuous need for studies that examine restorative and health-related benefits across different cultures and diverse demographic groups and among young children. It will also be important to discover which natural environments are especially effective in eliciting restorative or health-related responses.

Broadly speaking, research is needed to identify as yet unknown positive responses to nature. In this regard, the hypothesis that exposure to unthreatening natural settings may enhance creativity and high-order cognitive functioning is plausible and warrants evaluation. If the nature / creativity hypothesis were to receive empirical support, this research direction could evolve toward studies demonstrating that in certain situations natural environments can play an important role in generating significant economic value. For all types of biophilic responses, studies are needed to assess the extent to which biodiversity is related to the effectiveness of natural environments in eliciting positive reactions. Another question that warrants much research attention is whether the positive influences of viewing natural settings (restoration, creativity) are considerably attenuated—or not produced at all—by exposure to seriously damaged natural settings such as clear-cut forests or to urban environments lacking nature. While there is considerable knowledge about this issue in regard to aesthetic preferences, much more research is needed on restorative and health-related responses, as well as possible biophilic responses such as enhanced creativity. If built scenes and degraded natural settings are found to perform poorly in eliciting restorative and other positive responses, then one would often be justified in assigning markedly higher values to preserving nature in many decision-making contexts. As an example, if research indicates that views of old-growth forest stands have stress-mitigating influences that include lower blood pressure and reduced levels of stress hormones but views of clear-cut stands do not have these positive effects, then the potential loss of important public benefits relating to physiological well-being would constitute an argument against clear-cutting.

Finally, an important research issue concerns the effectiveness of *simulations* of natural environments (color photographs, videotapes), compared to real environments, in eliciting restorative and other positive responses. There is evidence that simulations can sometimes be at least partial substitutes for real nature in terms of eliciting short-term aesthetic liking and restoration. Studies are needed to evaluate the extent to which real settings may outperform simulations, for instance, in producing stress recovery during short-term exposure. Might simulations lose much of their effectiveness in long-term exposure contexts? It seems likely that, over a long-term exposure situation, real environments may be much more effective than simulations in sustaining positive responding owing to the ongoing visual changes and multisensory stimulation inherent in real environments (such as vegetation changes associated with seasons).

Research on biophilia is at a relatively early stage of development, and no findings have yet appeared that constitute convincing support for the proposition that positive responding to nature has a partly genetic basis. Perhaps the most persuasive findings currently available are the striking patterns across diverse groups and cultures revealing a preference for everyday natural scenes over urban scenes lacking nature, as well as the especially high preference for certain natural settings that presumably offered major survival-related advantages as habitats for early humans. While more cross-cultural and cross-group studies are needed for a number of reasons, this general approach can at best yield a provocative pattern of circumstantial support regarding the genetic hypothesis. Cross-cultural studies have limitations for resolving the genetic question—not only because of the control problems that typically accompany cross-cultural research, but mainly because such studies have not obtained genetic information which in turn can be linked directly to biophilic responding.

From the standpoint of efficacy and feasibility, behavior-genetic methods may offer the most promising approach for evaluating the genetic hypothesis. This approach to biophilia research would require an interdisciplinary team of investigators; it would be rather costly, but it could take advantage of the excellent twin registers that exist in different countries. Although the application of behavior-genetic methods to biophilia questions would be straightforward in most respects, sound research based on smaller twin samples would require focusing on a response characterized by wide variability among persons. In the case of aesthetic preferences for natural landscapes, the moderate to high similarity that usually characterizes different people's preferences would work against the statistical power of a behavior-genetic approach. This drawback, however, could probably be overcome by using a very large sample of twins. To fully exploit the considerable strengths of behavior-genetic methods, it would be advantageous first to perform studies that identify a biophilic response (or specify a particular aspect of a positive response) which exhibits high variability across individuals. Examples of such variables might include intensity of positive emotional responses, interest, and possibly differential rates of response habituation. It might be possible to develop questionnaires for screening people and assigning them to groups that are likely to vary widely in terms of a certain response. The development of such questionnaires might be based on the work of Kellert and others on group differences in certain emotion-saturated attitudes toward nature (Kellert 1980).

Another research approach that holds promise for addressing the genetic aspect of the biophilia hypothesis is *positive* in contrast to aversive Pavlovian conditioning. This laboratory method could be used to test the proposition that biophilic responses may be manifested in biologically prepared learning. Hence a positive conditioning experiment could evaluate, for instance, whether positive responding is conditioned more readily, and is more resistant to extinction, for certain premodern advantageous natural stimuli such as savanna-like views or settings with water than for neutral premodern stimuli or modern advantageous stimuli. Physiological techniques such as facial electromyography (EMG) could be used to measure the magnitude of conditioned positive emotional responses. Whereas positive conditioning studies hold promise and certainly should be attempted, this approach probably will be less efficacious for investigating biophilia than aversive conditioning has been for biophobia: producing an immediate positive stimulus in the laboratory is more problematic than providing a strong immediate negative stimulus.

A comprehensive program of biophilia research would also include studies of the responses of young children, including infants, to natural stimuli. The field of child development has generated an impressive repertoire of nonverbal methods for measuring responses such as attention/ interest and emotions. If very young children are shown slides of unthreatening natural stimuli, abstract stimuli, and modern stimuli that are comparable in information rate, brightness, and perhaps color, do the children evidence, for instance, greater perceptual intake for natural settings? If they do, the findings would constitute more convincing evidence of a partly genetic basis for biophilia than, say, cross-cultural studies of adults. Studies are also needed of children in different age groups—if genetic factors do play a role in biophilic responding, there is a distinct possibility of developmental characteristics in terms of an age-related proneness for acquiring a certain biophilic response. Here it is relevant to note that many other types of human behavior which apparently have a partly biological basis—such as phobias and the appearance of emotions during infancy— have clear age-related onset characteristics. In the event that one is indeed more likely to acquire biophilic responses at a certain age, deprivation of nature learning experiences (lack of direct contact, lack of vicarious learning, inadequate education) during such periods conceivably could be an important factor in the development of adults who are "bio-indifferent."

One other research approach that warrants consideration is the use of a high-resolution PET scanner (positron emission tomography—Druckman and Lacey 1989; Zappulla et al. 1991) to investigate the possible differential location of brain activity during processing of natural versus modern or built stimuli. Compared to the three research approaches discussed here—behavior-genetic, positive conditioning, and studies of young children—PET studies would be more of a long shot from the standpoint of shedding light on the genetic aspect of the biophilia hypothesis. Nonetheless, substantial differences might be found in the location of brain activity (metabolism, blood flow) as a function of viewing different stimuli that could be interpreted as evidence for a biological role in responses to nature. The efficacy of PET studies for addressing the genetic issue could increase if advances in technology are achieved that enable high-resolution scans of deep (and older) brain structures. Irrespective of the genetic issue, PET studies should eventually be performed because they are likely to yield important information regarding responsiveness to natural environments.

If biophilia does have a partly genetic basis, it might be speculated that the genetic contribution is roughly in the range of 20 to 40 percent and will vary for different biophilic responses. Moreover, there might well be age-related or developmental onset characteristics. Certain functional-evolutionary arguments imply the possibility of differences as a function of gender in the role of genetic factors. (Gender differences have not been addressed here because this would entail a discussion of research on a number of topics that are outside the scope of the chapter.) The theoretical propositions advanced here imply that partly genetic biophilic reponses should not appear spontaneously or in the absence of learning; rather, some learning or conditioning is necessary for acquiring biologically prepared positive responsiveness which is then marked by persistent retention. There is certainly no suggestion here that biophilia might be genetic in any deterministic or overriding sense. Learning is required for acquiring a positive response that is only partly predisposed by genetic factors, and the response is modified by conventional learning, experience, and culture.

If a behavioral scientist had argued seriously twenty years ago that humans have a partly genetic predisposition to respond positively to nature, the proposition would have been met with skepticism by most psychologists. Only a few years ago, if a researcher had advanced the theoretical proposition that genetic factors play a salient role in alcoholism among females, for instance, or that infants have a biological or innate capacity for mathematics, the ideas would have been received with derision by many social scientists. Recently, however, the mainstream theoretical orientation of the behavioral and brain sciences has been altered by a cascade of studies showing convincingly that biological or genetic factors play a role not only in alcoholism and math skills but in numerous other aspects of human behavior and response. For many important issues, the debate has shifted from bipolar nature/nurture distinctions to discussion of eclectic perspectives that recognize the crucial roles of both learning and genetics. In several key areas, the main question is no longer whether genetic factors play a role. Rather, the mainstream theoretical and research debate increasingly accepts the role of genetics but asks: is the genetic contribution 20 percent or 50 percent?

Against the background of this profound conceptual shift, the proposition that humans may have a partly genetic predisposition to respond positively to nature now seems plausible to many scientists. Natural settings provided the context of everyday experience throughout human evolution and were the source of critical advantages as well as challenges and risks. To propose that evolution might have left its mark on modern humans in the form of a partly genetic response disposition to nature is at least as plausible as expecting to find, for example, that genetics plays a role in cigarette use or personality. The need to expand our understanding of positive human responsiveness to nature, and to assess the question of its partly genetic basis, represents a major new direction for scientific research—one that can help us to learn more about ourselves as humans, to discover the key benefits that people derive from natural environments, and to gauge the losses in human benefits that result from the destruction of nature.

ACKNOWLEDGMENTS

The ideas and work discussed in this chapter have developed over a period of years and benefited from interactions with several colleagues and friends, especially Arne Öhman, Louis Tassinary, Ulf Dimberg, and Outi Lundén. I also thank Jon Rodiek and Russ Parsons for their helpful comments. The chapter also profited from my interaction with Professor Fernando G. Bernáldez of Universidad Autónoma de Madrid. His untimely death has deprived the scientific community of one of its leading researchers on biophilic responses to natural landscapes. Portions of the research discussed here were supported by NSF grant SES-8317803 and by a series of grants and cooperative agreements with the USDA Forest Service's Rocky Mountain Forest and Range Experiment Station.

REFERENCES

Anderson, L. M., and H. W. Schroeder. 1983. "Application of Wildland Scenic Assessment Methods to the Urban Landscape." *Landscape Planning* 10:219–237.

Appleton, J. 1975. *The Experience of Landscape*. London: Wiley.

Arthur, L. M. 1977. "Predicting Scenic Beauty of Forest Environments: Some Empirical Tests." *Forest Science* 23:151–159.

Asakawa, S. 1984. "The Effects of Greenery on the Feelings of Residents Towards Neighborhoods." *Journal of the Faculty of Agriculture, Hokkaido University* 62:83–97.

Balling, J. D., and J. H. Falk. 1982. "Development of Visual Preference for Natural Environments." *Environment and Behavior* 14:5–38.

Berlyne, D. E. 1971. *Aesthetics and Psychobiology*. New York: Appleton-Century-Crofts.

Bernáldez, F. G., and F. Parra. 1979. "Dimensions of Landscape Preferences from Pairwise Comparisons." In *Proceedings of Our National Landscape: A Conference on Applied Techniques for Analysis and Management of the Visual Resource*. USDA Forest Service General Technical Report PSW-35. Berkeley: Pacific Southwest Forest and Range Experiment Station.

Bernáldez, F. G., R. P. Abelló, and D. Gallardo. 1989. "Environmental Challenge and Environmental Preference: Age and Sex Effects." *Journal of Environmental Management* 28:53–70.

Brown, F., J. Harris, R. Leakey, and A. Walker. 1985. "Early *Homo Erectus* Skeleton from West Lake Turkana, Kenya." *Nature* 316:788–792.

Brush, R. O., and E. L. Shafer. 1975. "Application of a Landscape Preference Model to Land Management." In *Landscape Assessment: Values, Perceptions, and Resources*, edited by E. H. Zube, R. O. Brush, and J. G. Fabos. Stroudsburg, Pa.: Dowden, Hutchinson, and Ross.

Cacioppo, J. T., L. G. Tassinary, and A. J. Fridlund. 1990. "The Skeletomotor System." In *Principles of Psychophysiology: Physical, Social, and Inferential Elements*, edited by J. T. Cacioppo and L. G. Tassinary. New York: Cambridge University Press.

Chagnon, N. 1977. *Yanomamö: The Fierce People*. 2nd ed. New York: Holt, Rinehart & Winston.

Chokor, B. A., and S. A. Mene. 1992. "An Assessment of Preference for Landscapes in the Developing World: Case Study of Warri, Nigeria, and Environs." *Journal of Environmental Management* 34:237–256.

Civco, D. L. 1979. "Numerical Modeling of Eastern Connecticut's Visual Resources." In *Proceedings of Our National Landscape: A Conference on Applied Techniques for Analysis and Management of the Visual Resource*. USDA Forest Service General Technical Report PSW-35. Berkeley: Pacific Southwest Forest and Range Experiment Station.

Clamp, P. 1976. "Evaluating English Landscapes—Some Recent Developments." *Environment and Planning* 8:79–92.

Cook, E. W., R. L. Hodes, and P. J. Lang. 1986. "Preparedness and Phobia: Effects of Stimulus Content on Human Visceral Conditioning." *Journal of Abnormal Psychology* 95:195–207.

Cook, M., and S. Mineka. 1989. "Observational Conditioning of Fear to Fear-Relevant Versus Fear-Irrelevant Stimuli in Rhesus Monkeys." *Journal of Abnormal Psychology* 98:448–459.

———. 1990. "Selective Associations in the Observational Conditioning of Fear in Rhesus Monkeys." *Journal of Experimental Psychology: Animal Behavior Processes* 16:372–389.

Coss, R. G. 1990. "Picture Perception and Patient Stress: A Study of Anxiety Reduction and Postoperative Stability." Unpublished paper, Department of Psychology, University of California, Davis.

Coss, R. G., and M. Moore. 1990. "All That Glistens: Water Connotations in Surface Finishes." *Ecological Psychology* 2:367–380.

Costello, C. G. 1982. "Fears and Phobias in Women: A Community Study." *Journal of Abnormal Psychology* 91:280–286.

Daniel, T. C., and R. S. Boster. 1976. *Measuring Landscape Esthetics: The Scenic Beauty Estimation Method*. USDA Forest Service Research Paper RM-167. Ft. Collins, Colo.: Rocky Mountain Forest and Range Experiment Station.

Daniel, T. C., and J. Vining. 1983. "Methodological Issues in the Assessment of Landscape Quality." In *Human Behavior and Environment*, vol. 6: *Behavior and the Natural Environment*, edited by I. Altman and J. F. Wohlwill. New York: Plenum.

Darwin, C. 1877. "A Biographical Sketch of an Infant." *Mind* 2:285–294.

Dimberg, U. 1990. "Facial Electromyography and Emotional Reactions." *Psychophysiology* 27:481–494.

Druckman, D., and J. I. Lacey, eds. 1989. *Brain and Cognition: Some New Technologies*. Washington: National Academy Press.

Echelberger, H. E. 1979. "The Semantic Differential in Landscape Research."

In *Proceedings of Our National Landscape: A Conference on Applied Techniques for Analysis and Management of the Visual Resource.* USDA Forest Service General Technical Report PSW-35. Berkeley: Pacific Southwest Forest and Range Experiment Station.

Evans, G. W., and S. Cohen. 1987. "Environmental Stress." In *Handbook of Environmental Psychology,* edited by D. Stokols and I. Altman. New York: John Wiley.

Francis, C., and C. Cooper-Marcus. 1991. "Places People Take Their Problems." In *Proceedings of the 22nd Annual Conference of the Environmental Design Research Association,* edited by J. Urbina-Soria, P. Ortega-Andeane, and R. Bechtel. Oklahoma City: EDRA.

Frankenhaeuser, M. 1980. "Psychoneuroendocrine Approaches to the Study of Stressful Person-Environment Transactions." In *Selye's Guide to Stress Research,* vol. 1., edited by H. Selye. New York: Van Nostrand Reinhold.

Fyer, A. J., S. Mannuzza, M. S. Gallops, L. Y. Martin, C. Aaronson, J. M. Gorman, M. R. Liebowitz, and D. F. Klein. 1990. "Familial Transmission of Simple Phobias and Fears: A Preliminary Report." *Archives of General Psychiatry* 47:252–256.

Gabbay, F. H. 1992. "Behavior-Genetic Strategies in the Study of Emotion." *Psychological Science* 3:50–55.

Glass, D. C., and J. E. Singer. 1972. *Urban Stress: Experiments on Noise and Social Stressors.* New York: Academic Press.

Grahn, P. 1991. *Om Parkers Betydelse* [On the importance of parks]. Göteborg: Graphic Systems AB.

Hartig, T., M. Mang, and G. W. Evans. 1991. "Restorative Effects of Natural Environment Experiences." *Environment and Behavior* 23:3–26.

Heath, A. C., M. C. Neale, J. K. Hewitt, L. J. Eaves, and D. W. Fulker. 1989. "Testing Structural Equation Models for Twin Data Using LISREL." *Behavior Genetics* 19:9–35.

Heerwagen, J. H. 1990. "The Psychological Aspects of Windows and Window Design." In *Proceedings of the 21st Annual Conference of the Environmental Design Research Association,* edited by K. H. Anthony, J. Choi, and B. Orland. Oklahoma City: EDRA.

Herzog, T. R., and P. J. Bosley. 1992. "Tranquility and Preference as Affective Qualities of Natural Environments." *Journal of Environmental Psychology* 12:115–127.

Herzog, T. R., S. Kaplan, and R. Kaplan. 1982. "The Prediction of Preference for Unfamiliar Urban Places." *Population and Environment* 5:43–59.

Hockey, R., ed. 1983. *Stress and Fatigue in Human Performance.* New York: John Wiley.

Honeyman, M. K. 1992. "Vegetation and Stress: A Comparison Study of Varying Amounts of Vegetation in Countryside and Urban Scenes." In *The Role of Horticulture in Human Well-Being and Social Development*, edited by D. Relf. Portland, Ore.: Timber Press.

Hongxun, Y. 1982. *The Classical Gardens of China*. Translated by W. H. Min. New York: Van Nostrand Reinhold.

Hugdahl, K. 1978. "Electrodermal Conditioning to Potentially Phobic Stimuli: Effects of Instructed Extinction." *Behavior Research and Therapy* 16:315–321.

Hugdahl, K., and A. C. Karker. 1981. "Biological vs. Experiential Factors in Phobic Conditioning." *Behavior Research and Therapy* 19:109–115.

Hull, R. B., and I. D. Bishop. 1988. "Scenic Impacts of Electricity Transmission Towers: The Influence of Landscape Type and Observer Distance." *Journal of Environmental Management* 27:99–108.

Hull, R. B., and A. Harvey. 1989. "Explaining the Emotion People Experience in Suburban Parks." *Environment and Behavior* 21:323–345.

Hull, R. B., and G. R. B. Revell. 1989. "Cross-Cultural Comparison of Landscape Scenic Beauty Evaluations: A Case Study in Bali." *Journal of Environmental Psychology* 9:177–191.

Hull, R. B., and W. P. Stewart. 1992. "Validity of Photo-Based Scenic Beauty Judgments." *Journal of Environmental Psychology* 12:101–114.

Hultman, S.-G. 1983. "Allmänhetens Bedömning Av Skogsmiljöers Lämplighet För Friluftsliv" [Public judgment of forest environments as recreation areas]. Swedish University of Agricultural Sciences, Section of Environmental Forestry Report 27. Uppsala: Swedish University of Agricultural Sciences.

Hygge, S., and A. Öhman. 1978. "Modeling Processes in the Acquisition of Fears: Vicarious Electrodermal Conditioning to Fear-Relevant Stimuli." *Journal of Personality and Social Psychology* 36:271–279.

Isen, A. M. 1985. "The Asymmetry of Happiness and Sadness in Effects on Memory in Normal College Students." *Journal of Experimental Psychology: General* 114:388–391.

―――. 1990. "The Influence of Positive and Negative Affect on Cognitive Organization: Some Implications for Development." In *Psychological and Biological Approaches to Emotion*, edited by N. L. Stern, B. Leventhal, and T. Trabasso. Hillsdale, N.J.: Lawrence Erlbaum Associates.

Isen, A. M., M. M. S. Johnson, E. Mertz, and G. F. Robinson. 1985. "The Influence of Positive Affect on the Unusualness of Word Associations." *Journal of Personality and Social Psychology* 48:1413–1426.

Kaplan, R. 1983. "The Role of Nature in the Urban Context." In *Human Be-*

havior and Environment, vol. 6: *Behavior and the Natural Environment*, edited by I. Altman and J. F. Wohlwill. New York: Plenum.

Kaplan, R., and S. Kaplan. 1989. *The Experience of Nature*. New York: Cambridge University Press.

Kaplan, S., R. Kaplan, and J. S. Wendt. 1972. "Rated Preference and Complexity for Natural and Urban Visual Material." *Perception and Psychophysics* 12:354–356.

Kaplan, S., and J. F. Talbot. 1983. "Psychological Benefits of a Wilderness Experience." In *Human Behavior and Environment*, vol. 6: *Behavior and the Natural Environment*, edited by I. Altman and J. F. Wohlwill. New York: Plenum.

Katcher, A., H. Segal, and A. Beck. 1984. "Comparison of Contemplation and Hypnosis for the Reduction of Anxiety and Discomfort During Dental Surgery." *American Journal of Clinical Hypnosis* 27:14–21.

Kellert, S. R. 1980. "Contemporary Values of Wildlife in American Society." In *Wildlife Values*, edited by W. W. Shaw and E. H. Zube. Fort Collins, Colo.: Center for Assessment of Noncommodity Natural Resource Values.

Kendler, K. S., M. C. Neale, R. C. Kessler, A. C. Heath, and L. J. Eaves. 1992. "The Genetic Epidemiology of Phobias in Women." *Archives of General Psychiatry* 49:273–281.

Kennedy, C. B. 1989. "Vegetation in Tucson: Factors Influencing Residents' Perceptions and Preferences." Unpublished doctoral dissertation, Department of Geography and Regional Development, University of Arizona, Tucson.

Kennedy, S., R. Glaser, and J. Kiecolt-Glaser. 1990. "Psychoneuroimmunology." In *Principles of Psychophysiology: Physical, Social, and Inferential Elements*, edited by J. T. Cacioppo and L. G. Tassinary. New York: Cambridge University Press.

Klein, D. F. 1981. "Anxiety Reconceptualized." In *Anxiety: New Research and Changing Concepts*, edited by D. F. Klein and J. G. Rabkin. New York: Raven Press.

Knopf, R. C. 1987. "Human Behavior, Cognition, and Affect in the Natural Environment." In *Handbook of Environmental Psychology*, edited by D. Stokols and I. Altman. New York: John Wiley.

Lacey, J. I., and B. C. Lacey. 1970. "Some Autonomic–Central Nervous System Interrelationships." In *Physiological Correlates of Emotion*, edited by P. Black. New York: Academic Press.

Lambe, R. A., and R. C. Smardon. 1986. "Commercial Highway Landscape Reclamation: A Participatory Approach." *Landscape Planning* 12:353–385.

Lang, P. J., and M. K. Greenwald. 1985. "International Affective Picture System: Technical Report 1A." Gainesville: Center for Research in Psycho-physiology, University of Florida.

———. 1987. "International Affective Picture System: Technical Report 1B (Slide Set B)." Gainesville: Center for Research in Psychophysiology, University of Florida.

Latimer, D. A., H. Hogo, and T. C. Daniel. 1981. "The Effects of Atmospheric Optical Conditions on Perceived Scenic Beauty." *Atmospheric Environment* 15:1865–1874.

Leakey, M. 1980. "Early Man, Environment and Tools." In *Current Argument on Early Man*, edited by L.-K. Königsson. New York: Pergamon Press.

Loehlin, J. C., L. Willerman, and J. M. Horn. 1988. "Human Behavior Genetics." *Annual Review of Psychology* 39:101–133.

Lumsden, C. J., and E. O. Wilson. 1983. *Promethean Fire: Reflections on the Origin of Mind*. Cambridge: Harvard University Press.

McNally, R. J. 1987. "Preparedness and Phobias: A Review." *Psychological Bulletin* 101:283–303.

Marks, I. M. 1969. *Fears and Phobias*. New York: Academic Press.

Mednick, S. A. 1962. "The Associative Basis of the Creative Process." *Psychological Review* 69:220–232.

Mehrabian, A., and J. A. Russell. 1974. *An Approach to Environmental Psychology*. Cambridge: MIT Press.

Mineka, S., M. Davidson, M. Cook, and R. Keir. 1984. "Observational Conditioning of Snake Fear in Rhesus Monkeys." *Journal of Abnormal Psychology* 93:355–372.

Moore, E. O. 1982. "A Prison Environment's Effect on Health Care Service Demands." *Journal of Environmental Systems* 11:17–34.

Moran, C., and G. Andrews. 1985. "The Familial Occurrence of Agoraphobia." *British Journal of Psychiatry* 146:262–267.

Nasar, J. L. 1983. "Adult Viewers' Preferences in Residential Scenes: A Study of the Relationship of Environmental Attributes to Preference." *Environment and Behavior* 15:589–614.

———. 1988. *Environmental Aesthetics: Theory, Research, and Applications*. New York: Cambridge University Press.

Neale, M. C., A. C. Heath, J. K. Hewitt, L. J. Eaves, and D. W. Fulker. 1989. "Fitting Genetic Models with LISREL: Hypothesis Testing." *Behavior Genetics* 19:37–49.

Öhman, A. 1979. "Fear Relevance, Autonomic Conditioning, and Phobias: A Laboratory Model." In *Trends in Behavior Therapy*, edited by P. O. Sjödén, S. Bates, and W. S. Dockens. New York: Academic Press.

———. 1986. "Face the Beast and Fear the Face: Animal and Social Fears as

Prototypes for Evolutionary Analyses of Emotion." *Psychophysiology* 23:123–145.

Öhman, A., and U. Dimberg. 1984. "An Evolutionary Perspective on Human Social Behavior." In *Sociophysiology*, edited by W. M. Waid. New York: Springer Verlag.

Öhman, A., U. Dimberg, and L.-G. Öst. 1985. "Animal and Social Phobias: Biological Constraints on Learned Fear Responses." In *Theoretical Issues in Behavior*, edited by S. Reiss and R. R. Bootzin. New York: Academic Press.

Öhman, A., G. Erixon, and I. Löfberg. 1975. "Phobias and Preparedness: Phobic Versus Neutral Pictures as Conditioned Stimuli for Human Autonomic Responses." *Journal of Abnormal Psychology* 84:41–45.

Öhman, A., and J. J. F. Soares. 1993a. "On the Automatic Nature of Phobic Fear: Conditioned Electrodermal Responses to Masked Fear-Relevant Stimuli." *Journal of Abnormal Psychology*. In press.

———. 1993b. "Unconscious Anxiety: Phobic Responses to Masked Stimuli." *Psychophysiology*. In press.

Olmsted, F. L. 1865. "Preliminary Report upon the Yosemite and Big Tree Grove." Report to the Congress of the State of California. Reprinted in *The Papers of Frederick Law Olmsted*, vol. V: *The California Frontier, 1863–1865*, edited by V. P. Ranney, G. J. Rauluk, and C. F. Hoffman. 1990. Baltimore: Johns Hopkins University Press.

Orians, G. H. 1980. "Habitat Selection: General Theory and Applications to Human Behavior." In *The Evolution of Human Social Behavior*, edited by J. S. Lockard. New York: Elsevier North-Holland.

———. 1986. "An Ecological and Evolutionary Approach to Landscape Aesthetics." In *Meanings and Values in Landscape*, edited by E. C. Penning-Rowsell and D. Lowenthal. London: Allen & Unwin.

Parsons, R. 1991. "The Potential Influences of Environmental Perception on Human Health." *Journal of Environmental Psychology* 11:1–23.

Patey, R. C., and R. M. Evans. 1979. "Identification of Scenically Preferred Forest Landscapes." In *Proceedings of Our National Landscape: A Conference on Applied Techniques for Analysis and Management of the Visual Resource*. USDA Forest Service General Technical Report PSW-35. Berkeley: Pacific Southwest Forest and Range Experiment Station.

Penning-Rowsell, E. C. 1979. "The Social Value of English Landscapes." In *Proceedings of Our National Landscape: A Conference on Applied Techniques for Analysis and Management of the Visual Resource*. USDA Forest Service General Technical Report PSW-35. Berkeley: Pacific Southwest Forest and Range Experiment Station.

Rabinowitz, C. B., and R. E. Coughlin. 1970. "Analysis of Landscape Characteristics Relevant to Preference." *Regional Science Research Institute Discussion Paper Series*, no. 38. Philadelphia: Regional Science Research Institute.

Ribe, R. G. 1989. "The Aesthetics of Forestry: What Has Empirical Preference Research Taught Us?" *Environmental Management* 13:55–74.

Ródenas, M., F. Sancho-Royo, and F. G. Bernáldez. 1975. "Structure of Landscape Preferences: A Study Based on Large Dams Viewed in Their Landscape Setting." *Landscape Planning* 2:159–178.

Rose, R. J., J. Z. Miller, M.F. Pogue-Geile, and G. F. Cardwell. 1981. "Twin-Family Studies of Common Fears and Phobias." In *Twin Research 3: Intelligence, Personality, and Development*, edited by L. Gedda, P. Parisi, and W. E. Nance. New York: Alan Liss.

Ruiz, J. P., and F. G. Bernáldez. 1982. "Landscape Perception by Its Traditional Users: The Ideal Landscape of Madrid Livestock Raisers." *Landscape Planning* 9:279–297.

Rutherford, W., and E. L. Shafer. 1969. "Selection Cuts Increased Natural Beauty in Two Adirondack Forest Stands." *Journal of Forestry* 67:415–419.

Savolainen, R., and S. Kellomäki. 1984. "Scenic Value of the Forest Landscape as Assessed in the Field and the Laboratory." In *Multiple-Use Forestry in the Scandinavian Countries*, edited by O. Saastamoinen, S.-G. Hultman, N. E. Koch, and L. Mattsson. Communicationes Instituti Forestalis Fenniae, 120. Helsinki: Finnish Forest Research Institute.

Schroeder, H. W. 1986. "Psychological Value of Urban Trees: Measurement, Meaning, and Imagination." In *Proceedings of the Third National Urban Forestry Conference*. Washington: American Forestry Association.

———. 1989. "Environment, Behavior, and Design Research on Urban Forests." In *Advances in Environment, Behavior, and Design*, vol. 2., edited by E. H. Zube and G. T. Moore. New York: Plenum.

Schroeder, H. W., and L. M. Anderson. 1984. "Perception of Personal Safety in Urban Recreation Sites." *Journal of Leisure Research* 16:177–194.

Schroeder, H. W., and W. N. Cannon. 1983. "The Esthetic Contribution of Trees to Residential Streets in Ohio Towns." *Journal of Arboriculture* 9:237–243.

Seligman, M. E. P. 1970. "On the Generality of the Laws of Learning." *Psychological Review* 77:406–418.

———. 1971. "Phobias and Preparedness." *Behavior Therapy* 2:307–320.

Shafer, E. L, J. F. Hamilton, and E. A. Schmidt. 1969. "Natural Landscape Preferences: A Predictive Model." *Journal of Leisure Research* 1:187–197.

Shafer, E. L., and M. Tooby. 1973. "Landscape Preferences: An International Replication." *Journal of Leisure Research* 5:60–65.

Sheets, V. L., and C. D. Manzer. 1991. "Affect, Cognition, and Urban Vegetation: Some Effects of Adding Trees Along City Streets." *Environment and Behavior* 23:285–304.

Shepard, P. 1967. *Man in the Landscape: A Historic View of the Esthetics of Nature.* New York: Knopf.

Smardon, R. C. 1988. Perception and Aesthetics of the Urban Environment: Review of the Role of Vegetation." *Landscape and Urban Planning* 15:85–106.

Sonnenfeld, J. 1967. "Environmental Perception and Adaptation Level in the Arctic." In *Environmental Perception and Behavior*, edited by D. Lowenthal. Resource Paper 109. Chicago: Department of Geography, University of Chicago.

Taylor, J. G., E. H. Zube, and J. L. Sell. 1987. "Landscape Assessment and Perception Research Methods." In *Methods in Environmental and Behavioral Research*, edited by R. B. Bechtel, R. W. Marans, and W. Michelson. New York: Van Nostrand Reinhold.

Ulrich, R. S. 1977. "Visual Landscape Preference: A Model and Application." *Man-Environment Systems* 7:279–293.

———. 1979. "Visual Landscapes and Psychological Well-Being." *Landscape Research* 4(1):17–23.

———. 1981. "Natural Versus Urban Scenes: Some Psychophysiological Effects." *Environment and Behavior* 13:523–556.

———. 1983. "Aesthetic and Affective Response to Natural Environment." In *Human Behavior and Environment*, vol. 6: *Behavior and the Natural Environment*, edited by I. Altman and J. F. Wohlwill. New York: Plenum.

———. 1984. "View Through a Window May Influence Recovery from Surgery." *Science* 224:420–421.

———. 1986a. "Human Responses to Vegetation and Landscapes." *Landscape and Urban Planning* 13:29–44.

———. 1986b. "Effects of Hospital Environments on Patient Well-Being." Research Report 9 (55). Trondheim, Norway: Department of Psychiatry and Behavioral Medicine, University of Trondheim.

———. 1988. "Toward Integrated Valuations of Amenity Resources Using Nonverbal Measures." In *Amenity Resource Valuation: Integrating Economics with Other Disciplines*, edited by G. L. Peterson, B. L. Driver, and R. Gregory. State College, Pa.: Venture.

Ulrich, R. S., and D. Addoms. 1981. "Psychological and Recreational Benefits of a Neighborhood Park." *Journal of Leisure Research* 13:43–65.

Ulrich, R. S., U. Dimberg, and B. L. Driver. 1991. "Psychophysiological Indicators of Leisure Benefits." In *Benefits of Leisure*, edited by B. L. Driver, P. J. Brown, and G. L. Peterson. State College, Pa.: Venture.

Ulrich, R. S., U. Dimberg, and A. Öhman. 1993. "Spatially Restricted Versus Open Natural Environments as Conditioned Stimuli for Autonomic Responses." In preparation.

Ulrich, R. S., and O. Lundén. 1990. "Effects of Nature and Abstract Pictures on Patients Recovering from Open Heart Surgery." Paper presented at the International Congress of Behavioral Medicine, 27–30 June, Uppsala, Sweden.

Ulrich, R. S., and R. Parsons. 1992. "Influences of Passive Experiences with Plants on Individual Well-Being and Health." In *The Role of Horticulture in Human Well-Being and Social Development*, edited by D. Relf. Portland, Ore.: Timber Press.

Ulrich, R. S., R. F. Simons, B. D. Losito, E. Fiorito, M. A. Miles, and M. Zelson. 1991. "Stress Recovery During Exposure to Natural and Urban Environments." *Journal of Environmental Psychology* 11:201–230.

Verderber, S. 1986. "Dimensions of Person-Window Transactions in the Hospital Environment." *Environment and Behavior* 18:450–466.

Ward, L. M. 1977. "Multidimensional Scaling of the Molar Physical Environment." *Multivariate Behavioral Research* 12:23–42.

Wellman, J. D., and G. J. Buhyoff. 1980. "Effects of Regional Familiarity on Landscape Preferences." *Journal of Environmental Management* 11:105–110.

West, M. J. 1985. "Landscape Views and Stress Response in the Prison Environment." Unpublished master's thesis, Department of Landscape Architecture, University of Washington.

Wilson, E. O. 1984. *Biophilia*. Cambridge: Harvard University Press.

———. 1992. *The Diversity of Life*. Cambridge: Harvard University Press.

Wohlwill, J. F. 1976. "Environmental Aesthetics: The Environment as a Source of Affect." In *Human Behavior and Environment*, vol. 1, edited by I. Altman and J. F. Wohlwill. New York: Plenum.

———. 1983. "The Concept of Nature: A Psychologist's View." In *Human Behavior and Environment*, vol. 6: *Behavior and the Natural Environment*, edited by I. Altman and J. F. Wohlwill. New York: Plenum.

Yi, Y. K. 1992. "Affect and Cognition in Aesthetic Experiences of Landscapes." Unpublished doctoral dissertation, Department of Landscape Architecture and Urban Planning, Texas A&M University.

Zappulla, R. A., F. F. LeFever, J. Jaeger, and R. Bilder, eds. 1991. *Windows on the Brain: Neuropsychology's Technological Frontiers*. Annals of the New York Academy of Sciences, vol. 620, New York: New York Academy of Sciences.

Zube, E. H., R. O. Brush, and J. G. Fabos, eds. 1975. *Landscape Assessment: Values, Perceptions, and Resources*. Stroudsburg, Pa.: Dowden, Hutchinson, and Ross.

Zube, E. H., D. G. Pitt, and T. W. Anderson. 1975. "Perception and Prediction of Scenic Resource Values of the Northeast." In *Landscape Assessment: Values, Perceptions, and Resources*, edited by E. H. Zube, R. O. Brush, and J. G. Fabos. Stroudsburg, Pa.: Dowden, Hutchinson, and Ross.

Zube, E. H., D. G. Pitt, and G. W. Evans. 1983. "A Lifespan Developmental Study of Landscape Assessment." *Journal of Environmental Psychology* 3:115–128.

Zuckerman, M., R. S. Ulrich, and J. McLaughlin. 1993. "Sensation Seeking and Affective Reactions to Nature Paintings." *Personality and Individual Differences*. In press.

CHAPTER 4

Humans, Habitats, and Aesthetics

Judith H. Heerwagen and Gordon H. Orians

I N 1984 E. O. WILSON published his personal account of why viewing human beings as evolved organisms opens up new perspectives on many otherwise puzzling aspects of human behavior. Considering the deep-seated fear of snakes that is so widespread in human cultures, for example, Wilson suggested that "we need not turn to Freudian theory in order to explain our special relationship to snakes. The serpent did not originate as the vehicle of dreams and symbols. The relation appears to be precisely the other way around and correspondingly easier to study. Humanity's concrete experience with poisonous snakes gave rise to Freudian phenomena after it was assimilated by genetic evolution into the brain's structure" (1984:98).

Humans and Habitats

The study of landscapes and landscape aesthetics receives new insights as well when approached from an evolutionary perspective. Wilson notes that "the more habitats I have explored, the more I have felt that certain common features subliminally attract and hold my attention" (p. 106). Wilson does not claim that there is a hereditary program hard-wired into the brain, but he does suggest that our responses and learned reactions are biased in certain directions by our evolutionary history. For a number of years we have been exploring this uncertain terrain (Heerwagen and Orians 1986; Orians 1980, 1986; Orians and Heerwagen 1992). In this chapter we summarize our past efforts, extend our conceptual analysis, present some new hypotheses and tests, and look into the future.

Our basic approach is to look at human responses to landscapes from an adaptive problem-solving perspective. If we regard the human brain as an evolved organ especially designed to analyze and respond appropriately to the opportunities and constraints that existed in ancestral environments, we begin to look at human interactions with the natural world in a new way. (See Tooby and Cosmides 1992 for a detailed analysis of the adaptive character of the mind.) Our ancestors lived in environments devoid of modern comforts and conveniences. Their survival, health, and reproductive success depended on their ability to seek and use environmental information wisely. They had to know how to interpret signals from the animate and inanimate environments and how to adjust their behavioral response to the context at hand.

This problem-solving perspective suggests that our interactions with nature are likely to be quite complex. There are fear and loathing as well as pleasure and joy in our experiences with the natural world. Thus the real issue is not whether biophilia exists, but rather the particular forms it takes. We need to focus our attention on both positive and negative responses to the natural world and to discover what circumstances move people in one direction rather than another in their behavior and feelings about animals, plants, habitats, and ecosystems.

Habitat selection has been the conceptual focus of our theoretical and

Humans, Habitats, and Aesthetics

empirical studies of landscape aesthetics. A crucial step in the lives of most organisms, including humans, is selection of a habitat. If a creature gets into the right place, everything else is likely to be easier. Habitat selection depends on the recognition of objects, sounds, and odors to which the organism responds *as if* it understood their significance for future behavior and success. Responses are initially emotional feelings that lead to rejection, exploration, or a certain use of the environment. If the strength of these responses is a key to immediate decisions about where to settle, as empirical data suggest, then the ability of a habitat to evoke these emotional states should evolve to become positively correlated with the expected survival and reproductive success of an organism in it. Good habitats should evoke strong positive responses; poor habitats should evoke weak or even negative responses (Orians and Heerwagen 1992).

Although the conceptual foundation of an evolutionary approach to habitat selection and its important corollary, landscape aesthetics, is relatively simple, assessing habitats is a difficult process for organisms. The current state of the environment is usually important, but future states may be even more important. For this reason, many organisms evolve to use features of habitats that are good predictors of the future—how far into the future depends on the projected use. Many birds, for example, use general patterns of tree density and vertical arrangement of branches as primary settling cues (Cody 1985; Hildén 1965; Lack 1971) rather than attempting to assess food supplies directly, for the food that will support breeding often is not present when the habitats must be selected.

The use of features that are correlated with success but do not actually determine it characterizes human responses to landscapes as well. Human generation times are long, however, and offspring depend on their parents for more than a decade. Rarely did habitats occupied by humans during most of our evolutionary history provide reliable resources long enough to enable permanent occupancy of sites. Frequent moves through the landscape were the rule even though traditional sites might be revisited on an annual basis (Campbell 1985; Lovejoy 1981; Blumenschine 1987).

The needs of our ancestors were the same as ours: to find adequate food and water and to protect themselves from the physical environment, predators, and hostile conspecifics. We now seek these amenities in a much wider range of environments and by a broader array of means than our ancestors did. Nonetheless, from an evolutionary perspective, few generations have passed since we started to live in mechanized and urban environments. To the extent that we have evolutionary-based response patterns to landscapes, they should not have been modified by these new environments. Rather, these environments should be viewed from the perspective of an animal that has *modified* them according to preferences inherited from its distant past.

The History of Landscape Aesthetics

The study of aesthetics dates back to the philosophical explorations of Edmund Burke (1757), but during the succeeding two centuries attention has been directed almost exclusively to human artistic creations to the neglect of nature (Rose 1976). Indeed, in 1976 the philosopher R. W. Hepburn noted that contemporary writings on aesthetics deal mostly with the arts rather than natural beauty.

The neglect of nature is a by-product of the prevailing view that has dominated Western thinking for many centuries—namely, that cultural symbols and art forms create the aesthetic experience. Therefore, the study of aesthetics was viewed as the domain of artists and philosophers, and any attempt to explore the biophysical basis of aesthetic responses to the environment was regarded as both futile and ideologically dangerous. The prevailing view is well summarized by Cosgrove (1984:1–2): "Landscape is a way of seeing that has its own history, but a history that can be understood only as part of a wider history of economy and society." Cosgrove sees the development of the "landscape idea" as something unique to capitalistic societies within Western civilization between A.D. 1400 and 1900. Other societies or, for that matter, ancestral societies that preceded the rise of Western civilization, are deemed irrelevant to the study of landscape aesthetics.

No one could seriously doubt the strong influence of culture on the way humans see the environment and the symbolism we attach to environmental objects (Appleton 1990), but to assume a priori that evolutionary biology has nothing to contribute requires us, for example, to see serpent symbolism in Freudian terms rather than from the perspective proposed by Wilson.

Habitability Cues

The English word *aesthetic* is derived from the adjectival form of the Greek verb *aisthanomai*, which means "to perceive." Thus "aesthetic pleasure" means literally "pleasure associated with or deriving from perception." The sense of aesthetic pleasure and emotional enticement associated with nature is, in Wilson's view, the "central issue of biophilia."

The idea of beauty is a secondary association that has been attached to aesthetics comparatively recently (Appleton 1990). A danger arising from this association is that it may lead to the assumption that beauty is an intrinsic property of objects which we call "beautiful," comparable to such features as size, shape, texture, or weight, rather than being the product of an interaction between the traits of objects and the human nervous system. Investigators who believe that beauty is an intrinsic property of objects tend to seek statistical correlations between the characteristics of objects and aesthetic responses. Such an approach is currently popular because of the ease with which large quantities of data, often derived from questionnaires, can be stored and manipulated with computers.

Clearly questionnaires and statistical analysis are useful, but they should be carried out from the perspective that humans evaluate environments, not necessarily consciously, in terms of the opportunities they provide for pursuing activities that contribute positively to survival and reproductive success. This implies that the environment should be viewed in functional rather than morphological terms (Appleton 1975). Appleton identified three functional concepts—prospect, refuge, and hazard—upon which he built his theory of environmental aesthetics. His perspective has stimulated much activity among students of landscape and architecture (Hildebrand 1991; Jakle 1987), but it deals with only part of the complex environmental assessments humans make. Appleton's framework analyzes the initial evaluation and exploration of unfamiliar environments. It focuses on the opportunities to gather information and the safety with which environmental assessment can be accomplished. We expand that framework to include consideration of the types of information humans use to assess the habitability of environments.

Elsewhere (Orians and Heerwagen 1992) we have suggested that two complementary frameworks are useful in the study of human responses to landscapes. One framework is concerned with exploration of an unfamiliar landscape; the other involves the time frame over which information is relevant. We divided the process of exploration into three stages. Stage 1, accompanying an initial encounter with a landscape, is the decision to explore the landscape further or to ignore it and move on. Responses at this stage are known to be highly affective and almost instantaneous (Ulrich 1983; Zajonc 1980). If the response to the first stage is positive, then Stage 2, information gathering, follows. During this stage, cognition figures prominently and the process of exploration may last many days. Stage 3 concerns the decision to remain in the environment to carry out a certain set of activities. Depending on the relevant activities, the length of stay may be brief or last a lifetime.

Environmental information is relevant over highly varying time spans. Some environmental cues, such as weather, perception of prey, predators, enemies, or the arrival of a prospective mate, pertain to conditions that are transitory. These cues must be responded to quickly if the opportunity is to be seized or danger avoided. Other environmental cues signal changes that take place more slowly. Examples include seasonal changes in the vegetative and reproductive cycles of plants and the breeding activities of animals. Evaluating and responding to these seasonal cues is vital for successful functioning in the environment, but time is available for reflection and planning. At the other extreme are cues associated with relatively permanent features of the environment. Examples include topography, lakes, rivers, and vegetation structure. These features all change, but they do so very slowly.

For several reasons, human responses to all of these cues are highly variable. The needs of people change with age and the immediate situation. A family planning a picnic may view an approaching storm very differently from a farmer whose crops are wilting because of lack of rain, even though both may prefer to be in a dry place when the rain begins. A person's current physiological state influences the perceived importance of water, food, or

Humans, Habitats, and Aesthetics

shelter. Because over most of human history men have been hunters and women gatherers and child-tenders, one would expect men and women to evaluate the environment differently, including its prospect and refuge opportunities. Because of the high variability in response to environmental cues, no simple theory is likely to explain the complexity of human responses. But this does not render a search for patterns unprofitable. To focus our search, we now consider the key components of environmental habitability.

Resource Availability

The capacity of environments to provide food and water is vital to their habitability. At Stage 1, direct information may be available about these resources in the form of large mammals and birds, flowering and fruiting plants, and water. Indirect information is available in the form of topographic cues (a cliff may offer good viewing and may also have water at its base), signs of human occupation (if other people are living there, life-supporting resources must be available), and the presence of waterfowl or other animals (these may signal the presence of water or other resources).

More habitat information may be obtained during exploration. During Stage 2, information about future resources, particularly from a seasonal perspective, can be gathered. Flowering plants are potential sources of food (Peters and O'Brien 1981; Peters, O'Brien, and Box 1984). Furthermore, they signal future availability of fruits and honey while also providing cues about when and where the fruits can be found. Greening of grass and leafing of trees and shrubs can be used to predict future locations of large mammals. In tropical regions with pronounced dry seasons—the ancestral habitat of our species—approaching storms signal changes in the availability and location of food and water which can be used to make decisions about shifts in home bases and hunting trajectories.

Shelter and Predator Protection

Shelter from the physical environment and safety from predators, particularly at night, have been major factors in environmental assessment and valuation throughout human history. Caves and fires serve both purposes, and climbable trees offer safety from the major terrestrial predators that until recently were serious threats to humans. These environmental features, because they are relatively permanent, are important for safety both during exploration (Stage 2) and during occupancy (Stage 3). Indeed, Appleton's (1975) prospect/refuge theory postulates that evidence of safe places from which to explore is a key feature evoking initial positive responses to unfamiliar environments.

Hazard Cues

An environment may pose hazards during both exploration and occupancy. Some features, such as inclement weather, fire, presence of dangerous predators, and barriers to movement, may be perceived directly. Other hazards may be indicated by the behavior of animals or people. Alert and scanning antelope may have seen a predator, for example, whereas calm antelope that are grazing or resting indicate safety. A startled expression on another person's face, or body gestures that signal fear, can also alert other group members to dangers before they are actually seen. In contrast, relaxed and happy people indicate that all is well. Such indirect cues typically signal transient conditions, whereas many direct cues are relevant over much longer time spans. Also important are deficiency cues indicating that the environment is lacking in refuges, food, and water. In this regard, it is interesting to note that hell is typically painted as lacking basic amenities whereas heaven abounds in such resources. Paintings of heaven and hell show the extent to which painters use direct and indirect habitat cues to depict pleasure and pain, plenitude and deficiency, comfort and hazard.

Wayfinding and Movement

To be stimulated to explore an environment, a person must not only respond positively to its general features (Stage 1) but must also perceive pathways of movement. Before entering an unfamiliar environment—or imagining entering a landscape portrayed in a painting or photograph—a person typically decides on a route which will lead to places where information can be gathered without undue exposure to risk. These places are Appleton's attractive refuges that offer views (prospects) of the surrounding environment. If the route promises to open up vistas of portions of the environment invisible from the observer's current position, interest in exploration is likely to increase (Kaplan and Kaplan 1982). An environment that is difficult or dangerous to enter, even if it may contain significant resources, is likely to evoke negative responses at Stage 1.

Tests of Habitability and Aesthetics

The concepts we have just articulated are scientifically useful because they can be framed as testable hypotheses. The nature of the field of inquiry is such that no single test is, by itself, conclusive. Several tests, however, if they provide consistent results, can increase our confidence that the hypotheses have merit. Here we present some hypotheses about landscape aesthetics and perform tests using analyses of paintings, landscape architecture, selection for tree and leaf sizes and shapes under domestication, and human responses to, and manipulation of, flowers.

Contextual Factors

Decisions about where to go, how to get there, and what activities to engage in are highly influenced by personal factors, such as one's age or gender, as well as by situational events such as time of day or weather conditions. Because the ultimate goal of landscape preference is to guide environmental decisions in ways that enhance survival and fitness, the human psyche should be very responsive to cues that provide information relative to one's current and future needs and goals. The psychological mechanisms that guide adaptive functioning in environments may not even be conscious processes. All that is required for the mechanisms to be effective is that they produce the right behavior at the right time and place (Staddon 1987) and do so in situations that are likely to have had positive fitness consequences in ancestral environments (Tooby and Cosmides 1990).

One environmental cue to which our ancestors must have been highly receptive is time of day. Attention to events that signaled transition from day to night would have been highly adaptive for individuals of a species that regularly roamed long distances during daily gathering and hunting activities. Although there continues to be debate about the kinds of spaces our ancestors used regularly (social sites, butchering sites, home bases), there is general agreement that animal and vegetable foods were not totally consumed where they were located but usually were carried back to a safe place for later consumption and exchange among group members (Binford 1985; Isaac 1983). Sleeping and resting sites too were usually located away from foraging patches (Bourliere 1963).

Under these conditions, the setting sun would be a strong signal to return to a safe place for the night. Darkness has always been frightening for most people. The human visual system is adapted to daylight activities; night leaves us feeling vulnerable and helpless since we cannot rely on our primary means of perceiving the environment. Our vulnerability is all the more palpable when we consider that many of the primary human predators (hyenas, large cats, and wild dogs, as well as poisonous snakes) are nocturnal species. There are good reasons to fear the dark. Thus cues that signal the coming of night—such as the setting sun—should be highly motivating. The specific response to sunsets is likely to differ, however, depending on whether one is close to home or far away from it without any protection. The intense and lengthy shadows created by the setting sun provide a brief period during which depth perception is better than during midday, an advantage that would aid swift movement through the environment. Once the sun sinks below the horizon, the details and colors of the environment become less distinct and, thus, less readable.

Sunrise, on the other hand, does not create the same urgency for response. Normally the rising sun is viewed from a refuge in which the night was spent. Sunrise is a time of increasing light, signaling many hours of good visual perception to support movement through the environment and hazard detection.

Using Landscape Paintings to Test Hypotheses

Landscape painting, because it is meant to create an emotional response to a newly encountered place, is an ideal format in which to explore evolutionary hypotheses about environmental preferences and aesthetics. Viewers confront an unfamiliar landscape but imagine themselves in it and about to explore it. Clearly, most landscape paintings are designed to create positive emotional responses. Therefore, landscape paintings can be analyzed with respect to their prospect/refuge content and the opportunities for movement that they signal.

To test predictions about people's responses to signals of time of day, we conducted a content analysis of forty-six paintings of sunsets and sunrises. Paintings were selected from books based on exhibitions and represent a wide range of artists. One volume, however, was devoted exclusively to the paintings of Frederick Church, who painted numerous sunsets and sunrises. The following hypotheses were tested:

1. Landscape paintings of sunset should be high in refuge symbolism. Sunrise paintings may be much lower in refuge symbolism, provided the painter is able to clearly signal the time of day.

2. Paintings of sunsets with people should portray them in or very close to places they wish to spend the night if the scene is to be viewed as tranquil, particularly if the sun has gone below the horizon. People in sunrise settings may be shown in a greater variety of situations because the same time pressures do not exist. If a refuge is shown, people should be heading away from it.

3. Sunsets high in tension should portray people far from the refuge or engaged in activities that indicate lack of attention to signals of impending darkness.

In the content analysis, we scored each painting for prospect and refuge symbolism and noted the position and activity of people in the landscape. For refuge, scoring included the location, type, accessibility, and conspicuousness of the refuge. Prospect symbolism included the amount of horizon in the scene, the openness of the setting, and the presence of hills, mountains, towers, or other "secondary prospects" (Appleton 1975) that would afford an expansive view. In addition to the prospect and refuge symbolism, we noted the sky conditions, including the color of the sky and the presence of clouds.

Sunsets ($n = 35$) were much more common than sunrises ($n = 11$). In hindsight, this makes sense: the information provided by a sunset is much more valuable and requires more urgent attention than the information available in a sunrise. (On the other hand, we cannot rule out the possibility that artists may prefer to sleep well past sunrise if they work late into the night.)

As predicted, paintings of sunsets rate very high in refuge symbolism: 66 percent of the paintings score high on refuge compared with 9 percent of the sunrises (chi square = 10.89, $p = .004$). Analysis of the type and location of the refuge shows that sunset paintings had more built refuges (primarily houses but also some churches), with 46 percent of the buildings in the foreground or midground of the landscape. In contrast, 55 percent of the sunrises had no built refuges and none had a building in the foreground. The conspicuousness of the refuge in sunsets was also prominent. Of the paintings with a built refuge, 46 percent had a light in the window, 12 percent had smoke coming from a chimney, and 7 percent had smoke and a lighted window. When these signals were not present, the painter used high contrast to make the refuge stand out from its background (a white house against dark vegetation, for example).

To test the hypothesis regarding people, we noted the location and activity of each person in the scene and, where possible, identified the figure as male, female, or child. Not surprisingly, because it is a less desirable time to be outdoors, the sunsets had fewer people than the sunrises. Paintings of sunsets had a total of ninety-three people (the average number per painting was 2.6), and the sunrises had forty-eight people (the average was 4.4). All but one of the sunrise scenes (91 percent) had people compared with 70 percent of the sunsets. Furthermore, 63 percent of the sunsets had people in or near the refuge compared with 2 percent of the sunrises (chi square = 50,000, p = .0000).

Two of the three sunset paintings with people far from the refuge portray scenes that are hardly relaxing. Jerome Thompson's *Belated Party on Mansfield Mountain* shows three men and three women on a precipice high above the landscape as the sun disappears below the horizon and darkness masks one side of the hills. One of the men is pointing to his watch with a look of concern on his face. The description of the painting includes this comment: "A seated young man holds up his pocket watch to warn of the lateness of the hour (connoted also by the setting sun) and the need to descend before dark."

Lynn Meadows by Martin Johnson Heade shows three men clamming with no refuge in sight and the sun below the horizon. The accompanying text says: "The painting shows three men clamming in the middle ground of a deserted marsh, against an infinitely broad horizon. . . . These activities occur at the unexpected hour of sunset, when the drama in the sky can easily make such events seem insignificant. The two men digging in the muck are oblivious to their surroundings in a way that recalls the figure in *Lake George*, but, considering the events in the background, their level of ignorance seems profound."

The appeal of a refuge is influenced not only by its location and conspicuousness but also by its accessibility to the viewer. In some contexts, the refuge may deliberately be remote in order to discourage unwanted intrusions (as is the case with monasteries or fortresses located on treacherous peaks). If a painter is trying to elicit a feeling of tranquillity at sunset, however, the refuge should be easily reached. Several options exist: there may be a path or road leading to the refuge, or the ground may be open with low, easily traversed ground cover. We rated accessibility of the refuge as high, moderate, or low. High access was indicated by a path or open ground; moderate access meant that a potential barrier (a hill or woods) was surmountable; low access meant there was no indication of how to get to the refuge, or it was remote, or the terrain was difficult to cross. Some 92 percent of the refuges in sunset paintings were highly accessible. Similar results were found for sunrise paintings.

Gender Differences in Prospect and Refuge

Given substantial differences between the sexes with regard to reproduction and subsistence activities, males and females should use and assess environments differently. Although both males and females should find refuges attractive, for several reasons females should show a greater affinity for enclosure and protected places than do males. The continuous demands of pregnancy and childcare make females (and their dependent young) more vulnerable to predation; refuges reduce their exposure to these dangers. Pregnancy and lactation also increase energy demands on females; refuges, especially those affording enclosure above, offer relief from the sun and inclement weather, both of which can be serious energy drains. Refuges also allow for protection during childbirth, a time of acute vulnerability for both mother and infant.

Finally, because of their smaller physical size relative to males, females are at a disadvantage in encounters with predators. The wisest strategy under these circumstances is to avoid encounters in the first place by not venturing into predators' environments. In our evolutionary past, many of the primary predators on humans (the large cats and hyenas) hunted in open grasslands and savannas rather than in wooded areas affording refuge. In addition to animal predators, females are at risk for attack by human males seeking to gain a reproductive advantage. (See Smuts 1992 for a discussion of the evolutionary basis for male aggression against females; see Gowaty 1992 for women's use of "refugia" to protect themselves from sexual coercion by males.) Although highly enclosed places may be desirable for some activities, both females and males should prefer refuges that are penetrable and afford visual access to the surroundings to avoid entrapment or being taken by surprise. For females, "open" refuges would be decidedly advantageous when watching over young children who can readily get lost in dense bush (Hawkes 1987).

Another factor likely to influence male and female environmental preferences is the sexual division of labor that has persisted throughout human history. Despite the considerable literature on gender differences in subsistence and its complex relationship to social and reproductive behavior (Tooby and DeVore 1987), little attention has been given to the relationship between subsistence strategies and evolved responses to environments. Because men have been hunters and women gatherers throughout evolutionary time, environmental preferences should be consistent with subsistence roles. Specifically, males should show a greater affinity for open, prospect-dominant landscapes that in our evolutionary past supported the large game herds. Females should favor more heavily vegetated habitats that are likely to yield vegetables, fruits, and leafy greens.

Both landscape types existed on the African savannas in which early humans flourished. This habitat was a mosaic of open grassland, scattered copses, and denser woods near rivers and lakes (Butzer 1977; Blumenschine 1987). The open grasslands, with scattered trees, hills, and rock outcroppings, were an ideal habitat in which to hunt. The wide vistas provided the necessary space to plan distant moves, while the trees and prominences offered places from which to track moving animals. Such prospects also afforded visual surveillance of other human groups who were potential competitors for food and mates. If male-male competition played a crucial role in the evolution of human behavior, as some have proposed (Tooby and DeVore 1987), then the ability to see one's competition well in advance would be highly advantageous and allow for strategic planning.

We tested the gender hypotheses by conducting a content analysis of male and female landscape paintings. Although there are far fewer women painters than male painters, particularly in the heyday of landscape painting in the eighteenth and nineteenth centuries, we were able to compare samples of paintings by Irish and French women with paintings by Irish and French men of the same period. We gathered data on the prospect and refuge symbolism and on the location of people in the landscapes. A total of 108 paintings were analyzed: 52 by female painters and 56 by males. High prospect symbolism included an open landscape, view opportunities (hills, mountains, rock outcroppings), and a view of the horizon at least half the width of the painting. High refuge symbolism included houses or other buildings or vegetation cover, especially in the foreground (a tree canopy or woods).

Almost half of the women's paintings had high refuge symbolism compared with only 25 percent of the men's paintings (chi square = 6.89, p = .03). Men's paintings fell more frequently in the moderate category. Thus, as expected, men did not reject refuge entirely but found it less compelling than did women. Also, as predicted, men's paintings were more concerned with prospects: almost half of their paintings were in the high prospect category compared to only a quarter of the women's paintings (chi square = 12.07, p = .002). Almost three-quarters of the men's paintings were moderate or high in prospect symbolism compared to less than half of the women's paintings. Prospect orientation was also assessed by measuring the amount of horizon shown in the painting. The horizon covered at least half the width in 58 percent of the landscapes painted by men, whereas 75 percent of the women's paintings had no horizon or only a peephole and only 14 percent had broad views of the horizon.

In addition to the prospect / refuge symbolism, we assessed gender differences in the placement of people in the landscape. The paintings contained a total of 166 figures (21 percent men, 49 percent women, 11 percent children, and 19 percent too small or too far away to be identifiable). We predicted that female figures would more likely be positioned in refuge settings, whereas males would be placed in open spaces. The data support this prediction. Almost half of the male figures were in open spaces compared with only 14 percent of the female figures. The males were almost equally divided between prospect and refuge locations, whereas the vast majority of females (86 percent) were located within a refuge (chi square = 7.40, p = .006). Although we did not make any predictions about unidentifiable figures, they were found almost exclusively in men's paintings.

This gender bias is even stronger when paintings by males are compared with paintings by females. Women painters were more likely to place *both* male and female figures in refuge settings; only 6 percent of the female figures in women's paintings were in open spaces compared to 35 percent of the male figures (chi square = 9.00, p = .002). Men painters placed 62 percent of their male figures and 26 percent of their female figures in open spaces.

Analysis of the figures in the sunset and sunrise paintings also shows that more than half of the female figures were located in or close to the refuge compared with 29 percent of the male figures. Interestingly, there were only eleven female figures in the sunset paintings compared to fifty-eight men. This ratio suggests that night is not an appropriate time for women to be away from home.

Although the analysis of gender differences in landscape paintings supports our hypotheses, it is possible to interpret the results as arising from cultural factors rather than as evolved responses to the environment. Women's lives were much more restricted in the eighteenth and nineteenth centuries than they are now; thus their tendency to paint refuges may reflect the fact that they stayed much closer to home than did male painters. Nevertheless, it is very likely that women of that time ventured from home at least occasionally and undoubtedly experienced open spaces and panoramic views in parks and countryside. The fact that they did not choose to paint such scenes does not indicate lack of opportunity; it may simply represent a lack of interest.

Landscape Design

Building owners frequently hire landscape architects to make their grounds more aesthetically pleasing. As the designs are often undertaken at great expense, it is informative to look at the kinds of changes recommended by landscape designers. Humphrey Repton, an eighteenth-century British landscape architect, presented his customers with "before" and "after" drawings of their estates and bound these drawings and accompanying prose in red covers (Repton 1907). These "Repton Redbooks" are a valuable source with which to test evolutionary hypotheses. If humans have an intrinsic bias for certain landscapes and landscape elements, it should be possible to see this bias in the features that are added to environments to enhance their appeal. Repton is a particularly interesting case study because many of his manipulations are done solely for the sake of making the design more appealing to potential customers. Indeed, some of the changes were quite beyond Repton's powers to execute—for instance, he regularly added herds of sheep and deer to the estate grasslands and occasionally changed the shape of existing trees to make them more attractive.

With this in mind, we conducted content analyses of several Repton Redbooks. Although to conduct a thorough analysis one would have to travel to numerous libraries in the United States and England, we were able to obtain a sample of eighteen "before" and "after" designs to conduct an exploratory analysis. We tested the following hypotheses:

1. Landscapes are changed in ways that make them more savanna-like (for example, scattered trees and copses interspersed with open spaces).

2. Closed woods are opened up in ways that increase visual access and penetrability.

3. Distant views, especially to the horizon, are opened up.

4. Refuges are added if none presently exist.

5. Cues signaling ease of movement—paths, bridges, roads—are added.

6. Water and large, grazing mammals are added if they are not already present. Further, the animals should be shown in positions that indicate lack of tension or concern with hazards (lying down or grazing rather than alert and running).

As predicted, Repton added trees and copses to the open fields in almost half of his designs. In 26 percent of them copses were added at water's edge, a design feature that enhances the refuge character of the scene and protects people at a time when attention is turned to drinking, resting, or bathing. Repton also frequently broke up a straight pasture / wood edge by moving trees into the open space, creating an uneven edge. These features are all characteristic of savanna environments, the habitat in which humans lived for millions of years (Butzer 1977; Foley 1989). Repton also removed trees in 42 percent of the landscapes to open views to the horizon, and he opened up dense woods to make them more penetrable in four designs. He notes in his book, *The Art of Landscape Gardening* (1907:97), that too many trees "make a place appear gloomy and damp." His changes in water features are one of the more striking aspects of his designs. Repton added water; he expanded water features or made them more conspicuous; he enhanced visual access to water elements in 48 percent of the designs. His water features are in the open, often with a copse of trees at the edge. Moreover, he frequently added rocks to a brook to give it "a rippling, lively effect which is highly preferable to a narrow stagnant creek." Repton notes that "it is only by such deceptions that art can imitate the most pleasing works of nature" (1907:96).

Repton was notorious among his clients for adding animals and boats to his scenes to enhance their appeal. In the designs we analyzed, he added almost 200 large, grazing mammals. In four scenes he added boats. He notes that people often object to the addition of boats and animals, but he defends the practice by saying that "both are real objects of improvement, and give animation to the scene" (1907:42). His use of animals is especially interesting because these changes are simply tricks to enhance the appeal of the scene.

"Before" and "After" Analysis

Landscape painters frequently alter the scenes they paint to give them greater appeal or drama. John Constable, the renowned nineteenth-century English painter, is a particularly good source of data in this regard because he often composed sketches of scenes that can be compared with his final painting. We assume, of course, that the sketches approximate the actual scene.

We were able to locate nine pairs of sketches and paintings for which the drawing was detailed enough to permit a "before" and "after" analysis. We did a content analysis of the paintings and noted all the changes that were made to the landscape as depicted in the sketch. The results indicate that Constable's transformations of the landscape are highly consistent with the evolutionary interpretations we present in this chapter: the most frequently made changes involve the addition of houses, people, and animals, as well as alterations in vegetation that open views to the horizon or to the refuge and alterations in water features to make them more conspicuous. In seven of the nine pairs, Constable opened tree canopies to make the trunks and branching patterns conspicuous. He also added a magnificent rainbow in one scene and prominent clouds in two paintings.

Of the nine pairs, six included changes in refuge conditions. Constable added buildings in four paintings, opened up the refuge (by removing shrubbery in front of it) in two paintings, and made a refuge more conspicuous by painting it lighter in two cases. (This adds up to eight changes, not six, because Constable did multiple refuge alterations in two paintings.) The animals added to the scenes included a flock of sheep, six horses, several cows, two dogs, a large deer, and a flock of birds. A total of twenty people were added in six paintings.

Constable's trees are especially intriguing. The removal of leaves from the foreground trees exposed complex branching patterns that would not otherwise be open to view. He frequently sketched trees, often without foliage, to reveal trunk and branches. Constable's interest in tree structure is especially fascinating to us because our earliest research on human habitat selection was a project on aesthetic judgments of tree shape.

Analysis of Tree Shapes

Throughout most of human evolution, trees have played a central role in everyday life. They are sources of food and shade, they afford safe sleeping and eating places, and they offer vantage points for surveying the landscape (Bourliere 1963; Isaac 1983; Shipman 1986). As people became more proficient with technology, trees also provided materials for shelter, weapons, tools, and medicine (Lee 1979). Given the central role of trees in human life, it is worthwhile asking whether attractiveness in trees is related to these functional characteristics. Many studies indicate that trees figure largely in people's concepts of environmental quality, but few studies have addressed this question: are all trees equally effective in producing positive responses?

Using arguments from basic models of habitat selection, one of us (Orians 1980) proposed that tree shapes characteristic of environments providing the highest-quality resources for evolving humans should be more pleasing than shapes characterizing poor habitats. On the African savannas, these trees have canopies broader than they are tall, trunks that terminate and branch well below half the height of the tree, small leaves, and a layered branching system. Many tree species in drier African habitats lack trunks altogether and often have dense, convoluted branching patterns. By contrast, trees in wetter settings in East Africa have canopies that are taller than their width, high trunks, and branching patterns that are more uniform than layered.

In a preliminary test of the savanna hypothesis, Orians (1986) compared shapes of trees from high-quality African savannas with those selected for use in Japanese gardens. Two hypotheses were examined. The first was that the trees selected for garden use should resemble, in their natural growth forms, savanna trees more than trees not selected for use in gardens. The second hypothesis was that subsequent modifications of the growth forms of trees, whether by pruning or genetic manipulation, should tend to make them more like the savanna models than are the unmanipulated individuals. Support for the first hypothesis was found among maples (*Acer*), but chosen oaks (*Quercus*) did not differ in general growth form from those not used in gardens. Strong support for the second hypothesis was found among conifers used in Japanese gardens. Garden conifers are highly modified by pruning them to grow broader than tall; trunks are trained to branch close to the ground; foliage is trimmed to produce a distinct layering similar to that of a number of savanna species (Orians 1986).

We supplemented these preliminary results by determining people's responses to photographs of tree shapes. G. H. and E. N. Orians took a large number of photographs of *Acacia tortilis*, a tree characteristic of high-quality savannas in East Africa, showing the full range of shapes exhibited by that species. From the series we selected a set of trees that varied in height / width ratio, height at which the trunk bifurcated, and the extent of canopy layering. Because the species of tree was held constant, the importance of these variables could be assessed against a relatively uniform background. We tested four hypotheses:

1. Trees with lower trunks should be more attractive than trees with high trunks.
2. Trees with moderate canopy density should be more attractive than trees with low or high canopy density.
3. Trees with a high degree of canopy layering should be more attractive than trees with low or moderate degrees of layering.
4. The broader the tree canopy, relative to its height, the more attractive the tree should be.

The stimuli used in the study were black and white photographs printed in a photoquestionnaire. All of the tree photos were taken at approximately the same distance and under similar sky conditions. Although it was impossible to isolate the trees completely, we made a deliberate attempt to focus on individual trees. We eliminated all photos with obvious mountains or hills in the background because these features are known to affect preference judgments.

We selected photos so that we could vary each major category (trunk pattern, canopy density, canopy layering) while holding the others as constant as possible. For example, we held canopy layering and overall shape constant while varying trunk patterns. Each of the trees considered for the final questionnaire was rated independently by both of us on each of the major dimensions. Only trees for which there was complete agreement were selected for study. The primary dependent measure in this research was attractiveness rating. We deliberately chose to use the dimension attractive / unattractive rather than the more common like / dislike dimension because we were specifically looking at aesthetic judgments. There are many reasons why someone might like a particular tree that have little to do with its judged attractiveness: it could look like a tree in their backyard, for instance, or it might remind them of a particular place they had been.

We also measured each tree on four dimensions: canopy width, canopy height, tree height, and trunk height. The measures were made from the photographs using a millimeter ruler. The data were then transformed into canopy width / canopy height ratio, canopy width / tree height ratio, and trunk height / tree height ratio. These measures provided an objective way of assessing canopy features and trunk height. Subjects' ratings were made on a six-point scale: 1 to 2 = "unattractive," 3 to 4 = "moderately attractive," and 5 to 6 = "very attractive." The scale included division points between each interval so that a subject could, for instance, rate a tree as 3.4, rather than 3 or 4.

A total of 102 people responded to the survey: 72 were drawn from in front of the University of Washington Bookstore and 30 from a restaurant on campus. Ages ranged from eighteen to sixty. People were approached and asked if they would mind filling out a questionnaire on the role of trees in environmental aesthetics. If they agreed, they were given a survey. The instructions on the survey read: "In the attached photoquestionnaire, we ask you to rate the relative attractiveness of a number of trees. There are six trees on each page, with the rating scale under each. Circle the point on the scale which best corresponds to your opinion of the tree's attractiveness." They were asked to rate the trees on each page before they continued on to the next and not to turn back once they had completed a page.

Our analysis indicates that trunk height, canopy layering, and the canopy width / tree height ratio significantly influenced attractiveness scores. The canopy width / canopy height ratio did not, contrary to our predictions, have a significant effect on ratings of attractiveness. We were not able to assess the impact of canopy density because the final version of the photoquestionnaire printed dark and made some of the canopies appear much denser than they were in the original photographs.

In addition to examining the mean scores for all trees, we also compared the seven most attractive trees with the seven least attractive. As can be seen in Table 4.1, the most attractive trees had highly or moderately layered canopies and also had lower trunks and a higher canopy width / tree height ratio than the least attractive trees. Visual analysis of the trees indicates that several of the least attractive trees had broken branches, deformed trunks, and highly asymmetrical canopies.

Briefly, the results of this study indicate that: some tree shapes are clearly more attractive than others; aesthetic quality is based on trunk height, canopy layering, and the ratio of the tree's canopy width to tree height; the canopy width / canopy height ratio did not significantly influ-

Humans, Habitats, and Aesthetics

TABLE 4.1. *Comparisons of Most and Least Attractive Trees*

	Seven Most Attractive	Seven Least Attractive	t	p
Mean attractiveness score	3.91	2.9	12.58	.000
Trunk height / tree height ratio	0.17	0.33	8.24	.000
Canopy width / tree height ratio	1.93	1.53	5.89	.000
Canopy width / canopy height ratio	3.63	3.56	0.20	.83

ence attractiveness scores; signs of resource depletion (defoliated branches) or unhealthiness (dead branches, deformation) negatively affected the attractiveness scores. We did not make specific predictions about the healthiness of the trees and, in fact, tried to hold tree condition constant. We could not entirely eliminate such factors as broken or defoliated branches, however, because of intense foraging pressures by elephants related to drought conditions and habitat loss. Despite our efforts to make the tree sample uniform with respect to tree condition, subjects readily perceived the imperfections.

If trees have been important in our evolutionary past for protection and visual scanning of the environment, as well as for food, then the features most likely to influence these needs are the trunk pattern, canopy shape, and health indicators. A low trunk is easier to climb than a high one; a broad, umbrella-like canopy affords greater refuge from sun or rain than a narrow, high canopy. The results from our preliminary survey are in accordance with a functional-evolutionary perspective on the relationship between trees and humans.

Artificial Selection of Plant Characteristics

For millennia people have practiced artificial selection of plants to increase their value as sources of food, fiber, and aesthetic pleasure. Most research has concentrated on changes designed to increase the quality and quantity of edible parts of plants, the principal component of plant domestication. Nevertheless, the aesthetic role of plants in human life is sufficiently important that considerable effort has been expended to modify, either by genetic selection or pruning, the shapes and sizes of entire plants and their leaves and flowers. As an example of such selection, we present an analysis of the changes produced in cultivars of a Japanese maple (*Acer palmatum*), the species used most extensively in formal Japanese gardens.

Because most trees of high-quality African savannas have small, compound leaves, we predicted that selection would increase the number of lobes on the leaves and the depth of lobing. Although we were unable to test the prediction that selection would tend to make leaves smaller—insufficient information was available in Vertrees' compendium (1978) to estimate leaf sizes—we note that all cultivars in the dwarf series have leaves much smaller than wild *A. palmatum*. We also predicted that artificial selection would favor trees with spreading crowns that were broader than they were tall.

Our predictions about selection for coloration of leaves are more complicated. In tropical environments where humans evolved, bright coloration is generally restricted to newly expanding leaves. Flushing of new leaves is often associated with the beginning of the rainy season, a time of increasing resources. Moreover, the locations of certain tree and shrub species may be more readily determined at that time than later, when all plants have green leaves. Conspicuous colors are generally lacking in leaves of tropical plants when the leaves are senescing and falling. Bright coloration is characteristic of most trees when they flower because the vast majority of tropical trees have animal-pollinated flowers and many of them flower en masse so that the entire tree is ablaze with color, but only for a few days. In regions with pronounced dry seasons, many trees flower during the dry season when they are leafless, making the individuals especially conspicuous.

Flowering trees should be especially appealing because they have been major food sources for people throughout evolutionary time. Indeed, because flowers typically lack many of the toxins plants sequester in their leaves to deter grazing, they are often especially desirable foods. Flower tissue also contains high levels of water and nitrogen in the form of pollen, nectar, and developing ovules (Strong et al. 1984; Scriber and Slansky 1981; Bazzaz et al. 1987). When concentrated as honey, moreover, flowers were, until recently, the only major source of natural sugar. Beekeeping is an ancient art practiced by human cultures for thousands of years. Even if they are not eaten, flowers play a role in signaling the future availability of fruits and nuts, many of which are edible. Therefore, paying attention to flowering plants should have contributed positively to survival among our ancestors. And because flowering periods may be very brief, continual attention to flowering is necessary to track current and future resources.

Maples, unlike most tropical trees, have inconspicuous wind-pollinated flowers that are poor targets for selection. Leaves, however, which do contain red and yellow pigments even when they are masked by chlorophyll, can be selected to make those colors conspicuous during seasons when they would not normally be visible. Thus we predicted that, in lieu of flowers, Japanese maples would be selected for bright coloration of summer leaves.

Our statistics are drawn from the extensive compendium of Japanese maples assembled by Vertrees (1978). "Wild type" *A. palmatum* has seven lobes, green leaves, and a moderate depth of lobing. Vertrees divides the maple cultivars into groups labeled "palmate," "dissectum," "deeply divided," "linearilobium," and "dwarf." For named cultivars in each of these groups, we recorded the color of the summer leaves, the number of lobes, and the depth of lobing. Depth of lobing was scored on a scale of 1 to 5, where 1 = very shallow lobing, 3 = lobed approximately halfway to the base of the leaf, and 5 = lobed to the base so that the leaf is effectively compound. "Wild type" *A. palmatum* has a lobing score of 2 to 3; lobing varies both within and among individuals, but leaves with lobing greater than 3 are rare in nature.

The results of our analysis are shown in Table 4.2. They reveal that there has been strong selection for reddish coloration of leaves. More than half the cultivars have nongreen leaves. Many have leaves that are reddish when young but become greener when they mature. Vertrees also describes an additional forty cultivars (not in the table) with variegated leaves, most of which have red in them. We emphasize that Table 4.2 scores the color of summer leaves; nearly all the cultivars have red or orange leaves in autumn.

Contrary to our prediction, selection on the number of lobes has been weak except in the "dissectum" group, where the deep division tends to separate the basal lobes into two, thus yielding a 7–9 score. The depth of lobing, however, has been under strong selection. Very few varieties are as shallowly lobed as "wild type" *A. palmatum*, and more than half the vari-

TABLE 4.2. *Characteristics of* Acer palmatum

Group	No. of Culti- vars	Leaf Color				No. of Lobes					Depth of Lobing[b]			
		Grn.	Red	Ppl.	Other[a]	5	6	7	8	9	2	3	4	5
Palmate	44	20	11	12	1	4	15	20	4	1	4	17	15	5
Dissectum	27	9	7	7	4			13	10	4				27
Deeply divided	39	11	9	13	6	1	5	29	3				3	36
Linearilobium	10	5	4		1	7	3						1	9
Dwarf	20	12	2	1	5	16	2	2			1	7	8	4
Total	140	57	33	33	17	28	22	67	17	5	5	24	27	81

Note: "Wild type" *A. palmatum* has seven lobes, green leaves, and depth of lobing of 2–3.
[a]"Other" includes variegated and cases where young leaves differ in color from mature leaves.
[b]Scale: 1 = shallow to 5 = very deep (almost to base).
Source: Data from Vertrees (1978).

eties are lobed all the way to the base of the leaves. These leaves appear to be palmately compound. Moreover, the lobes of many of the "dissectum" cultivars are themselves deeply dissected (hence the name), making the leaves appear to be doubly compound.

Selection for pendulant, even prostrate, forms has been very strong. All the "dwarf" cultivars mature at a height of 1 meter or less. Many of the taller forms have been selected to be very shrubby, to be broader than tall when mature, and to have drooping branches. This is the case for nearly all the "dissectum" and many of the "palmate" cultivars. In contrast, most of the "deeply divided" cultivars are erect in form.

Flowers as Resource Signals

Domestication of flowers is a major human enterprise. Visits to sick people are inevitably accompanied by flowers, and flowers are often brought to hosts by guests at dinner parties. Indeed, people annually spend millions of dollars on flowers. If we are correct in our view that flowers evoke strong positive feelings because they have long been associated with food resources, then selection on flowers should increase those traits that signal quantity of resources.

Amount of nectar is positively correlated with flower size and flower zygomorphism—that is, asymmetry in shape. Zygomorphism evolves, in part, to restrict a pollinator's entry to a single route, thereby resulting in precise placement of pollen on the pollinator's body. Zygomorphism may also limit entry to a restricted set of potential pollinators. Pollinators must learn how to enter and exploit these complex flowers, but once they know how to use them the rate of harvesting may be high. Indeed, to favor specialization on the part of pollinators, plants must provide larger than average quantities of resources in zygomorphic flowers. Otherwise pollinators would not find it profitable to visit these flowers.

We have not yet carried out an analysis of changes in flowers under domestication, but it is evident from a cursory examination that flowers are generally selected to increase their size and multiply the number of floral parts (especially sepals and petals). Both of these changes would signal larger resource amounts. It is less clear that zygomorphic flowers are more attractive than symmetrical ones. Orchids, the ultimate in zygomorphic flowers, are highly attractive, but they are also usually large and have the desirable feature of lasting for many days after being picked. Nonetheless, it is our impression that flowers with unusual shapes are regarded as especially attractive and desirable. Selection for flower shapes will be the subject of a future investigation.

Attraction vs. Aversion

For those of us interested in the biophilia phenomenon, it is often the obvious that provokes our curiosity. Many of the fears and pleasures associated with nature may seem, at first glance, to be unworthy of scientific investigation because they are so commonplace. Upon reflection, however, it is even more curious that the real hazards of modern life—guns, bombs, drugs, polluted water—do not generate nightmares or intense fears as frequently as do hazards from our evolutionary past (snakes, predators, darkness).

The same might be said of things and places that bring us pleasure. Although flowers, grazing deer, and climbable trees are no longer as crucial to survival as they once were, our emotional attachment to these amenities has persisted across the gulf of time and space that now separates us from our hunting and gathering ancestors. We modern humans spend inordinate sums of money on "dream vacations" that transport us back to our tropical beginnings. We deliberately seek out environments lush with water, green trees, flowers, and grass, adorned by spectacular sunsets, and capped off by sumptuous meals—all experienced from comfortable refuges in the company of close companions. Our emotional attachment to nature and its processes is as much a part of human life as are the modern technologies that make our journey to paradise possible.

It is especially telling to examine people's emotional responses to the recent devastation of southern Florida by Hurricane Andrew. A multitude of stories in the national media have described people's feelings of sadness and loss born of experiencing their lush environment now bereft of trees, flowers, and animals. Even in the face of immense survival pressures that come with the loss of homes, stores, and workplaces, people grieve for the loss of natural amenities that make life pleasurable.

Since humans spent 99 percent of their history as hunter-gatherers, the answer to why we are attracted to water and trees and are afraid of snakes or darkness must lie, at least in part, in the day-to-day relationship between nature and our hominid ancestors. Here we have shown some ways in which landscape preferences and aesthetics can be studied from a functional-evolutionary perspective. Unlike many current approaches to landscape aesthetics, we do not assume that everyone will respond similarly to a certain environment. In fact, an evolutionary approach postulates quite the opposite view. Variability in people's preferences and assessments is expected, but that variability is not random. Rather, it is a function of such biologically relevant factors as age, gender, familiarity, physical condition, and presence of others. For instance, we based our gender-related predictions on differential vulnerability to environmental hazards and predation. If our assumptions are correct, then people's preferences for spatial enclosure should move back and forth along a prospect-refuge continuum in accordance with their social, physical, and emotional feelings of vulnerability. Thus children and elderly people, as well as those who are physically ill or depressed, should prefer spaces offering refuge rather than open spaces where they can readily be seen by others.

Although an evolutionary approach to human interactions with the environment is based on the assumption that responses evolve in ways that enhance fitness, we have not attempted to measure fitness directly, nor do we believe it is profitable to do so. Our objective has been to discover problems faced by humans in their interactions with the natural world and to uncover the basic architecture of the problem-solving process—including the types of information used to make decisions and the ways in which context influences responses. Our approach uses a "computational" framework initiated by Marr (1982) in his studies of the visual system and most recently expanded by Tooby and Cosmides in a major theoretical discussion of the adaptive functioning of the human mind (1992).

A functional-evolutionary approach to human problems has attracted attention in a number of fields related to the environment. Researchers in psychiatry (Nesse 1990), medicine (Williams and Nesse 1991), and nutrition (Harris and Ross 1987) have begun to look at modern diseases from the perspective of hunter-gatherer lifestyles during the Paleolithic period. Nesse's psychiatric work, which focuses on the adaptive functions of emotions, is especially relevant to environmental preferences and assessments. He argues that negative emotional states such as fear, anxiety, or depression may be our mind's way of telling us to attend to the situation at hand. Medication may, in some circumstances, be truly maladaptive because it enables us to escape from negative situations rather than understanding and learning from them. Although Nesse deals primarily with the social environment, our work suggests that physical and biological features of the environment also have consequences, not currently well understood, for both positive and negative emotional states.

The few studies of the relationship between environment and human response suggest that nature has a more powerful impact on our emotional and physical health than has been appreciated to date (Chapter 3 in this volume; Ulrich et al. 1992; Ulrich 1984). The benefits do not even require direct contact with nature: they can be experienced from the passive viewing of posters and slides, as well as window views. Our own studies of visual decor in office settings (Heerwagen and Orians 1986) show that people in windowless offices used more nature posters and photos to decorate their work spaces than did people in windowed offices where views of nature were available (perhaps as a way to reduce work stress and enhance their psychological functioning). The widespread appeal of nature posters is evident in many public buildings these days—particularly medical offices and hospitals—where pictures of landscape scenes, flowers, and animals abound in stairwells, hallways, and waiting rooms. And though little research has been done on the effects of indoor plants, they too should have positive effects. A post-occupancy evaluation of seven office buildings in the Pacific Northwest, for instance, found that half of the 260 people surveyed had added plants or flowers to enhance their work space (Heerwagen 1991).

Most of the recent work on people and nature has focused on beneficial outcomes, but nature can also produce fear and tension. Work by Öhman (1986), Dimberg (1989), and Coss (1968) is especially illuminating. All three researchers have studied emotional and physiological response to animals or hostile humans. Coss, for instance, has found that people react most negatively to large, threatening animals (such as lions who look hungry) and respond more positively to small animals or those seen in profile. Öhman and Dimberg, using classical conditioning paradigms, were able to elicit and maintain fear responses to spiders, snakes, and hostile human faces but not to "neutral" stimuli such as flowers or to such modern hazards as guns. Indeed, Dimberg's studies have found that people do not react neutrally to flowers; they tend, instead, to exhibit positive emotions, an outcome that is highly understandable from our framework.

As we have illustrated, human response to features of the natural environment is complex and highly variable. Expectations of simple relationships between environmental variables and aesthetic judgments are certain to be frustrated. Nonetheless, we believe we have demonstrated that interesting response patterns can be detected if appropriate hypotheses are posed and tested. These hypotheses and tests, however, do little more than touch the surface of a much richer research agenda. We have not begun to explore changes associated with age, for example. During a lifetime, a person changes from a creature unable to move and entirely dependent upon its parents, through various stages of mobility and independence, to arrive eventually at old age with restricted sensory capacities and mobility. Because needs change during life's journey, so too should one's preferences and emotional response to environments.

Nor have we considered the potentially powerful applications of this approach to the built environment. From the perspective of habitat selection, buildings and urban environments provide both prospect and refuge and, moreover, expose inhabitants to a multitude of hazards both physical (stairways) and social (human predators). The attributes of the physical environment have implications for human behavior that are not adequately understood or utilized by designers and environmental planners. Recent work by Grant Hildebrand (1991) is an exception. Hildebrand's use of Appleton's (1975) prospect/refuge theory to analyze the consistent appeal of Frank Lloyd Wright's houses is an example of how designers could use the evolutionary approach to understand human responses to built environments.

Evidence pointing to the profound extent to which people are emotionally and physiologically bound to the natural world is steadily increasing. As Wilson pointed out in *Biophilia*, these ties have strong implications for the preservation of biodiversity. A biologically impoverished planet will not only reduce humanity's economic options, it will diminish our emotional lives as well. And it is a loss from which recovery will be virtually impossible. As Wilson puts it: "The one process now going on that will take millions of years to correct is the loss of genetic and species diversity by the destruction of natural habitats. This is the folly our descendants are least likely to forgive us" (1984: 121). The study of humans as evolved animals can, perhaps, help us avoid committing this ultimate folly.

ACKNOWLEDGMENTS

We would like to thank the participants of the Biophilia Symposium, especially Edward Wilson and Stephen Kellert, for invigorating and provocative discussions of the human/nature relationship. We would also like to thank Roger Ulrich, John Tooby, and Leda Cosmides for their insightful comments on the ideas addressed here and in our previous work. We are grateful also to the Landscape Research Group for their support of our research on trees.

REFERENCES

Appleton, J. 1975. *The Experience of Landscape*. London and New York: Wiley.
———. 1980. "Landscape Aesthetics in the Field." In *The Aesthetics of Landscape*, edited by J. Appleton. Didcot, England: Rural Planning Services, Ltd.
———. 1990. *The Symbolism of Habitat*. Seattle: University of Washington Press.

Balling, J. D., and J. H. Falk. 1982. "Development of Visual Preference for Natural Environments." *Environment and Behavior* 14:5–28.

Bazzaz, F. A., N. R. Chiarello, P. D. Coley, and L. F. Pitelka. 1987. "Allocating Resources to Reproduction and Defense." *BioScience* 37:58–67.

Binford, L. R. 1985. "Human Ancestors: Changing Views of Their Behavior." *Journal of Anthropological Archaeology* 4:292–327.

Blumenschine, R. J. 1987. "Characteristics of an Early Hominid Scavenging Niche." *Current Anthropology* 28(4):383–407.

Bourliere, F. 1963. "Observations of the Ecology of Some Large African Mammals." In *African Ecology and Human Evolution*, edited by F. C. Howell and F. Bourliere. Chicago: Aldine.

Burke, E. 1757. *Philosophical Enquiry into the Origin of our Ideas of the Sublime and Beautiful*.

Butzer, K. W. 1977. "Environment, Culture, and Human Evolution." *American Scientist* 65:572–584.

Campbell, B. 1985. *Human Evolution*. 3rd ed. New York: Aldine.

Cody, M. L. 1985. "An Introduction to Habitat Selection in Birds." In *Habitat Selection in Birds*, edited by M. L. Cody. New York: Academic Press.

Cosgrove, D. 1984. *Social Formation and Symbolic Landscape*. London: Croom Helm.

———. 1986. "Critiques and Queries." In *Landscape Meanings and Values*, edited by E. C. Penning-Rowsell and D. Lowenthal. London: Allen & Unwin.

Cosmides, L., and J. Tooby. 1987. "From Evolution to Behavior: Evolutionary Psychology as the Missing Link." In *The Latest on the Best: Essays on Evolution and Optimality*, edited by J. Dupre. Cambridge: MIT Press.

Coss, R. G. 1968. "The Ethological Command in Art." *Leonardo* 1:273–287.

Dimberg, U. 1989. "Facial Electromyography and Emotional Reactions." *Psychophysiology* 27:481–494.

Foley, R. 1989. "The Ecological Conditions of Speciation: A Comparative Approach to the Origins of Anatomically-Modern Humans." In *The Human Revolution*, edited by P. Mellars and C. Stringer. Edinburgh: University of Edinburgh Press.

Gowaty, P. 1992. "Battles of the Sexes and the Evolution of Mating Systems: Female Resistance to Male Control Selects for Frequency Dependent Reproductive Patterns." Paper presented at the Human Behavior and Evolution Society conference, Albuquerque, N.M., 22–26 July.

Harris, M., and E. B. Ross, eds. 1987. *Food and Evolution*. Philadelphia: Temple University Press.

Hawkes, K. 1987. "How Much Food Do Foragers Need?" In *Food and Evolu-*

tion, edited by M. Harris and E. B. Ross. Philadelphia: Temple University Press.

Heerwagen, J. 1991. "Post Occupancy Evaluation of Energy Edge Buildings." Final Report. Portland, Ore.: Bonneville Power Administration Project.

Heerwagen, J., and G. H. Orians. 1986. "Adaptations to Windowlessness: A Study of the Use of Visual Decor in Windowed and Windowless Offices." *Environment and Behavior* 18:623–639.

Hepburn, R. W. 1968. "Aesthetic Appreciation in Nature." In *Aesthetics in the Modern World*, edited by H. Osborne. London: Thames & Hudson.

Hildebrand, G. 1991. *The Wright Space*. Seattle: University of Washington Press.

Hildén, D. 1965. "Habitat Selection in Birds." *Annales Zoologici Fennici* 2:53–75.

Isaac, G. 1978. "Food Sharing and Human Evolution." *Journal of Anthropological Research* 34:311–325.

———. 1983. "Bones in Contention: Competing Explanations for the Juxtaposition of Early Pleistocene Artifacts and Faunal Remains." In *Animals and Archaeology*, Part I: *Hunters and Their Prey*, edited by J. Clutton-Brock and C. Grigson. Oxford: B.A.R. International Series.

Jakle, J. J. 1987. *The Visual Elements of Landscape*. Amherst: University of Massachusetts Press.

Kaplan, S., and R. Kaplan. 1982. *Cognition and Environment: Functioning in an Uncertain World*. New York: Praeger.

Lack, D. 1971. *Ecological Isolation in Birds*. Oxford: Blackwell.

Lee, R. B. 1979. *The !Kung San: Men, Women, and Work in a Foraging Society*. Cambridge and New York: Cambridge University Press.

Lee, R., and I. DeVore, eds. 1976. *Kalahari Hunter-Gatherers*. Cambridge: Harvard University Press.

Lovejoy, C. O. 1981. "The Origin of Man." *Science* 211:341–350.

Marr, D. 1982. *Vision: A Computational Investigation into the Human Representation and Processing of Visual Information*. San Francisco: Freeman.

Nesse, R. 1990. "Evolutionary Explanations of Emotions." *Human Nature* 1(3):261–289.

Öhman, A. 1986. "Face the Beast and Fear the Face: Animal and Social Fears as Prototypes for Evolutionary Analyses of Emotion." *Psychophysiology* 23(2):123–145.

Orians, G. H. 1980. "Habitat Selection: General Theory and Applications to Human Behavior." In *The Evolution of Human Social Behavior*, edited by J. S. Lockard. New York: Elsevier.

———. 1986. "An Ecological and Evolutionary Approach to Landscape Aesthetics." In *Landscape Meanings and Values*, edited by E. C. Penning-Rowsell and D. Lowenthal. London: Allen & Unwin.

Orians, G. H., and J. H. Heerwagen. 1992. "Evolved Responses to Landscapes." In *The Adapted Mind: Evolutionary Psychology and the Generation of Culture*, edited by J. Barkow, L. Cosmides, and J. Tooby. Oxford and New York: Oxford University Press.

Peters, C. R., and E. M. O'Brien. 1981. "The Early Hominid Plant-Food Niche: Insights from an Analysis of Plant Exploitation by *Homo*, *Pan*, and *Papio* in Eastern and Southern Africa." *Current Anthropology* 22:127–140.

Peters, C. R., E. M. O'Brien, and E. O. Box. 1984. "Plant Types and Seasonality of Wild-Plant Foods, Tanzania to Southwestern Africa: Resources for Models of the Natural Environment." *Journal of Human Evolution* 13: 397–414.

Potts, R. 1987. "Reconstructions of Early Hominid Socioecology: A Critique of Primate Models." In *The Evolution of Human Behavior: Primate Models*, edited by W. G. Kinzey. Albany: State University of New York Press.

Repton, H. 1907. *The Art of Landscape Gardening*. Boston and New York: Houghton Mifflin.

Rose, M. C. 1976. "Nature as Aesthetic Object: An Essay in Meta-Aesthetics." *British Journal of Aesthetics* 16:3–12.

Scriber, J. M., and F. Slansky. 1981. "The Nutritional Ecology of Immature Insects." *Annual Review of Entomology* 26:183–211.

Shipman, P. 1986. "Scavenging or Hunting in Early Hominids: Theoretical Framework and Tests." *American Anthropologist* 88:27–40.

Smuts, B. 1992. "Male Aggression Against Women: An Evolutionary Perspective." *Human Nature* 3(1):1–44.

Staddon, J. 1987. "Optimality Theory and Behavior." In *The Latest on the Best: Essays on Evolution and Optimality*, edited by J. Dupre. Cambridge: MIT Press.

Strong, D. R., J. H. Lawton, and R. Southwood. 1984. *Insects on Plants: Community Patterns and Mechanisms*. Cambridge: Harvard University Press.

Tooby, J., and L. Cosmides. 1990. "The Past Explains the Present: Emotional Adaptations and the Structure of Ancestral Environments." *Ethology and Sociobiology* 11:375–424.

———. 1992. "The Psychological Foundations of Culture." In *The Adapted Mind*, edited by J. H. Barkow, L. Cosmides, and J. Tooby. Oxford and New York: Oxford University Press.

Tooby, J., and I. DeVore. 1987. "The Reconstruction of Hominid Behavioral Evolution Through Strategic Modeling." In *The Evolution of Human Behavior: Primate Models*, edited by W. G. Kinzey. Albany: State University of New York Press.

Ulrich, R. S., 1983. "Aesthetic and Affective Response to Natural Environment." In *Human Behavior and Environment*, vol. 6: *Behavior and the Nat-*

ural Environment, edited by I. Altman and J. F. Wohlwill. New York: Plenum.

———. 1984. "View Through a Window May Influence Recovery from Surgery." *Science* 224:420–421.

———. 1986. "Human Response to Vegetation and Landscapes." *Landscape and Urban Planning* 13:29–44.

Ulrich, R. S., et al. 1992. "Stress Recovery During Exposure to Natural and Urban Environments." *Journal of Environmental Psychology* 11:201–230.

Vertrees, J. D. 1978. *Japanese Maples*. Forest Grove, Ore.: Timber Press.

Washburn, S. L., ed. 1963. *Social Life of Early Man*. Chicago: Aldine.

Williams, G. C., and R. M. Nesse. 1991. "The Dawn of Darwinian Medicine." *Quarterly Review of Biology* 66(1):1–21.

Wilson, E. O. 1984. *Biophilia*. Cambridge: Harvard University Press.

Zajonc, R. 1980. "Feeling and Thinking: Preferences Need No Inferences." *American Psychologist* 35:151–175.

CHAPTER 5

Dialogue with Animals:
Its Nature and Culture

Aaron Katcher and Gregory Wilkins

T HE THREADS OF discourse in Edward Wilson's *Biophilia* (1984) include an apologia, a major hypothesis about human perception and motivation, and a moral or political agenda. This kind of heuristic weaving is integral to the writings of moral, religious, and political leaders, as well as some scientists, who wish to change the way in which the world is perceived or acted upon. Although these works are an efficacious means of persuasion and recruitment, they can be an impediment to testing a central hypothesis about reality. The possibility of disconfirmation seems to threaten the moral agenda and, in some instances, the significance of the apologia.

The coupling of moral agendas to hypotheses either by their defenders or opponents can obscure, politicize, or deaden the investigation of scientific concepts. One need only think of the conflation of sociobiology with Social Darwinism, the stigma of racism which almost precludes study of the genetic basis of human intelligence or aggression, the juncture of ideas about the cause of the Jurassic extinctions and ideas about nuclear winter, and the linkage of the evaluation of the ability of primates to use signs with the need to assert or deny the unique character of the human mind. Moreover, the moral agenda central to biophilia—preservation of habitat and species—would be just as imperative if there were no innate tendency to relate to other kinds of life or if human beings were genetically programmed to be hostile to or ignore other life-forms. In the latter case, it would be necessary to consider what cultural patterns could oppose and reverse those tendencies.

Examination of the biophilia hypothesis requires particular caution because of a highly pervasive trend in modern political and moral thinking: the attempt to counter the anxiety generated by the enormous technological and political changes of this century by infusing new value into the past, actual or mythical, to create a new moral authority. Ayatollahs and Serbian Nationalists, Solzhenitsyn and TV Evangelists, Conservationists, Archaeologists, and Antiabortionists, the Sierra Club and the Platform Committee of the Republican Party are all searching for authority in history.

Our specific agenda for the preservation of wilderness, wetlands, and rain forest and the species they contain is to some extent a manifestation of this general desire to preserve and seek instruction from the past. In the case of the rain forest we seek to preserve the intricate record of interdependent biological evolution, a record in large part unstudied and unrecorded and hence vulnerable to loss. When we look for the source of a love for other life-forms in our genetic inheritance we are searching our past for the authority to act on that love. That bias should cause us to question our own assumptions so that we restrain our human tendency to see things as we want them to be.

If the term biophilia is stripped of ethical connotation so that it denotes only the hypothesis that human beings have an innate tendency to "focus on living things," then its value can be assessed independently of a moral agenda. Our studies of children with conduct disorders and hyperactivity illustrate a salient benefit of thinking about biophilia as only a tendency to focus on living things without a preconception about the benign or destructive aspect of any subsequent action (Katcher and Wilkins 1992). Cruelty to animals is one of the defining symptoms of conduct disorder, but an intense *interest* in animals, not cruelty, was common to almost all children in our study. Some children, however, told us that previous to the education program they had liked to hunt and kill animals, but now they enjoyed taking care of them. To quote one student directly: "Helped me not to do something to animals, not to kill 'em, like mice. 'Cause if I see a mice run like that I got a shoe in my hand, it's dead. And they don't want me to do that. I used to set traps for 'em, try to kill 'em. It helped me not to do stuff like that."

Children who throw stones at birds and children who feed birds are both responding to what may be an innate tendency to focus their attention on living things. The choice of behavior used to engage the animal in the interaction is different and it is a *learned* behavior. The study of cruelty to animals as well as the study of pet keeping would profit from the recognition that both activities are alternative and learned means of dealing with an innate attraction to animals.

One way of studying how human beings pay attention to nature is to observe the physiological consequences of such attention. Roger Ulrich (Ulrich 1979, 1983; Ulrich and Simons 1986; Chapter 3 in this volume) has described a large and consistent body of data on the stress-reducing effects of viewing still pictures of natural settings. The same physiological changes can be observed when people contemplate arrays of moving animals, specifically tropical fish swimming in a tank. A very simple experiment (Katcher et al. 1983) demonstrated that watching an aquarium resulted in significant decreases in blood pressure below the resting level in both hypertensive and normal subjects. In a second experiment (Katcher, Segal, and Beck 1984) a more complex design was used to examine the influence of aquarium contemplation on patients about to undergo oral surgery. The subjects were randomly assigned to one of five half-hour pretreatments prior to surgery: (1) contemplation of an aquarium, (2) contemplation of a poster of a sylvan waterfall, (3) contemplation of the same poster after a hypnotic induction, (4) contemplation of an aquarium after a hypnotic induction, and (5) a control condition in which the subject was instructed to sit in a comfortable chair and relax. During the subsequent surgery, extraction of a third molar, their response to the procedure was recorded with

TABLE 5.1. *Effects of Hypnosis and Contemplation on Patient Comfort*

Group	Contemplation	Hypnosis	Mean	Standard Error	Significant Contrasts
1	Aquarium	no	40.3	3.0	Group 2****, 5***
2	Poster	no	26.5	3.1	Group 1****, 3***, 4***
3	Poster	yes	37.7	3.0	Group 2***, 5*
4	Aquarium	yes	38.3	3.0	Group 2***, 5*
5	None	no	29.9	2.7	Group 1***, 3**, 4*

Note: Higher values = more comfort and relaxation.
**** $p < .001$; *** $p < .01$; ** $p < .05$; * $p < .06$.

three methods: responses were scored by both an observer and the oral surgeon, who were unaware of the nature of the pretreatment, and after surgery the patient filled out a questionnaire about his relative comfort during the procedure.

Results were the same for all three outcome measures, although the largest changes were seen in the patient's self-evaluation (see Table 5.1). The two hypnotic conditions were not significantly different from aquarium contemplation without hypnosis, and all three conditions produced significantly more relaxation and comfort than poster contemplation without hypnosis and the control treatment. Thus aquarium contemplation was as effective as hypnosis in relaxing subjects and hypnosis did not improve the effects of aquarium contemplation.

These results have been confirmed in other studies in the United States (DeSchriver and Riddick 1990) and Europe (Bataille-Benguigui 1992) but have not received even limited cross-cultural testing. They did, however, suggest a testworthy hypothesis related to biophilia. There is a whole class of events in nature that have two properties: first, a Heraclitean motion that is always changing but always remaining the same; second, a psychological association with relative safety or comfort. Aquarium fish are one example. Others are waterfowl swimming, horses or cows grazing, the pattern of light and shade created by broken cloud cover, fire in a fireplace, the undulation of a breeze-blown field of hay or wheat, waves lapping on a shore, or the shuttling of feeding birds. All these events have been associated with tranquillity or absence of danger. They are also dramatically contrasted with events that do signal danger—the erratic motion of injured or dying fish, birds taking flight or animals breaking into a run to escape a predator, the black clouds of an approaching storm, fire burning out of control in a forest, the patterns produced by wind or waves during a storm. Certainly, sensitivity to these distinctive patterns of safety or danger could be increased by selection over the long course of years. Moreover, the relaxed attentional state produced by these natural signs of safety may facilitate a more creative, less stereotyped pattern of thought (Bachelard 1956; Crook 1991). Thus increased sensitivity to phenomena like the quiet motion of pond life could offer secondary advantages through the benefits of the physiological relaxation or, perhaps more important, through better problem solving.

The two characteristics we have been discussing—Heraclitean motion and association with safety—are separable and can be studied independently. Yi Fu Tuan (1984), for example, has called attention to the singular characteristic of fountains: water running uphill, an event which, with the rare exception of thermal geysers, does not occur in nature and can have no history of either benign or dangerous associations. Yet the motion of water in fountains seems to have the same hypnotic attraction as water flowing downhill in a waterfall. In similar fashion, the experimental observation of the physiological consequences of observing the quiet and repetitive motion of predators—sharks in an aquarium, circling birds of prey, or the stalking movements of wolves or large felines—which combine Heraclitean movement with potential danger would be an interesting test case. The growing availability of computer-generated images would offer the opportunity to contrast the impact on attention and arousal level of different patterns of movement independent of any consciously identifiable event. Thus a harmless tropical fish could be given the swimming pattern of a shark or the motion imparted to a tree by a severe storm could be simulated by objects not recognizable as trees and with patterns of light not characteristic of storms. It would also be possible to search for general characteristics of the patterns in nature that produce relaxation. Exploring the ability of computer-generated fractal structures to entrain subjects' attention and induce calm could be a promising approach, as well, since waves, flames, and clouds can be duplicated by fractals. Fractal structures could also relate the physiological and cognitive effects of both natural phenomena such as waves and cultural artifacts like music. Without such a dissection of complex events it is premature to attribute the calming influence of aquarium contemplation to any innate tendency to respond to living things.

Another method of exploring the biophilia hypothesis is to examine the responses of diverse patients to recreational or therapeutic programs using animals. If there is a very similar pattern of responses among patients with diverse functional and organic pathology, the pattern may be innate. In the absence of cross-cultural data, however, the evidence can only be suggestive. One can illustrate this approach by describing two contrasting groups of patients: adults with chronic brain syndrome secondary to cerebral arteriosclerosis or Alzheimer's disease and autistic children.

One of the more striking changes in the care given to the institutionalized aged has been the increase in frequency of resident animals as well as animal visitation programs in which groups of volunteers and their pets regularly visit the inmates. This growth of animal visitation followed the report of Corson and Corson in 1977 describing the positive social responses of depressed and asocial patients to a visiting animal. Since that time there have been hundreds of clinical reports of the powerful impact of animal contact on patients with chronic brain syndrome who were unresponsive to their environment, negative toward their caretakers, or even mute. Uniformly, these socially withdrawn patients are said to:

· Focus their attention on the animals.
· Interact with the animals, holding, stroking, and hugging them.
· Smile and laugh.
· Talk to the animals and volunteers.

Not every patient responds, of course, but the majority of patients visited do experience this kind of social facilitation.

Ange Condoret (1983) reported on the history of a mute and asocial autistic girl who began to make clinical progress after her attention was captured by a dove brought into the classroom. Since that publication there have been a series of clinical reports of autistic children who responded to animals though they still ignored their human therapists. These clinical reports were tested in a controlled study by Redefer and Goodman (1989), who observed that when a dog was introduced into outpatient therapy sessions with autistic children there was a significant decrease in autistic behavior (self-stimulation) and in increase in social responses—first directed at the animal and then at the therapist and the animal. As sessions progressed there was an increased frequency of appropriate speech, sustained interaction with animal and therapist, and more intricate social behavior.

We have confirmed Redefer and Goodman's work and demonstrated that dogs increased social responding even in more severely autistic children who were resident in a treatment facility (Katcher and Campbell 1990). Half of our sample were mute and used only signs to communicate. We also noted that children responded differentially to specific animals or combinations of an animal and a human handler. Our findings support the clinical literature which notes that autistic children may ignore a variety of different animals before directing their attention to a particular one (Condoret 1983). Press reports of autistic children miraculously responding to the presence of captive dolphins (Rousselet-Blanc and Mangez 1992) are the result of uncontrolled clinical studies and the suggestive influence of the cult that has evolved around dolphins and other cetaceans. Equally spectacular results have been noted with cats, dogs, birds, and even a small turtle.

One of the patients in our study used speech only for self-stimulation, endlessly repeating nonsense syllables or fragments of television commercials. He was one of the more passive patients in the study, and for the first two sessions he studiously ignored a well-trained dog and its owner. On the third session that animal was replaced with a hyperactive adolescent dog, Buster, taken from a local humane society. Apparently this dog too was ignored. At the next session, however, without any other change in regimen, the patient eagerly ran into the therapy room and within minutes said his first new words in six months: "Buster Sit!" The words were used appropriately and the patient learned both to play ball with the animal and give him food rewards. He began the next session by spontaneously searching out the ball to initiate play with the dog. In that session and subsequent ones with Buster, he continued to play appropriately with the animal and approached both Buster and the volunteer for comfort.

In patients with chronic organic brain damage and autistic children with congenital brain dysfunction, therefore, the entry of animals into a purely human environment resulted in focused attention, increased social responding, positive emotion, and, critically, speech. The results for a wide variety of functional mental disorders (Beck, Saradarian, and Hunter 1986; Levinson 1969; Corson et al. 1977; Peacock 1984; McCulloch 1981) are the same as those described for organic disorders. These results with patients should be interpreted against the background of the normal response to animals in our society. People have an inveterate tendency to speak discursively to animals. Moreover, in American and European populations speaking to animals is associated with less sympathetic nervous system activation, that is, lower blood pressure and pulse rate, than speech directed at human partners (Katcher 1981; Baun et al. 1984; Grosberg and Alf 1985; Friedmann et al. 1983). Dialogue with animals is accompanied by stereotyped vocal characteristics: lower volume, higher pitch, rising inflection at the end of speech segments, and the creation of a pseudodialogue by the insertion of pauses for the animal to "reply" (Katcher and Beck 1989). These vocal characteristics are also accompanied by typical facial expressions: smooth brow, lowered eyelids, and a faint rather than forced smile. The speech resembles that used with infants who do not talk—motherese (Hirsh-Pasek and Treiman 1982). These physiological and behavioral characteristics of speech to animals, being less stressful than dialogue with humans, could explain why animals tend to facilitate talk. There is also evidence from the United States and Europe that the presence of an animal increases social interaction. Normal adults or handicapped children are much more likely to be approached by strangers if they are in the presence of an animal (Hoyt and Hudson 1980; Messent 1983; Hart, Hart, and Bergin 1987). The tendency of animals to facilitate approaches to other people is confirmed by studies of the positive influence of animals on the social attractiveness of human figures in projective tests (Lockwood 1983) or photographs of human subjects (Katcher and Beck 1989). Thus all the desirable responses to animals in therapeutic situations reflect the influence of interaction with companion animals within the general population.

Recently we examined the influence of education structured around animal contact with boys between the ages of nine and fifteen who were in residential treatment at the Brandywine campus of the Devereux Schools in Pennsylvania (Katcher and Wilkins 1992). Most of the students shared the central characteristics of attention-deficit hyperactive disorders (ADHD) and more than half were classified as having conduct disorders or oppositional defiant disorders (CD and ODD). Current conceptualizations of ADHD and its core problems emphasize biologically based deficiencies in the regulation of behavior by rules and consequences (Barkley 1990). Children with this disorder have deficits in behavioral regulation that may stem from a diminished sensitivity to behavioral consequences, lessened control of behavior by partial reinforcement schedules, and poor rule-governed behavior. Barkley (1990:71) reports: "These deficiencies give rise to problems with inhibiting, initiating, or sustaining responses to tasks or stimuli and adherence to rules or instructions, particularly in situations where consequences for such behavior are delayed, weak, or nonexistent." One of the hallmarks of the condition is the rapid extinction of the influence of rewards and punishments. ADHD children seem to require highly novel forms of reinforcement and rapid alternation between different kinds of reinforcements rewards if behavioral control is to be maintained.

As a result of their inability to inhibit impulsive behavior and direct their attention consistently, children with ADHD are, from early life on, subjected to conflictual and unrewarding social interaction with parents, siblings, peers, and teachers. The co-morbidity of CD with ADHD occurs most frequently in families and social milieus where behavior is controlled by violence and aggressive coercion. With this experience children are prone to make negative attributions about human motivation and act upon them. They are suspicious and mistrustful and project their own angry and retaliatory impulses onto others. One of the defining symptoms of CD is cruelty to animals.

We set up a controlled crossover experimental design in which fifty children were randomly assigned to one of two voluntary experiences complementing their regular school curriculum. The control procedure was an outward bound course where the children learned the rudiments of rock climbing, canoeing, and water safety. The experimental treatment was a nature education program for five hours of the school week. After six months, the outward bound group was taken into the nature education program and the students in the original experimental group were returned to their regular school program. They were, however, permitted to visit their animals in their free time. The reason for the partial crossover was our belief that it was unethical for us to separate the children from their pets.

The focus of the nature education program was a 14- by 32-foot building housing a collection of small animals: rabbits, gerbils, hamsters, mice, chinchillas, iguanas and other lizards, turtles, doves, and chicks. One of the two nature educators had a dog that was present most of the time.

The children were given only two general rules. They had to be gentle with the animals and that included talking softly while in the zoo. They had to respect the animals and each other and avoid speech that devalued each other or the animals. These rules were designed to favor behavior that demanded motor inhibition and impulse control: speaking softly, being gentle, and focusing attention on the animals' rather than personal needs. The use of the word respect was not an attempt to define the animals anthropomorphically but a means of getting the children to think reflexively with the animal as an intermediate—that is, thinking about their own feelings by reasoning about the animals' feelings. The teachers in the program defined their tasks as caring for the animals in the zoo and the places visited by the children as well as helping the children learn.

The first task given the students was learning the general requirements for care of animals and the proper means of holding them. The second task was learning the biology and care requirements for one animal: the animal the child chose to adopt as his pet. After adoption there were some twenty-one other knowledge and skill areas that a child could learn. Skill areas included learning how to weigh and measure their pet, chart its growth, and compute its food and bedding requirements, learning how to breed their pet and care for the mother and young, and learning how to show a pet to children in other special education classes at Devereux or adults in rehabilitation hospitals. In one term the average child progressed through eight such areas completely and finished parts of three or four more. Once the child adopted his pet, there were no other direct rewards for learning.

The program also included hikes through the Devereux campus, fishing and camping trips, and visits to local state parks, to pet stores, to farms, and to a veterinarian's office. On these trips children learned about water cycles, general aspects of wetlands, pasture, and woodlot, and identification of indigenous birds, trees, reptiles, and small mammals. The zoo had spaces for guest wildlife such as found insects, amphibians, and reptiles. These animals were housed in the zoo for a few days for identification and then returned to the wild. In addition to their regularly scheduled times, the children in the program could visit the zoo to care for the animals or play with them on their free time after breakfast, at lunch, and after school.

The relevant results could be summarized as follows. The animals caught and held the children's attention. There were significant differences in attendance between the zoo and the outward bound groups. The average zoo attendance for the summer was 93 percent and in the outward bound 71 percent ($t = 3.4$; $p < .001$). In the fall term the average attendance for the zoo program was 89 percent and for the outward bound 64 percent ($t = 2.8$; $p < .01$). When the outward bound group shifted to the zoo program, there was a significant increase in attendance. The same children who had a 67 percent attendance in outward bound now attended the zoo program 87 percent of the time ($t = 2.9$; $p < .01$).

Using a variety of criteria such as the number of skill areas mastered, scores on an objective knowledge examination, and results of a weekly progress review, we observed that 80 percent of the children made a good clinical response to the zoo education program. Some students who had made no progress in the regular school program for as long as four years rapidly accomplished learning tasks in the zoo. The capacity of the animals to engage the children's attention was evident in the frequency of their visitation during their free time even when they were not being given time off from school to attend the program. A high sustained level of zoo visitation during free time was highly correlated with a good clinical response to the zoo program. Impulse control was consistently better in the zoo than in either the school or the residences. Children frequently had to be restrained

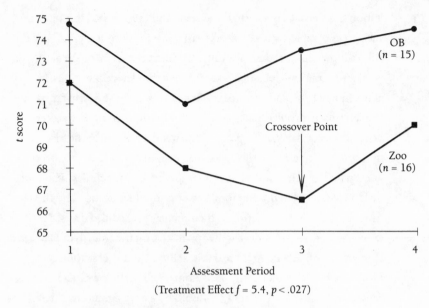

FIGURE 5.1. *Total Symptom Level: Achenbach Teacher Rating Form*

and medicated because of explosive outbursts in both the residences and the classroom. No child ever had to be restrained in the zoo, although from the frequency of restraints during the school day we would have anticipated twenty-four incidents. Teachers were able to use a visit to the zoo as a therapeutic intervention to calm children who otherwise would have had to be restrained or medicated.

We used the Achenbach Child Behavior Checklist (a well-standardized and highly reliable instrument) to measure the severity of the child's symptoms. The symptom inventory was completed by both the residential counselors and the teachers in the child's regular school program. The teachers noted a significant decline in symptom level in the zoo group in comparison to the controls (see Figure 5.1). The effect of animal contact on attention and impulsivity was evident only in the structured environment of the school during the first six months of the program. During that time behavior of both the experimental and control group worsened in the residences during the afternoon and evening. Only after students had been in the zoo program for six months did we note any improvement in their behavior in the residences.

The following conclusions can be drawn from all the clinical data:

1. Animals brought into a human context are powerful reinforcers of human attention and behavior. Their effects are persistent even in populations in which the effects of other reinforcers rapidly attenuate.

2. When the child is given an opportunity to interact with the animal as well as watching it, there are more positive changes in behavior and they are more persistent.

3. Human speech and the nonverbal expression of emotion are facilitated by the presence of animals.

4. Children affiliate with the therapists and teachers who are associated with the animals and responsible for their care. In the case of the children with ADHD, rules structured around the care of animals are accepted as rational when other social expectations are rejected as arbitrary.

5. The response to animals seems to remain intact even when social and emotional responses to other humans are compromised by a variety of structural or functional disorders. This finding suggests that the "programs" mediating responses to animals may have a highly redundant representation within the central nervous system and thus may be more resistant to disease or injury.

One can develop a simple hypothesis about the evolutionary role of biophilia based on certain assumptions about human evolution and the observations of the social relationships between modern humans and their animals we have just cited. This exploration is based on two salient observations: the tendency of animals to evoke speech from people and the propensity of people to consider animals as kin. The argument posits a twofold elaboration of an original use of language to refer to sources of food or danger that were not visible to the self or others: one was the use of social intelligence to make inferences about animal behavior as if animals were purposefully signaling the way humans do (anthropomorphic inference); the other was the ascription of kinship to certain animals.

Dialogue with Animals: Its Nature and Culture

The growth of hunting and gathering, the need to transport food to a home site, and the patterns of reciprocal feeding that characterized evolving humans (Potts 1991) demanded a greater appreciation of the kinds of information that could be obtained from the environment. Evolving humans would have profited greatly from the ability to tell each other about the location of game and other edibles. Indeed it is possible to make up a "Just So" story—we should follow Kroeber's example and label any hypothesis about prehistoric mind or social structure a "Just So" story rather than a hypothesis—positing that language became advantageous as an adjunct to hunting and gathering. Thus human speech may be a reinvention of the dance of the bees, but with a need for more complexity since bees forage for nectar while human beings harvested a broad array of plants and animals. If a major function of early language, either a language of gestures or vocalizations, was referential—indicating the presence of plants or animals that were not visible—then the naming of animals was a critical event in human evolution. This conjunction of speech and animals marks the biblical account of humankind's first task (Genesis 2:18): "So out of the ground the Lord God formed every beast of the field and every bird of the air, and brought them to the man to see what he would call them; and whatever the man called every living creature, that was its name. The man gave names to all cattle, and to the birds of the air, and to every beast of the field . . ."

Naming animals may have made the coordination of mutual activity around hunting and gathering easier, but making correct inferences about animal behavior demanded a more sophisticated use of language. Intelligence is not a unitary phenomenon, and the ability of primates to make inferences in the social domain is not matched by their intelligence in other areas. Monkeys do use alarm calls to warn of the presence of predators and thus have a limited vocabulary of "names" for animals (Cheney and Sayfarth 1990). Monkeys are, however, highly limited in the kinds of inferences they draw from the environment—for example, the sight of a simulated leopard kill in a tree, the track of a python, or the sight of lions feeding do not alarm these primates. Certainly our ancestors started their trek toward humanity with the same highly developed social intelligence. Once they had learned to name animals, then viewing those same animals, indeed all of nature, as sentient and capable of engaging in dialogue would greatly enlarge their abilities to learn from their environment. When living things are thought to be purposefully signaling to us, then all of the environment becomes a social environment.

Early humans could use all of their highly developed social intelligence to understand the behavior of animals, vegetation, or any feature of the world. The creation of the pathetic fallacy and anthropomorphic inference was one of the central events in human evolution. The first time one of our ancestors said "If I were a turtle where would I hide my eggs?" or "The sky looks angry, I'd better go back to my cave!" he began to use a whole new pattern of inference in environmental investigation. Anthropomorphic inference may deserve its bad reputation within the narrow confines of Cartesian science, but it is a highly useful tool in negotiating difficult and sometimes unfriendly natural environments. Lockwood (1989:49) has described "applied anthropomorphism" as "the use of our own personal perspective on what it's like to be a living being to suggest ideas about what it is like to be some other being of either our own or some other species. This process is a form of projection, and it is a process that makes our life on earth as social beings possible."

Using social intelligence to make inferences about the environment facilitated hunting, gathering, avoidance of predators, and prediction of weather. At some time, however, the animals became more than sentient. They were adopted into human society. It is at this point in time that Rousseau places the origin of language in metaphor. Animals became a mirror reflecting the human social world. Social relationships and strategies of social attribution were mapped out using the distinct forms of living things to make abstract social or psychological attributes concrete and obvious. This superimposition of human and natural worlds in the essence of totemism (Lévi-Strauss 1974). Creating a kinship with animals did more than illuminate the complexity of human social relationships, however; it made the world a more comfortable place by reducing human isolation. The presence of animals, their touch, their attention, became part of the pattern of social dialogue that helps to maintain human health.

Dialogue with Animals: Its Nature and Culture

The prolonged care of infants, as well as the developing patterns of interdependence between adults that paralleled human evolution, were facilitated by the development of physiological, psychological, and social rewards for nurturing behavior. It is obvious that the survival of children depends on the quality of their nurturing. Nurturing comprises more than the provision of food and protection from danger, however. Tactile comfort and appropriate social interaction are also critical. Indeed, friendly social interaction is now recognized to be necessary for the health of adults as well. Social support, family composition, and friendship have powerful influences on the health and well-being of adults. The strong links between life expectancy and illness rates and a person's social support network were reviewed in *Science* by House, Landis, and Umberson (1988). They concluded: "The evidence on social relationships is probably stronger, especially in terms of prospective studies, than the evidence which led to the certification of the Type A behavior pattern as a risk factor for coronary heart disease. The evidence regarding social relationships and health increasingly approximates the evidence in the 1964 Surgeon General's report that established cigarette smoking as a cause or risk factor for mortality and morbidity from a range of disease. The age-adjusted relative risk ratios are stronger than the relative risks [for all causes of] mortality reported for cigarette smoking."

If tactile comfort and companionship make people healthier, then animal companionship—especially contact with animals treated as kin, that is, pets—should increase well-being. There is even some direct evidence that the companionship of animals makes a difference in health (Friedmann et al. 1980; Siegel 1990; Anderson et al. 1992). Certainly there is indirect evidence suggesting that having a dialogue with animals or considering animals as kin could influence health through stress reduction (Katcher 1983). It is not known when human beings began to keep pets; nor do we know the temporal relationship between thinking about animals as kin and actually using them as companions. From the presence of pets among preagricultural people, it can be inferred that the use of animals as companions may have preceded the evidence of domestication by a very long period of time.

Affect and Aesthetics

If animals are so woven into the history, and perhaps the neural structure, of our social dialogue, why has the living environment suffered so from unrestrained destructive human behavior? If we have a predisposition to treat at least some animals as kin, why have we exterminated so many of them and why are we so indifferent to their loss? Why is biophilia, if it exists, so weak a determinant of human behavior?

When animals became our kin, the metaphor implies that they will continue to nourish their human relatives eternally. Prey and predator are bound, like family, into a never-ending dialogue. Individuals perish, but the two kinds of beings are assumed to be linked in perpetuity. The hunter kills the individual, but the hunter's tribe depends on the hunted species' continued life. If we were related to the natural world by an unending dialogue demanding the continuance of every partner in the dialogue, we would not be threatened by today's environmental disasters. But use of the category "kin" to describe our reciprocal relationships with animals or plants implies the category "not kin." And "not kin" can devolve into the category "vermin" and remorseless destruction can follow.

The category of vermin implies that certain animals or plants (weeds are plant vermin) can or must be exterminated. There is no longer any concern for their persistence. Extinction becomes a desirable end. Our tendency to view an animal or plant as utterly bad, or unnecessary, or valueless is accentuated by the metaphorical use of animals and nature that is so fundamental a part of all languages. In metaphor a being is stripped of its behavioral complexity and refined into the reflection of a single attribute, even one as broad as "good" or "bad." The complexity of any being which reflects the totality of its behavior, both good and bad, is distilled into a single image: either all good or all bad. Humans simplify their maps of complex social situations by reducing the complex images of people and animals to metaphorical icons which reflect only a fragment of reality. The use of such icons to reduce complexity is an important mental mechanism of both conscious and unconscious thought. The usefulness of animals in metaphor and allegory depends, in part, on their great utility in representing such images. Thus we can be "brave as a lion" though that beast is frequently a carrion eater and "wise as an elephant" though that animal blunders around de-

Dialogue with Animals: Its Nature and Culture

stroying its own habitat. When Kipling used wolves and panthers to stand for orderly predation and wild dogs and a tiger to stand for disorderly predation, he was using animal images to make a simple contrast between good and bad behavior. Our willingness to exterminate animals and destroy habitat are reflections of this same universal tendency to reduce the complex roles played by animals to simple images defined by human interest or need. Coyotes and wolves are always bad; sheep and cattle are always good.

We have now come almost full circle. Our explication of what may be an innate tendency to affiliate with animals has recognized the power of the cultural categorization which can change any animal's status from kin to vermin—or, more recently, from kin to thing. To complete the circle it is necessary to consider how culture and human nature interact and reintroduce the moral or political agenda for the preservation of biodiversity.

The precarious state of the world's dwindling reservoirs of wilderness provides overwhelming evidence that no matter how deeply the tendency to engage other kinds of life is rooted in our brains, our culture determines how our actions impinge on our common natural world. Any doubts about the power of cultural instruction to threaten our fragile environment should have been quenched by Mr. Patrick Buchanan's speech on the first night of the 1992 Republican Convention. Environmentalism was explicitly linked with radical feminism, atheism, homosexuality, and all the other evils of the Democrat Party. It was opposed to Freedom and the Judeo-Christian Ethic and the Right of Working Men to Have Jobs. Although the cultural and political opposition to environmental conservation is beyond the domain of this volume, we might wish to consider how the science instruction received by our children could hinder or contribute to the development of an ethic of biophilia.

Our current relationship with animals and the living environment is historically unique. Animals have become marginal in our lives, but only within this century. In a magnificent essay entitled "Why Look at Animals?" John Berger (1980) describes the progressive marginalization of animals in industrial society. Although the philosophical justification for that exile was framed by Descartes in the seventeenth century, the trivialization

Affect and Aesthetics

of animals in industrial societies began in the nineteenth century and has greatly accelerated in ours. The causes are many: the migration of progressively larger fractions of the population to cities, the replacement of draft animals, the banishment of production animals from cities, the mechanization of farming and subsequent loss of the farm workforce, and the progressive destruction of forest and wilderness. Berger (1980:1) writes: "The 19th century, in Western Europe and North America, saw the beginning of a process, today being completed by 20th century corporate capitalism, by which every tradition which has previously mediated between man and nature was broken. Before this rupture, animals constituted the first circle of what surrounded man. Perhaps that already suggests too great a distance. They were with man at the center of his world."

The consequences of all of these changes on our relationship with nature have been amplified by the character of modern science and in particular the way science is taught to our children. Modern science has, like the worm in an apple, undermined the value of appearance. It has devalued the world created by the senses and parted company from both the common sense that we use to construct a world from our senses and the speech which holds us together in the body politic (the same body politic that biologists must sway if the moral agenda of habitat and species preservation is to be accomplished). Hannah Arendt (1958:283, 4) describes this relationship between modern science, our common world, and our common dialogue as follows:

> Cartesian reason is entirely based "on the implicit assumption that the mind can only know that which it has itself produced and retains in some sense within itself." Its highest ideal must therefore be mathematical knowledge as the modern age understands it, that is, not the knowledge of ideal forms given outside the mind but of forms produced by a mind which in this particular instance does not even need the stimulation—or, rather, the irritation—of the senses by objects other than itself. This theory is certainly what Whitehead calls it, "the outcome of common-sense in retreat." For common sense, which once had been the one by which all other senses, with their intimately private sensations, were fitted into the common world, just as vision fitted man into the vis-

ible world, now became an inner faculty without any world relationship. This sense now was called common merely because it happened to be common to all. What men now have in common is not the world but the structure of their minds. . . .

For the sciences today have been forced to adopt a "language" of mathematical symbols which, though it was originally meant only as an abbreviation for spoken statements, now contains statements that in no way can be translated back into speech.

Obviously the solution for this world alienation is not a Luddite uncoupling of mathematics and science. Instead there needs to be a juncture between education and direct sensory experience of life and nature. Children's time in school is spent with text, not the experience of events, social or natural. They learn science by reading words, not by observation of their own place in the world. After years of excellence in science, students enter medical and dental school unable to distinguish reliably between veins and arteries. Our children should be educated with natural history rather than the science of the invisible as it is now taught. The ability of living things to hold children's attention should be used to help them learn how to understand the living environment, respect the appearance of the world, and take responsibility for its care. Natural history, even in its modern forms, works on the surface of the world amid the bustling activities of living things. The study of animal populations does profit from mathematical formulations, of course, but one can learn a great deal about the world in the observable and speakable realms of taxonomy, ecostructure, and animal behavior before having to deal with unspeakable mathematical formulations.

It is sometimes difficult for those who experienced childhood without television to recognize how difficult it is for modern children to see the world. Bill McKibben (1992), in *The Age of Missing Information*, explicitly contrasts the kinds of information that can be extracted from television versus the experience of a hike on an Adirondack mountain. Children raised on television are exposed to vast amounts of information but fail to learn very much about their immediate environment. Too much is learned from a small two-dimensional representation of global events and too little from direct exploration of their own place in the world:

The definition of television's global village is just the contrary—it's a place where there's as little variety as possible, where as much information as possible is wiped away to make "communications" easier. . . .

The mountain says you live in a particular place. Though it's a small area, just a square mile or two, it took me many trips to even start to learn its secrets. Here there are blueberries, and here there are bigger blueberries. The swamp is impenetrable here, but easier to skirt on the other side. You pass a hundred different plants along the trail—I know maybe twenty of them. One could spend a lifetime learning a small range of mountains, and once upon a time people did. [McKibben 1992:52]

The hyperactive children who were taught how to care for animals used that experience to learn biology and to learn how to learn. They acquired method, fact, and a new moral orientation toward nature. If the curriculum was effective, it is because they were taught by people who cared for animals and nature and reflected that concern in their demeanor and instruction. Nature was not merely subject matter; it was the world common to teacher and student for which both had to assume responsibility. With that kind of teaching, the innate tendency to continue a dialogue with other kinds of life can be joined to the moral agenda of preserving that life.

If biophilia exists, then it most probably exists as a disposition to attend to the form and motion of living things and, for animals at least, incorporate them into the social environment. The tendency may be so general that it extends to elements in the environment that display patterned movements like flowing water or the motion of clouds. Moreover, cultural instruction may well be the primary determinant of how that general disposition is incorporated into behavior and affects the environment. If, as is probable, our present environmental predicament indicates that any innate biophilic tendency has such a weak influence on human behavior, why explore an innate mechanism when cultural instruction is so powerful? One reason is the peculiar nature of social theory in the modern world. Social theories act reflexively and are one of the means by which human behavior is shaped, though in unpredictable ways. Positing the existence of biophilia will most likely facilitate the ongoing social campaign to protect biodiversity and slow down the continuing degradation of the global environ-

Dialogue with Animals: Its Nature and Culture

ment. The concept of biophilia may have its most powerful influence through changes in the way science is taught, however, rather than through programs of research seeking to demonstrate strong effects attributable to biophilia.

When the teaching of biological science is infused with responsibility for the object of study, then the moral agenda advanced in E. O. Wilson's *Biophilia* will be well begun. Hannah Arendt (1963:196) describes the relationship between education and responsibility as follows:

> Education is the point at which we decide whether we love the world enough to assume responsibility for it and by the same token save it from that ruin, which, except for renewal, except for the coming of the new and young, would be inevitable. And education, too, is where we decide whether we love our children enough not to expel them from our world and leave them to their own devices, nor to strike from their hands their chance of undertaking something new, something unforeseen by us, but to prepare them in advance for the task of renewing a common world.

REFERENCES

Albert, A., and K. Bulcroft. 1987. "Pets and Urban Life."*Anthrozoos* 1:9–25.

Anderson, W., C. Reid, and G. Jennings. 1992. "Pet Ownership and Risk Factors for Cardiovascular Disease." *Medical Journal of Australia* 157:298–301.

Arendt, H. 1958. *The Human Condition*. Chicago: University of Chicago Press.

———. 1963. *Between Past and Future*. Cleveland: World Publishing Co.

Bachelard, G. 1956. *The Psychoanalysis of Fire*. Translated by C. C. M. Ross. Boston: Beacon Press.

Barkley, R. 1990. *Attention Deficit Hyperactivity Disorder*. New York: Guilford Press.

Bataille-Benguigui, M. 1992. "Man-Fish Relationship in the Therapy of Conflict." *Proceedings of the International Conference on Science and the Human-Animal Relationship*. Amsterdam.

Baun, M., N. Bergstrom, N. Langston, and I. Thoma. 1984. "Physiological Effects of Petting Dogs: Influences of Attachment." In *The Pet Connection: Its Influence on Our Health and Quality of Life*, edited by R. B. Anderson, B. Hart, and A. Hart. St. Paul: Grove Publishing.

Beck, A. M., L. Saradarian, and G. F. Hunter. 1986. "Use of Animals in the Rehabilitation of Psychiatric Inpatients." *Psychological Reports* 58:63–66.

Berger, J. 1980. *About Looking*. New York: Pantheon Books.

Cheney, D., and R. Seyfarth. 1990. *How Monkeys See the World*. Chicago: University of Chicago Press.

Condoret, A. 1983. "Speech and Companion Animals: Experience with Normal and Disturbed Nursery School Children." In *New Perspectives on Our Lives with Companion Animals*, edited by A. Katcher and A. Beck. Philadelphia: University of Pennsylvania Press.

Corson, S. A., and E. Corson. 1977. "The Socializing Role of Pet Animals in Nursing Homes: An Experiment in Non-Verbal Communication Therapy." In *Society, Stress, and Disease*, edited by L. Levi. London: Oxford University Press.

Corson, S. A., E. Corson, P. Gwynne, and E. Arnold. 1977. "Pet Dogs as Nonverbal Communication Links in Hospital Psychiatry." *Comprehensive Psychiatry* 18:61–72.

Crook, J. 1991. "Consciousness and the Ecology of Meaning: New Findings and Old Philosophies." In *Man and Beast Revisited*, edited by M. Robinson and L. Tiger. Washington: Smithsonian Publications.

DeSchriver, M., and C. Riddick. 1990. "Effects of Watching Aquariums on Elders' Stress." *Anthrozoos* 4:44–48.

Friedmann, E., A. Katcher, J. Lynch, and S. Thomas. 1980. "Animal Companions and One-Year Survival of Patients Discharged from a Coronary Care Unit." *Public Health Reports* 95:307–312.

Friedmann, E., A. Katcher, S. Thomas, J. Lynch, and P. Messent. 1983. "Social Interaction and Blood Pressure: Influence of Animal Companions." *Journal of Nervous and Mental Disease* 171:461–465.

Giddens, A. 1990. *The Consequences of Modernity*. Stanford: Stanford University Press.

Grossberg, J., and E. Alf. 1985. "Interaction with Pet Dogs: Effects on Human Cardiovascular Response." *Journal of the Delta Society* 2:20–27.

Hart, L. A., B. L. Hart, and B. Bergin. 1987. "Socializing Effects of Service Dogs for People with Disabilities." *Anthrozoos* 1:41–44.

Hirsh-Pasek, K., and R. Treiman. 1982. "Doggerel: Motherese in a New Context." *Journal of Child Language* 9:229–237.

House, J., K. Landis, and D. Umberson. 1988. "Social Relationships and Health." *Science* 241:540–545.

Hoyt, L. A., and J. W. Hudson. 1980. "Dog-Guides or Canes: Effects on Social Interaction Between Sighted and Unsighted Individuals." *International Journal of Rehabilitation Research* 3:252–254.

Katcher, A. 1981. "Interactions Between People and Their Pets: Form and Function." In *Interrelations Between People and Pets*, edited by B. Fogle. Springfield: Thomas.

———. 1983. "Man and the Living Environment: An Excursion into Cyclical

Time." In *New Perspectives on Our Lives with Companion Animals*, edited by A. Katcher and A. Beck. Philadelphia: University of Pennsylvania Press.

Katcher, A., E. Friedmann, A. Beck, and J. Lynch. 1983. "Looking, Talking and Blood Pressure: The Physiological Consequences of Interaction with the Living Environment." In *New Perspectives on Our Lives with Companion Animals*, edited by A. Katcher and A. Beck. Philadelphia: University of Pennsylvania Press.

Katcher, A., H. Segal, and A. Beck. 1984. "Comparison of Contemplation and Hypnosis for the Reduction of Anxiety and Discomfort During Dental Surgery." *American Journal of Clinical Hypnosis* 27:14–21.

Katcher, A., and A. Beck. 1989. "Human-Animal Communication." In *International Encyclopedia of Communications*, edited by E. Barnow. London: Oxford University Press.

———. 1991. "Animal Companions; More Companion Than Animal." In *Man and Beast Revisited*, edited by M. Robinson and L. Tiger. Washington: Smithsonian Publications.

Katcher, A., and C. Campbell. 1990. "Social Interaction with Animals." Paper presented to the 1990 Meeting of the Pavlovian Society, Philadelphia.

Katcher, A., and G. Wilkins. "A Controlled Trial of Animal Assisted Therapy and Education in a Residential Treatment Unit." Paper presented to the Sixth International Conference on Human Animal Interactions, Montreal.

Levinson, B. 1969. *Pet Oriented Child Psychotherapy*. Springfield: Thomas.

Lévi-Strauss, C. 1974. *Le Totemisme Aujourd'hui*. Vendôme: Presses Universitaires de France.

Lockwood, R. 1983. "The Influence of Animals on Social Perception." In *New Perspectives on Our Lives with Companion Animals*, edited by A. Katcher and A. Beck. Philadelphia: University of Pennsylvania Press.

———. 1989. "Anthropomorphism Is Not a Four Letter Word." In *Perception of Animals in American Culture*, edited by R. Hoage. Washington: Smithsonian Publications.

McCulloch, M. 1981. "The Pet as Prosthesis: Defining Criteria for the Adjunctive Use of Companion Animals in the Treatment of Medically Ill, Depressed Outpatients." In *Interrelations Between People and Pets*, edited by B. Fogle. Springfield: Thomas.

McKibben, B. 1992. *The Age of Missing Information*. New York: Random House.

Messent, P. 1983. "Social Facilitation of Contact with Other People by Pet Dogs." In *New Perspectives on Our Lives with Companion Animals*, edited by A. Katcher and A. Beck. Philadelphia: University of Pennsylvania Press.

Peacock, C. 1984. "The Role of the Therapist's Pet in Initial Psychotherapy Sessions with Adolescents: An Exploratory Study." Doctoral dissertation, Boston College Graduate School of Arts and Sciences.

Potts, R. 1991. "Untying the knot: Evolution of Early Human Behavior." In *Man and Beast Revisited*, edited by M. Robinson and L. Tiger. Washington: Smithsonian Publications.

Redefer, L., and J. Goodman. 1989. "Brief Report: Pet-Facilitated Therapy with Autistic Children." *Journal of Autism and Developmental Disorders* 19:461–467.

Rousselet-Blanc, V., and C. Mangez. 1992. *Les Animaux Guerisseurs*. Paris: J. C. Lattès.

Siegel, J. 1990. "Stressful Life Events and Use of Physician Services Among the Elderly: The Moderating Role of Pet Ownership." *Journal of Personality and Social Psychology* 58:1081—1086.

Tuan, Y. F. 1984. *Dominance and Affection: The Making of Pets*. New Haven: Yale University Press.

Ulrich, R. 1979. "Visual Landscapes and Psychological Well-Being." *Landscape Research* 4:17–23.

———. 1983. "Aesthetic and Affective Response to the Natural Environment." In *Behavior and the Natural Environment*, edited by I. Altman and J. Wohlwill. New York: Plenum.

———. 1984. "View Through a Window May Influence Recovery from Surgery." *Science* 224:420.

Ulrich, R., and R. Simons. 1986. "Recovery from Stress During Exposure to Everyday Outdoor Environments." *Proceedings of the Seventeenth Annual Conference of the Environmental Design Research Association*. Washington: EDRA.

Wilson, E. 1984. *Biophilia*. Cambridge: Harvard University Press.

Part Three

CULTURE

Searching for the Lost Arrow: Physical and Spiritual Ecology in the Hunter's World

Richard Nelson

J UST BELOW THE Arctic Circle in the boreal forest of interior Alaska; an amber afternoon in mid-November; the temperature minus twenty degrees; the air adrift with frost crystals, presaging the onset of deeper cold. Five men—Koyukon Indians—are leaning over the carcass of an exceptionally large black bear. For two days they have traversed a sprawling tract of the Koyukuk River valley, searching for bears that have recently entered hibernation dens. The animals are in prime condition at this season, but extremely difficult to find. Den entrances, hidden beneath eighteen inches of powdery snow, are betrayed only by the subtlest of clues—patches where no grass protrudes above the surface because it has been clawed away for insulation; faint concavities hinting of footprint depressions in the moss below.

Earlier this morning the hunters made their first kill, a yearling. It was discovered by the group's leader, Moses Sam, who has trapped in this territory since childhood, following trails established long before by his father. Moses is in his early sixties. He is known for his detailed knowledge of the land and his extraordinary skill and success as a bear hunter. *"No one else has that kind of luck with bears,"* I am told. *"Some people are born with it. And he always takes good care of his animals—respects them. That's how he keeps his luck."*

A few minutes later, Moses pulls a small knife from his pocket, kneels beside the bear's head, and carefully slits the clear domes of its eyes. Vitreous fluid glistens on his fingers. *"Now the bear won't see,"* he explains softly, *"if one of us makes a mistake or does something wrong."*

In Koyukon tradition, there are hundreds of rules for proper treatment of killed animals. A bear's feet should be removed first, for example, to keep its spirit from wandering. And certain parts are eaten away from the village at a kind of funeral feast attended by men and boys. Koyukon hunters know that an animal's life ebbs slowly, that it remains aware and sensitive to how people treat its body. This is especially true for the potent and demanding spirit of the bear.

Speaking of the black bear's larger relative, a Koyukon elder once told me: *"Every hair on a brown bear's hide has a life of its own . . . so it can't keep still; it can't keep its temper. It takes a few years for all that life to be gone from a brown bear's hide. That's the kind of power it has."*

Perhaps the most important observation to be made about this episode is that contemporary Euro-Americans are likely to find it exotic. Yet over the long run of history, stories like this have been utterly commonplace, the essence of our interactions with the natural world, the very crux of human experience. For 99 percent of our history, as anthropologists Richard Lee and Irven DeVore (1968:3) have pointed out, human beings lived exclusively as hunter-gatherers. On a relative time scale, agriculture has existed only for a moment and urban societies scarcely more than a blink.

From this perspective, much of the human lifeway over the past several million years lies beyond the grasp of urbanized Western peoples. And if we hope to understand what is fundamental to that lifeway, we must look to traditions far different from our own.

I can imagine nothing for which this is more true than the human relationship to life-forms other than ourselves. Probably no society has been so deeply alienated as ours from the community of nature, has viewed the natural world from a greater distance of mind, has lapsed to a murkier comprehension of its connections with the sustaining environment. Because of this, we are greatly disadvantaged in our efforts to understand the basic human affinity for nonhuman life.

Here again, I believe it's essential that we learn from traditional societies, especially those in which most people experience daily and intimate contact with the land—above all, those who harvest a livelihood from the wild environment: the hunters, fishers, trappers, and gatherers. In such communities, we find knowledge similar to that achieved by our own scientific disciplines. And we are given important insights about how humans engage themselves with the living process, insights founded on a wisdom that we had long forgotten and are now beginning to rediscover.

In the discussions to follow, I draw mainly from the teachings of Native American peoples, with whom I have worked as an ethnographer. This work includes four periods of intense study in northern Alaska: something over a year with Inupiaq Eskimo in the arctic coastal village of Wainwright (1964–1966, 1971, 1981); a year in the Gwich'in Indian community of Chalkyitsik (1969–1971); six months in the Kobuk River Inupiaq villages of Ambler and Shungnak (1974–1975); and several accumulated years with the Koyukon Indian people of Huslia and Hughes (1971–1984).

All of this work focused on human relationships to the environment: knowledge of animals and plants, methods of subsistence and survival, religious and spiritual beliefs centered on the natural world, ecological concepts and conservation practices, affiliations with place, and the moral or ethical dimensions of a hunting lifeway.

Searching for the Lost Arrow

At the outset, I want to emphasize my own limitations and shortcomings. Although the anthropologist may be regarded as an authority at home, people in the communities where he has studied are likely to feel otherwise. His experience with their culture seldom goes beyond a year or two; as in our own fields of study, this is not enough to advance beyond the stage of an apprentice. (I use the male pronoun as a convenience and as a reference to myself.)

Moreover, it's difficult to find words that summarize the intricate interplay between hunter-gatherers and the natural communities in which they hold membership. Our books and essays reduce to a few pages the nearly infinite complexities of a human lifeway. We might best hope to create a metaphor, an image that hints of something much larger and deeper than we've been able to penetrate.

Here is an example. A Navajo elder named Claus Chee Sonny recited texts from the Deer Huntingway religious tradition which were recorded by ethnographer Karl Luckert (1975:39). Among the instructions given to hunters is this statement attributed to the divine Deer-people: "*Animals are our food. They are our thoughts.*"

Reading this statement is like walking through a doorway into a wild and illimitable terrain: it opens in all directions. For me, these few words epitomize the pervasiveness of animals—and the natural world as a whole—in the lives and minds and cultures of hunting-gathering peoples like those who inhabited much of North America before the first Europeans arrived.

Among these communities, the natural environment finds expression in virtually all aspects of culture and tradition. In a vast and intricate accumulation of knowledge about animals, plants, and what Western people define as the nonliving environment. In a bewildering array of methods used to harvest a livelihood and process what the natural surroundings provide. In the complex of tools, weapons, and other technology associated with subsistence and survival. In adaptive patterns such as community size, mobility, and annual cycle. In elements of social organization—family and kinship structure, residence patterns, social systems, economic institutions, legal concepts, politics. In a range of customs—from communal activities and sharing to deeply ingrained patterns of personality. In language

Culture

and aesthetic expression—stories, poetry, music, and visual arts. In a complex of beliefs and rituals that may be viewed collectively as one of the earth's great religious traditions. In ways people conceptualize the surrounding world and see themselves in relationship to it. And in intellectual traditions that parallel our Western concepts of ecology, resource management, environmental ethics, and philosophy.

Among hunting-gathering peoples, the intricate weaving together of nature and culture is like the exchange between living cells and their surroundings: the vital breathing in and out, the flux of water and nutrients, the comminglings of outer world and inner flesh.

In the following pages, I discuss a few of these cultural patterns as they are expressed in communities where I have studied. My focus is on relationships between people and animals. The first section, drawing from the example of the Inupiaq Eskimos, deals with empirical or scientific knowledge on which traditional hunting is based. The second section describes religious knowledge, environmental ethics, and conservation practices in the traditions of Koyukon Athabaskan Indians. And the final section considers how non-Western intellectual traditions might contribute to our search for a more harmonious and sustainable relationship to life on earth.

Knowledge: The Hunter's Eye

Eskimos who live on the arctic coast of Alaska call themselves Inupiaq, which means "the Real People." In keeping with this name, they regard themselves proudly. And well they might: Eskimos are among the most exquisitely adapted of all human groups, and they have long inhabited some of the most challenging environments on earth. Their homeland stretches from Western Alaska and adjacent Siberia, across the entire North American continent to Greenland, encompassing latitudes from the bitterest high arctic to mild subarctic coasts. Due to the scarcity of plant life, Eskimos are more dependent than any other people on hunting. Before the present century, some groups lived almost exclusively on mammals, birds, and fish.

Of course, Eskimos are famous for the cleverness and complexity of their technology, which includes kayaks, skin clothing, harpoons, snow houses, and dog teams. But I believe their greatest genius, and the foundation of their adaptive success, lies in the less tangible realm of the intellect—the nexus of mind and nature. What struck me above all else in my work with north Alaskan Eskimos was their knowledge of the environment, which is remarkable for its depth, detail, and accuracy (see Nelson 1969).

Several times, when an Inupiaq hunting companion did something particularly clever, he pointed to his head and declared: "*You see . . . Eskimo scientist!*" At first I took it as hyperbole, but as time went by I realized he was telling the truth. Scientists had often visited his community and he recognized a familiar commitment to the empirical method. (*Note*: My accounts of Inupiaq hunting emphasize men because they are the primary hunters; women catch fish, take small game, and are responsible for most processing of the larger animals.)

Traditional Inupiaq hunters spend a lifetime acquiring knowledge—from other members of their community and from their own experience and observation. This interest is not surprising, because their survival depends on possessing a large body of absolutely dependable information. When I first went to live with Eskimo people, I often doubted things they told me or had difficulty taking them seriously. Somehow I had learned that Western knowledge, embedded in our own scientific tradition, carried a more substantial weight of truth than what was often termed "folk knowledge." But the longer I stayed, the more I trusted their assertions, because experience so frequently showed them right.

For example, hunters say the behavior of ringed seals surfacing in open leads is a reliable way to forecast the weather. And because a sudden gale can set people adrift on the pack ice, accurate prediction can be a matter of life and death. When seals raise chest-high in the water, snout pointed skyward, acting as if they're in no hurry to go anywhere, it indicates stable weather conditions. But if they surface briefly, head low, snout parallel to the water, and tend to show themselves only once, a storm may be approaching. These indicators are most important when combined with others, such as frequent howling by the sled dogs, stars twinkling erratically, and current running from the south. My own experiences with seals and storms through the course of winter affirmed what the Eskimos had said.

Many times I found that the seemingly mystical abilities of Inupiaq hunters rested on the keen edge of intimate knowledge. Like a young Inupiaq in training, I grew less skeptical and began consistently applying what I was told. Had I ever been rushed by a polar bear, for example, I would have jumped away to the animal's *right* side. Inupiaq elders say polar bears are left-handed, so you have a slightly better chance to avoid their right paw, which is slower and less accurate. I'm pleased to say that I never had the chance for a field test. But when we're judging assertions like this, we should remember that Eskimos have had close contact with polar bears over a period of several thousand years.

A hunter named Migalik recalled spotting a polar bear that was waiting beside a seal's breathing hole in a large area of flat ice. He couldn't hope to approach it undetected, so he watched from the concealment of a nearby ice ridge. As the hours passed, the animal took turns lifting first one foot, keeping it up for a while, easing it quietly down, and then lifting another. From this, Migalik concluded that the soles of a polar bear's feet are sensitive to the moist, salty, frigid surface of young sea ice. When at last the bear gave up and headed for rough ice, Migalik predicted its direction, circled out ahead, and waited until it came within easy hunting range.

During the winter, ringed and bearded seals maintain tunnel-like breathing holes in ice many feet thick. These holes are often capped with an igloo-shaped dome created by water sloshing onto the surface when the animal enters from below. Inupiaq elders told me that polar bears are clever enough to excavate around this dome, leaving it perfectly intact but thin enough so a hard swat will shatter the ice and smash the seal's skull. I couldn't help wondering if this were true; but then a younger man described tracking a bear for many miles, and along the way it had excavated one seal hole after another, waiting unsuccessfully at each before moving on to try again.

When I lived in the village of Wainwright, the most respected hunter was Igruk, a man in his seventies. He had a remarkable sense for animals—a genius for understanding and predicting their behavior—as if he could penetrate their minds. Although he was no longer quick and strong, he joined a crew hunting bowhead whales during the spring migration, his main role being that of advisor. Igruk had hunted bowheads since childhood and probably knew them better than anyone else. Yet each time he spotted a whale coming from the south, he counted the number of blows, timed how long it stayed down, and noted the distance it traveled along the open lead, until it vanished toward the north. He never tired of studying whales and passing along what he had learned to the other hunters.

Knowledgeable elders like Igruk are held in great respect; they are the masters of vital skills, the learned intellectuals in their communities, the living libraries of tradition.

These stories illustrate the way hunters accumulate knowledge through patient observation and close interaction with their surroundings. Among the Inupiaq, as among ourselves, communication is a vital part of the learning process. Throughout the year, hunting activities and environmental phenomena are constant topics of discussion. They often dominate the conversation among families at meals, among friends visiting each other's homes or staying in camps, among groups gathered at community buildings, among people working together in various contexts, and among those who hunt, trap, fish, or travel together. In this way, information is exchanged between active hunters, passed down by the elders, and taught to the young. As the seasons change, so do the subjects discussed and the intensity of people's interests. During one season in a whaling camp, for example, I seldom heard a sustained conversation that wasn't related in some way to hunting.

Outsiders tend to underestimate the knowledge of indigenous people like the Eskimos. Little of this knowledge has been written down, so it remains largely invisible and inaccessible. Yet I believe the expert Inupiaq hunter possesses as much knowledge as a highly trained scientist in our own society, although the information may be of a different sort. Volumes could be written on the behavior, ecology, and utilization of arctic animals—polar bear, walrus, bowhead whale, beluga, bearded seal, ringed seal, caribou, musk ox, and others—based entirely on Eskimo knowledge.

Comparable bodies of knowledge existed in every Native American culture before the time of Columbus. Since then, even in the far north, Western education and overall culture change have taken a serious toll. Children now spend most of their time in school, not out on the tundra and sea ice. Television has come to the villages. Imported food diminishes the subsistence imperative. And introduced technology, such as firearms and snowmobiles, reduces the amount of knowledge and the range of techniques necessary for successful hunting. (The same is true in our own culture—farming and commercial fishing are examples. As technology becomes more sophisticated and complex, there is a corresponding decrease in the sort of knowledge we're discussing here.)

Reflecting on a time before Europeans arrived, we can imagine the whole array of North American animal species—deer, elk, black bear, wolf, mountain lion, beaver, coyote, Canada goose, ruffed grouse, passenger pigeon, northern pike—each known in hundreds of different ways by tribal communities: the entire continent sheathed in intricate webs of knowledge. Taken as a whole, this comprised a vast intellectual legacy born of intimacy with the natural world. Sadly, not more than a hint of it has ever been recorded.

Over the past 500 years, we have concentrated on promoting Western secular and religious education in Native American communities and have almost completely ignored what they could teach us in return. As a result, we've lost the opportunity for a balanced exchange and coalescence of our two great intellectual traditions.

Like other Native Americans, the Inupiaq acquired their knowledge through gradual accretion of naturalistic observations—year after year, lifetime after lifetime, generation after generation, century after century. In Western science, we often use other techniques—specialized full-time observation, controlled experiments, captive animal studies, and technological devices such as radio collars or electronic monitoring. These methods allow us to gather similar information, but much more quickly. I have not forgotten that disciplines such as molecular biology and cellular physiology have no equivalent among hunter-gatherers. But I also believe there is much in the native traditions that we have yet to learn.

I'll risk carrying the interchange between Eskimos and animals a step further to suggest that people have not only learned *about* animals, but also *from* them. We've already seen how polar bears hunt seals by waiting at their winter breathing holes, although I omitted details about keeping silent, staying downwind from the hole, and knowing when a seal comes up for air.

Polar bears use a completely different method to hunt seals in the spring, when they crawl up on the ice to bask in the sun. This tactic involves stalking over the ice until they're close enough to catch the seal before it can slip into a hole or crack. Here again, the bear's success depends on keeping silent, avoiding detection of its scent, moving at appropriate times, using concealment, and being still whenever the seal looks around. (According to elders, the approaching bear will even use a paw to cover its conspicuous black nose.)

Eskimo methods for hunting seals, both at breathing holes and atop the spring ice, are essentially identical to those of the polar bear. Is this a case of independent invention? A kind of convergent evolution? Or is it possible that ancestral Eskimos learned the techniques by watching polar bears, who had perfected an adaptation to the sea ice environment long before humans arrived in the arctic?

The hunter's genius centers on knowing an animal's behavior so well he can turn it to his advantage. For instance, an Inupiaq method of hunting polar bears is to mimic a seal lying atop the ice, enticing the bear to stalk within shooting range. This may be the only way to take a bear that's on the far side of an open lead or in a large flat area where it would see anyone trying to approach.

Here is another example. A polar bear will occasionally charge someone, especially if the bear has been wounded. It's hard to imagine a situation where the incorrect response would have a more immediate and purely Darwinian result. Inupiaq elders warn that bullets sometimes glance off a bear's skull, so you should wait until the animal raises its head and then aim for the neck. Failing this, shoot for the bulging hindquarter, because a bear hit in that spot will turn and bite the wound, exposing its neck for a deadly shot. One man said that when he was charged by a bear, he stood up and ran straight at the animal, certain that it would stop—as it did—giving him a clear, unmoving target.

Earlier I mentioned the great hunter Igruk, who understood animals so well he almost seemed to enter their minds. In April 1971, I was in a whaling camp several miles off the Wainwright coast. Onshore winds had closed the lead that whales usually follow in migration; but one large opening remained, and here the men placed their camp. For a couple of days there had been no whales, so everyone stayed inside the warm tent, talking and relaxing. The old man rested on a soft bed of caribou skins with his eyes closed.

Then, suddenly, he interrupted the conversation: "*I think a whale is coming, and perhaps it will surface very close . . .*"

To my amazement, everyone jumped into action, although none had seen or heard anything except Igruk's words. Only he stayed behind, while the others rushed for the water's edge. I was last to leave the tent. Seconds after I stepped outside, a broad, shining back cleaved up through still water near the opposite side of the opening, accompanied by the burst of a whale's blow.

Later, when I asked how he'd known, Igruk said, "*There was a ringing inside my ears.*" I am surely not a mystic; I have no explanation other than his; and I can only report what I saw. None of the Inupiaq crew members even commented afterward, as if nothing out of the ordinary had happened.

It's important to say that Eskimos are not strict empiricists; they have a complex religious tradition that imbues all of nature with spiritual powers. This aspect of Inupiaq life was mostly inaccessible to me, perhaps because Christian beliefs have supplanted many of the ancestral ways. But elders sometimes talk about animal abilities that we would call supernatural. Men speak carefully about their hunting plans, for example, because animals are said to hear and understand, even from afar.

An Inupiaq hunter is taught never to boast of his skills or say demeaning things about animals, especially large and dangerous ones like the polar bear, whale, and walrus. Offended animals may take revenge or shun those who are disrespectful. "*When you hunt walrus,*" I was advised, "*you must not act like a man. Do not be arrogant; be humble.*" Sometimes a few walrus are shot on an ice floe, but others from the herd refuse to leave, surfacing at the edge, snorting and glaring. The men stand facing them, extend their opened arms, and beg them to go away.

One Inupiaq elder recalled the ideals he was taught as an apprentice hunter: "*When I was young, the old-timers always told me to respect all the birds and animals, and don't ever kill any unless you want it. That way you can live a long time.*"

Wisdom: The Hunter's Heart

This brings us to the religious and spiritual dimensions of hunting, another elaborate and important realm of traditional knowledge. To illustrate, I turn to the Koyukon Indians, who inhabit a vast wildland along the Yukon and Koyukuk rivers in Alaska. The Koyukon belong to a family of peoples called the Northern Athabaskan, which includes many distinct groups throughout the Alaskan interior and across subarctic Canada from British Columbia to Hudson Bay.

In the following pages, I describe the spiritual concepts that guide Koyukon people in their relationships to nature (see also Nelson 1983). As in our own society, there are variations in the extent to which individuals conform to these guidelines in their daily behavior. But as a whole, Koyukon society is bound to the natural community through these principles, and I believe their usefulness is reflected in the health of the surrounding environment.

Like the Inupiaq, Koyukon people possess a large body of empirical knowledge encompassing every aspect of their boreal forest homeland. This scientific information is interwoven with an equally voluminous knowledge from a realm that lies beyond the senses. For the Koyukon, animals, plants, and elements of the physical world possess qualities that are both natural and supernatural. The environment is inhabited by watchful and potent beings who feel, who can be offended, and who should be treated with respect.

Consider the Koyukon hunters I described earlier having taken a black bear in its den. From the beginning, they knew a great power had revealed itself to them. No one discovers a bear by skill and cleverness alone—or purely by accident—because in the Koyukon world animals *give* or *withhold* themselves. A hunter's "luck" with a particular species depends on the respect he has shown toward it, which keeps him in a state of harmony or grace. When Koyukon people talk about luck in hunting, they might say *bik'ohnaatltonh*, "something took care of him." This "something" is the animal's spiritual power.

A bear hunter, confronting this power, should carefully follow the many rules for proper behavior toward the animal. Earlier I mentioned slitting its eyes and taking off its feet as the first step in butchering. Fresh bear meat should also be kept strictly away from women, whose own feminine spirituality could alienate all bears from the hunter. Although Koyukon women hunt and trap, they are prohibited from taking bears and certain other spiritually powerful animals, such as wolves and otters.

When he comes home, a successful bear hunter should wait a while and then make a cryptic comment like : "*I found something in a hole.*" This avoids any semblance of boasting or reveling in the animal's demise, a breach of etiquette that could case bad hunting luck, sickness, even death to a hunter or a member of his family.

Several days or more after a bear is killed, men and boys gather away from the village—exchange stories, talk about bear hunting, and cook cerain parts of the animal over an open fire. Elders say this is done to honor the bear, protect those who hunt for it, and assure that bears will offer themselves in the future. Sitting beside the fire, an old man once told me: "*What we eat here is the main part of the bear's life.*"

According to Koyukon people, each individual animal has its own spirit, but offending one can alienate all others of the species. Violations against nature can bring every sort of bad luck or personal harm, especially when the most spiritually potent animals are involved—black bear, brown bear, wolverine, lynx, otter, and wolf. But mistreating or disrespecting any creature can alienate its spirit. Even a redback vole or a ruby-crowned kinglet is a power to be recognized. No one is ever alone in the wild: no one is ever outside the bounds of moral restraint. "*There's always something in the air that watches us,*" a village elder said.

Some animals, like the bear, are surrounded by dozens of rules or taboos; for others there are only a few. But the basic rules of respect apply to everything in nature—animals, plants, earth, water, sky. Here is a scattering of examples: People should not point at a mountain or at the stars. Explaining this, a woman told me: "*You should never point at something that's so much greater than you are.*" After peeling its bark, you should bury a birch log under the snow, not leave it naked and exposed to the cold air. It is taboo to scrape moose hide at night; to wear clothing made from lynx if you are a woman; to eat meat from a loon if you are a youth or a fertile woman; to intentionally trap a porcupine (though it can be hunted); to leave a wolverine's carcass near children (the emanating power of its spirit can cripple a child); to put fresh bear meat where it might be eaten by dogs; to discard the bones of game animals with household refuse. A complete list would include hundreds of rules like these.

The necessary killing of animals and harvesting of plants is not considered disrespectful. The natural order—established in a time beyond memory—dictates that humans and other animals must sustain themselves by taking other lives. But people should do everything possible to prevent unnecessary suffering; they should never take more than they need or waste what has been given them; and they should treat all remains according to the traditional rules.

The spiritual interactions between people and nature are expressed in many ways. For example, animals sometimes give signs or omens. If a hawk owl flies past a hunter from behind, it foretells good luck; but if it crosses his trail or comes toward him, the hunt will be poor. "*I really hate to see one of those things do that,*" a man lamented. "*When it does I still go out, thinking I can beat him! But I never did yet; he's always right. If you see that, you might as well go back home.*" Hearing a red squirrel chatter in the night is an omen of impending death. On the other hand, people can bring themselves good luck by catching a boreal owl, fastening a tiny bit of dried fish to its back feathers, and releasing the bird.

A hunter seeing a raven fly overhead may shout: "*Tseek'aał, [Old Grandfather], drop your pack to me!*" If the bird tucks a wing and rolls in flight, as if to drop a loaded packsack, it means good luck. Because ravens have a great, benevolent power, people often ask them for help and protection. A Koyukon woman described praying to a raven when she was desperately sick, then explained: "*It's just like talking to God; that's why we pray to ravens.*"

In Koyukon tradition, it was Dotson'sa, the Great Raven, who created the earth and guided it through dreamlike transformations to the world we know today. These events are recounted in elaborate stories from the Distant Time—a Koyukon equivalent of Genesis or Darwin—explaining how certain animals, plants, and elements of the physical environment came to be as they are (see Attla 1983, 1990). Some story cycles take many evenings for a complete telling; written versions are hundreds of pages long. Similar stories, called by the wholly inadequate terms folklore and myth, are a part of all Native American traditions. They form the basis for an entire view of the world and the ideological foundation for human relationships to the natural community.

This traditional worldview carries us beyond the idea that humans have a tendency to affiliate with other life . . . to the possibility that our fellow creatures also have a tendency to affiliate with us. Among people like the Koyukon, awareness of the interweaving of human and nonhuman life runs so deep that distinctions between the two seem to fade. There is but one living community in which all organisms—including humankind—hold full and functional membership.

In a telling reversal of Western thought, Koyukon stories explain that humans were created by an omnipotent *animal* figure during the Distant Time. As Raven designed and transformed all of nature, distinctions arose in the two orders of being. Yet our worlds are still closely unified—not so much because humans possess animal qualities but because animals possess human qualities. In the Distant Time, it is said, animals *were* humans: we lived together in one society and spoke a common language.

Animals can still understand what people say, regardless of distance. This power has positive repercussions—as when they answer prayers or requests—or it can be negative, as when someone brags about hunting and offends the species involved. Animals also have humanlike personalities, a legacy from their behavior in the Distant Time. The sucker fish was a thief, for example, and things he stole became an assortment of oddly shaped bones in the fish's head. Such qualities—good or bad—can be somewhat contagious. Because of this, a man told me: "*In spring, sometimes we run short of food . . . but if we catch a sucker in the net, I just can't eat him.*"

It's difficult to express the power and substantiality of this worldview. People like the Koyukon experience a natural environment in which two dimensions—the material and spiritual—are fully expressed and profoundly important in normal daily life. The raven soaring above is not just a bird but a potent spiritual being with abilities far beyond those possessed by humans. A brown bear at the forest edge is both a temperamental beast and a preternatural force. The white-crowned sparrow sings words that tell of its life as an animal-person in the Distant Time.

Among the Koyukon, humans and animals are bound together in ways that challenge an outsider's comprehension. All creatures, no matter how small and inconspicuous, carry the luminescence of power.

I was with an old hunter named Grandpa William when a strange bird (I later concluded it was young northern shrike) perched in a nearby tree-top and started babbling melodically. For Grandpa William, encountering this unknown and loquacious creature was deeply unsettling, because an unusual event like this can be a dangerous omen. After a few minutes, he looked up at the bird and spoke in a soft, pleading voice: *"Who are you? And what are you saying to us?"* He paused, watched, then added, *"Wish us good luck, whoever you are. . . . Wish us well, and surround us—your grandchildren—within a circle of protection."*

I had few experiences with Koyukon people that more forcefully revealed the difference between their world and mine: a man speaking to a bird, humbling himself as a child to an elder, imploring mercy and protection. It filled me with wonder and amazement . . . and I longed for an understanding that seemed always beyond reach.

Here is a natural world pervaded with dimensions unknown among contemporary Western peoples, specifically denied in our modern traditions of science, philosophy, and religion. Through cultural learning, we acquire a particular worldview that becomes the bedrock of our minds: as deep inside us as language, as far beyond conscious thought as the act of breathing, as potent as our involuntary emotional responses.

In Western cultures, we have learned to perceive a natural world whose dimensions are entirely material and open to discovery through empirical methods. But among people like the Koyukon, we encounter the possibility of yet another dimension in nature, as if someone were telling us about the far side of the moon. If we can never experience or comprehend this dimension as they do, we might at least acknowledge the possibility of its existence.

And we might recognize the wisdom it embodies. When a Koyukon woman catches fish in her net, she treats them in ways that recognize their sensitivity and awareness. When a man finds an otter in his trap, he accepts a constellation of moral responsibilities toward that animal. For Koyukon people, nature has the greatest power, and humans must humble themselves to it. The key to this humility is respect—through innumerable gestures of politeness and through adherence to rules or taboos. A strict code of morality extends beyond the enclave of human society to include the entire community of life.

In recent times, this code has also influenced Koyukon villagers' attitudes toward biological research. Studies in nearby wildlands have involved live capture of small mammals, translocation of fish, and use of radio collars on caribou. In each case, they felt the methods were dangerously offensive to the animals involved and likely to make them shun people or move elsewhere.

Killing an animal for food is one thing; capturing, manipulating, and releasing it is another. As I understand it, these intrusions violate a creature's inherent right to live with dignity. Both Koyukon and Inupiaq people objected most strenuously to the use of radio collars, and some said they would try to kill such an animal so it wouldn't have to go on living that way.

Among the Koyukon and many other traditional peoples, what Westerners define as the physical world is considered a part of the living community. In Koyukon tradition, elements of the "nonliving" environment—earth, mountains, rivers, lakes, ice, snow, storms, lightning, sun, moon, stars— all have spirit and consciousness. The soil underfoot is aware of those who bend to touch it or dig into it. Certain localities are alive with power, sometimes dangerous and sometimes benevolent. Winter cold has a mind of its own, which people may anger or assuage.

When the river ice breaks up each spring, people speak to it, respectfully acknowledging its power. Elders make short prayers, both Christian and traditional Koyukon, asking the ice to drift easily downstream without jamming and causing floods. By contrast, some years ago the U.S. Air Force bombed an ice jam on the Yukon River to prevent inundation of communities. Far from approving, some villagers blamed subsequent floods on this arrogant use of physical force. In the end, nature will assert the greater power. The proper role of humans is to move gently, humbly, pleading or coercing, but always avoiding belligerence.

For the Koyukon and people with similar traditions, an embracing concept of affinity toward life and living processes would therefore include the physical universe. An anthropologist named A. Irving Hallowell (1976:362) questioned an Ojibwa elder: *"Are all the stones we see about us here alive?"* After thinking about it, the man replied: *"No! But some of them are."*

Perhaps it's human nature to embrace the absoluteness of one culturally defined worldview. In Western society, we rest comfortably on our inherited truths about the nature of nature. Our burgeoning environmental literature, for example, contains a nearly endless variety of statements about the absence of mind in nature. The environment is numb to a human presence—blind, deaf, inert, insentient, compassionless, sometimes brutal in its raw, random power. I've seen many statements like this one: "However much I love these mountains, they are indifferent to my presence and have no concern for my fate."

We may be elevated by the beauty of nature, cling to it, crave to protect it; but we cleave to the coldness of stone, the storm that carries us away without knowing, the waters that kill without reason. We live alone in an uncaring world of our own creation.

Despite our certainty on this matter, the anthropological literature indicates that most of humankind has concluded otherwise. Perhaps our scientific method does follow the path to a single, absolute truth. But there may be wisdom in accepting other possibilities and opening ourselves to different views of the world. The most important reason for this acceptance lies in our behavior toward the environment. Although we can never know if there is sensitivity and awareness in nature, we might govern our actions as if it were so, following the basic principles of humility, respect, and moral concern that bind Koyukon people to their encompassing natural community.

When I asked a Koyukon man about the personality of the Canada goose, he described it as a gentle and good-natured animal. *"Even if it had the power to knock you over,"* he concluded, *"I don't think it would do it."* For me, these words carried a deep metaphorical wisdom. They epitomized the Koyukon people's own restraint toward the world around them. And they offered a contrast to Western culture, in which our power to overwhelm the environment has long been sufficient justification for its use.

There is a further and thoroughly pragmatic dimension to Koyukon environmental relationships, a dimension having to do with ecological dynamics, human use, and the maintenance of productivity. Almost every animal and edible plant species significant in the Koyukon economy undergoes marked changes of abundance or availability. Some species have fairly regular fluctuations, such as the approximately ten-year population cycle of the snowshoe hare. Others, like caribou and beaver, fluctuate irregularly. Before imported foods were available, Koyukon people experienced times of scarcity or starvation when important food species reached simultaneous population lows. Some villagers remember spring cold snaps when stored supplies were gone and people survived by gathering songbirds that froze to death under the trees.

Given these uncertainties, it's not surprising that Koyukon tradition includes strong prohibitions against waste. If an animal is killed it should be carefully butchered, stored where it will not spoil or be defiled by scavengers, and used as fully as possible. To do otherwise will offend its spirit, bringing bad luck or sickness. Meat is a sacred substance, still permeated with the animal's spirit. As a matter of respect, one woman advised that a platter of meat should be covered with a cloth before carrying it outside to a neighbor's house.

Hunters should also do everything possible to avoid losing wounded game, lest they be punished for wasting an animal or causing it to suffer unnecessarily. If someone kills a diseased or starving animal for humane reasons, it should be symbolically butchered and covered with brush to appease the spirit. "*Otherwise it would look like you just killed it for nothing.*"

Avoidance of waste is a pervasive theme in Koyukon environmental ethics. Another theme is intentional limitation of harvests to help maintain plant and animal populations. The underlying principle is essentially identical to our own concept of sustained yield management. Koyukon people are keenly aware of ecological processes. During their lifetimes, elders have observed population changes in most of the economically significant species. They have felt the weather "growing old," as winters lose their former intensity. They have seen the country become drier—lakes changing to meadows, meadows to thickets, thickets to forests. They have seen floods revitalize sterile lakes with increasing populations of fish. They have watched successional changes in vegetation and animal communities after fires. And they have observed the effects of both overharvesting and fallowing.

Building on this awareness, Koyukon villagers attempt to manage their uses of animals and plants. For example, they advise simple, common-sense practices like cutting large trees for firewood and leaving the smaller ones. People who seine whitefish in the fall prefer wide-meshed nets that allow younger fish to escape, although all sizes are equally valuable for feeding sled dogs. Trappers regulate their take of furbearers, hoping for the best long-term yields. Special trap sets are made to catch only large beaver, for instance, and the traps are usually removed after two are taken. This leaves a nucleus of young beavers in each house. One man criticized himself for taking too many otters from his trapline in a single year. Another said proudly that he trapped the same area for most of his life, and the country is as rich today as when he started.

Hunters try to limit their moose takes according to need. They also avoid local overharvest by hunting in dispersed areas and foster reproduction by making selective kills. Men from one village decided against taking black bears in the spring so that more would be available during fall when bears are in peak condition. For obvious reasons, people follow these practices most strictly when food is readily available and less so during hard times such as the spring shortages earlier in this century.

Even before Western technology became available, Koyukon people had the capacity to overexploit certain species. Traditional snares and deadfalls are highly effective, for example, and I believe their unrestrained use could have a serious impact on species as large as moose or as small as beavers. Today, the same prudent principles have been applied to firearms, snowmachines, steel traps, cable snares, and other modern devices.

Conservation practices like these are based partly on knowledge of ecological dynamics, partly on moral principles and spiritual beliefs. They emerge from a worldview that strongly opposes unrestrained exploitation of an environment that is not only finite and changeable but also aware. "*The country knows,*" an elder told me. "*If you do wrong things to it, the whole country knows. It feels what's happening to it. I guess everything is connected together somehow, under the ground.*"

The subject of Native American conservation practices has been controversial: some analysts have doubted the effectiveness of such practices and others have questioned their very existence. I can only speak with assurance about my experiences and research among the Koyukon, for whom a conservation ethic and sustained yield management are indeed basic elements of environmental relationships. Ethnographic accounts strongly suggest that similar traditions exist (or existed) among other Native American peoples. (See, for example, Vecsey and Venebles 1980; Hunn and Williams 1982.)

In judging the evidence, it's important to remember what I mentioned earlier: indigenous communities are no different from any others; some individuals violate even the strictest laws or moral edicts. Among the Koyukon, there are puritans and sinners, conformists and lawbreakers, and all shades between. Even orthodox people can recall occasions when they disobeyed the code of respect toward an animal, offending its spirit and bringing themselves bad luck. Moreover, whole societies may adhere strictly to some of their own ideals, less strictly to others; and the degree of adherence can vary from one time to another.

It would be surprising, indeed, to find that Native Americans had never been guilty of waste, overharvest, or environmental damage. But it would also be a mistake to conclude that such breaches nullify the entire existence of conservation ethics and practices in these cultures.

In my opinion, the ethnographic record supports the existence of a widespread and well-developed tradition of conservation, land stewardship, and religiously based environmental ethics among Native Americans. I believe that Aldo Leopold's eloquent and insightful formulation of a land ethic (1949:201–226) is a fascinating example of convergence with Native American thought. Again, I speak with assurance about the Koyukon people, for whom the land ethic is a founding principle of religious belief, ideology, and behavior toward the environment.

In all likelihood, both the Native American land ethic and the one espoused in Leopold's famous essay rest upon a similar fusion of scientific knowledge and environmental philosophy. The one is lavishly and intricately expressed in a multitude of cultural contexts; the other is contained in a credo remarkable for its simplicity and wisdom. But taken together, they represent a powerful statement about human relationships to the natural world.

As Europeans explored and settled North America, they compiled voluminous accounts of the continent's natural history. What rings clear is the vastness of unbroken forest, the unfettered expanses of prairie, the phenomenal abundance of wildlife—the richness and diversity of natural communities.

Even today, we speak of this continent as a pristine wilderness when the first Europeans arrived. Yet for an enormous span of time—variously estimated at 12,000 to 30,000 years—Native American people had inhabited and intensively utilized the land; had gathered, hunted, fished, settled, and cultivated; had learned the terrain in all its details, infusing it with meaning and memory; and had shaped every aspect of their lifeways around it. That humans could sustain membership in a natural community for such a protracted time, without profoundly degrading it, fairly staggers the Western imagination.

I do not mean to imply that Native Americans lived without altering their environments. This would be impossible for any human group. Hunting affected game populations; gathering had impacts on plant communities; settlements and agricultural fields caused more visible and significant changes. Perhaps most important was wholesale manipulation of the environment, on a continental scale, through intentional use of fire. For example, fires were used to improve habitat for game and edible plants in the California chaparral, the eastern tall-grass prairie, the southeastern longleaf pine forests, and the Canadian subarctic forests. Studies such as those by Henry Lewis (1977, 1982) and Omar Stewart (1954) reveal not only the significant environmental effects of Indian fires, but also the complex ecological knowledge that guided their application and allowed people to anticipate the beneficial results.

The basic point is this: although no human society can exist without affecting its surroundings, Native American people inhabited the continent for a prodigious span of time, yet it remained in a condition that Europeans identified as "wilderness." This is strong testimony to the adaptation of mind—the braiding together of knowledge and ideology—that linked North America's indigenous people with their environment.

A Koyukon elder, who took it upon himself to be my teacher, was fond of telling me: *"Each animal knows way more than you do."* He spoke as if it summarized all that he understood and believed, as if he wanted to bury it somewhere inside my consciousness. For me, this statement epitomizes relationships to the natural world among many Native American peoples, as they can be understood through the ethnographic literature. And it goes far in explaining the diversity and fecundity of life on our continent when the first sailing ship approached these shores.

Confluence: Finding the Lost Arrow

In the brief passage of five centuries, European culture and technology have radically overturned an order that prevailed in North America for thousands of years. Perhaps, at some level, Euro-Americans also carry inside them a deep affinity for life; if so, it has been subservient to other values and motivations. Compared with their Native American predecessors, Euro-Americans have exploited the living community with an utter disregard for restraint or moral concern.

Of course, there are signs of change. Ecological perspectives that emerged during the past century are now spreading beyond the scientific enclave into the population at large. Environmental ethics and related bodies of thought are also diffusing outside the academic sphere. But for the most part, our society remains embedded in the Western worldview, which isolates us from the natural community and leaves us spiritually alienated from nonhuman life. We have created for ourselves a profound and imperiling loneliness.

Based on the anthropological literature, it's reasonable to suggest that most of humanity—over the span of our species' history and evolution—has conceptualized the natural world according to principles much like those of the Koyukon and other Native American peoples.

Anthropologist Robert Redfield cites three basic characteristics of this worldview. First, humanity, nature, and the sacred are thoroughly conjoined. We cannot speak of humans *and* nature, because humans are *in* nature, wholly a part of it. Nor should this approach to sacredness be called mysticism, for "mysticism implies a prior separation of man and nature and an effort to overcome the separation" (Redfield 1953:105). Second, relationships between humans and environment are based on orientation rather than confrontation: people do not aspire to control or master their surroundings; rather they seek to work with them through placation, appeal, or coercion. And third (p.106): "Man and Not-Man are bound together in one moral order. The universe is morally significant. It cares. . . . When man acts practically toward nature, his actions are limited by moral considerations."

According to Redfield, abandonment of this primary worldview by contemporary societies like our own represents one of the greatest transformations of the human mind (p.108): "Man comes out from the unity of the universe . . . as something separate from nature and comes to confront nature as something with physical qualities only, upon which he may work his will. As this happens, the universe loses its moral character and becomes to him indifferent, a system uncaring of man."

I do not mean to idealize traditional peoples or to imply that they live in an elysian world of harmony and perfection—they do not. But in communities like those of the Koyukon, ideological constraints on human behavior and uses of technology (both traditional and modern) create a truly sustainable relationship between humans and environment. In this relationship, people are nourished by what the natural community provides while the diversity and fecundity of nature are nourished in turn.

Cases like that of the Koyukon offer little support for the widely held view that humans are by definition a blight; that we cannot exist without destroying our environment; that we have no rightful place on earth. These self-accusations may not reflect a *human* condition so much as a cultural condition brought about by agriculture and domestication—what anthropologist Hugh Brody has called "the neolithic catastrophe."

The worldview that has emerged among industrialized agricultural societies has brought us to the edge of ecological collapse. If we cannot avert a cataclysm, Richard Lee and Irven DeVore (1968:3) suggest that "interplanetary archaeologists of the future will classify our planet as one in which a very long and stable period of small-scale hunting and gathering was followed by an apparently instantaneous efflorescence of technology and society leading rapidly to extinction."

What I am suggesting is that biophilia—a deep, pervasive, ubiquitous, all-embracing affinity with life—lies at the very core of traditional hunting-fishing-gathering cultures. That people like the Koyukon manifest biophilia in virtually every dimension of their existence. That connectedness with nonhuman life infuses the entire spectrum of their thought, behavior, and belief.

Indeed, it might be impossible for such people to stand far enough outside themselves to imagine a generalized concept of biophilia and give it a name. As a prerequisite, they might need a measure of disaffiliation from nonhuman life, a remoteness from other organisms that is impossible for those so intimately bound with the natural world.

Perhaps, like the curved edge of earth, biophilia only becomes visible from a distance.

Yet an affinity for other life may be as vital to us as water, food, and breath; may be so deep in us that only by a centuries-old malaise of drifting away have we come to the point of thinking about it. At the conclusion of *Biophilia*, Wilson (1984:185) asks: "Is it possible that humanity will love life enough to save it?" Surely there is no more important question in the latter twentieth century. But it seems nearly certain that throughout most of history, humans did love life. Every aspect of culture and mind was permeated with biophilia.

The essential question may not be whether biophilia is an innate and universal human tendency, but why a very recent branch of human culture has veered away from it. And how long we can survive in its absence. As we work to resurrect a sustainable human lifeway, I believe we have much to learn from traditional cultures like the Inupiaq and Koyukon. Even now, we are finding our way back to the same principles that have guided their long and successful membership in the natural community.

If we can recover that lost wisdom, the physical and spiritual affinity with life celebrated by the concept of biophilia might so deeply pervade our worldview that we would no longer apply a name to it.

Among my strongest memories from the years in Alaskan villages are two offhand comments uttered in very different contexts. Ending a conversation about the incursions of Euro-American culture into his arctic world, an old Inupiaq hunter said: *"The white man is a genius alright, but . . ."* His voice trailed off, and in a thoroughly Eskimo way he let silence carry the message: Western culture lacks something of great importance.

Years later, I was trekking through the forest with a young Koyukon man. In the midst of our casual banter, he teased: *"Dick, you're smart but not wise."* Over the following days, his words came back to me again and again. I had never before thought of the distinction, yet it was clear even to a teen-

Searching for the Lost Arrow

ager in this community, where people recognize that the young can know a great deal but only the elders can be wise. I've always suspected his comment was not meant only for me, personally, but for the culture to which I belong.

I believe people like the Koyukon and Inupiaq have a far better sense than modern Euro-Americans about the relative importance of knowledge and wisdom. In Western society, we emphasize the paramount value of knowledge: information, facts, that which can be discovered through our empirical disciplines; the palpable, material truth.

In traditional societies of my acquaintance, people recognize the vital importance of knowledge; but I suspect they would judge wisdom to be even more important. My old dictionary defines wisdom as "the power of true and right discernment; also, conformity to the course of action dictated by such discernment." And the entry for *wise* reads: "seeing clearly what is right and just; having sound judgment . . . prudent, sensible . . . having great learning . . . versed in mysterious things."

Is it possible that wisdom has been more important than knowledge as a basis for the long and successful habitation of North America by its indigenous people? Could it be wisdom that explains why the first European travelers found here a vast and untrammeled beauty, an extraordinary wealth and diversity of wild species, an array of intact natural communities?

Perhaps our imbalance with the environment and our loss of affinity with life reflect a single-minded pursuit of knowledge and a diminished regard for wisdom. And the greatest promise of Western science may be less in the knowledge it brings us than in its ability to reveal a wisdom similar to that so pervasive in Native American traditions.

There is a riddle, one of many told by Koyukon people "to help shorten the winter":

Wait, I see something. I am looking everywhere for a lost arrow.
Answer: *The search for a black bear's den.*

For me, this riddle stands both as an illustration of Koyukon people's relationship to nature and as a metaphor. The lost arrow represents our abandoned sense of physical connectedness to the natural world; the bear symbolizes our need to rediscover a deep, perhaps spiritually based, affiliation with life. Scientific knowledge is our way of finding tracks under the snow, watching for places where the grass is scratched away, leading us toward the hidden den. But knowledge may not suffice without the balance of harmony, without that state of grace through which the animal reveals itself.

In the end, it is the wisdom of humility and respect that brings a Koyukon hunter to the bear.

ACKNOWLEDGMENT

I am grateful to the people of Wainwright, Ambler, Chalkyitsik, Huslia, and Hughes, Alaska, for sharing their lives and traditions with me over the years. I know only enough of Inupiaq and Athabaskan Indian traditions to realize how little I have learned, so I beg my instructors' understanding for the errors and shortcomings in my work. Personal names mentioned in the text are pseudonyms.

REFERENCES

Attla, Catherine. *As My Grandfather Told It: Traditional Stories from the Koyukuk.* Fairbanks: Alaska Native Language Center, 1983.

———. *K'etetaalkkaanee: The One Who Paddled Among the People and Animals.* Fairbanks: Alaska Native Language Center, 1990.

Hallowell, A. Irving. *Contributions to Anthropology.* Chicago: University of Chicago Press, 1976.

Hunn, Eugene, and Nancy Williams (eds.). *Resource Managers: North American and Australian Hunter-Gatherers.* Boulder: Westview Press, 1982.

Lee, Richard, and Irven DeVore (eds.). *Man the Hunter.* Chicago: Aldine, 1968.

Leopold, Aldo. *A Sand County Almanac.* New York: Oxford University Press, 1949.

Lewis, Henry T. "Maskuta: The Ecology of Indian Fire in Northern Alberta." *Western Canadian Journal of Anthropology* (1977) 7:15–52.

———. "Fire Technology and Resource Management in Aboriginal North America and Australia." In E. Hunn and N. Williams (eds.), *Resource Managers: North American and Australian Hunter-Gatherers.* Boulder: Westview Press, 1982.

Luckert, Karl. *The Navajo Hunter Tradition.* Tucson: University of Arizona Press, 1975.

Nelson, Richard. *Hunters of the Northern Ice.* Chicago: University of Chicago Press, 1969.

———. *Make Prayers to the Raven.* Chicago: University of Chicago Press, 1983.

Redfield, Robert. *The Primitive World and Its Transformations*. Ithaca: Cornell University Press, 1953.

Stewart, Omar C. "The Forgotten Side of Ethnography." In F. Spencer (ed.), *Method and Perspective in Anthropology*. Minneapolis: University of Minnesota Press, 1954.

Vecsey, Christopher, and Robert W. Venebles (eds.). *Indian Environments: Ecological Issues in Native American History*. Syracuse: Syracuse University Press, 1980.

Wilson, E. O. *Biophilia*. Cambridge: Harvard University Press, 1984.

CHAPTER 7

The Loss of Floral and Faunal Story: The Extinction of Experience

Gary Paul Nabhan and Sara St. Antoine

The new knowledge the human race is acquiring does not compensate for the knowledge spread only by direct oral transmission, which, once lost, cannot be regained or retransmitted: no book can teach what can be learned only in childhood if you lend an alert ear and eye to the song and flight of birds and if you can find someone who knows how to give them a specific name.

ITALO CALVINO (1983)

BIOPHILIA HAS BEEN described as the innate human need for contact with a diversity of life-forms (Wilson 1984). Therefore, the expression of biophilia should be possible in any individual. Yet if this phenomenon called biophilia is the *genetic* affinity for other life, why is it expressed in some people and cultures more than others? At least three answers to this question are possible.

First, perhaps biophilia is not genetically determined but is a set of learned responses. Or, second, biophilia could be a set of behavior based on a number of genes, for which any particular individual may have some but not all of the genes. In other words, some human genetic lineages may have been selected for biophilic responses more than others. The third possibility, which we believe is consistent with Wilson's original hypothesis, is that a child's learning environment greatly conditions the expression of any genetic basis for biophilia. Unless the appropriate environmental triggers are present in a certain cultural/environmental context, biophilia is unlikely to be fully expressed.

Biophilia, then, may take on different shapes and shades in its expression within various cultures and individuals, depending on their degree of exposure to and engagement with other organisms. Obviously, biophilia must have evolved during the hundreds of thousands of years of our existence as hunter-gatherers, but the very different conditions prevailing in modern societies today may not be sufficient to trigger similar phenotypic expressions of our affinity for other life-forms. One way to assess this possibility is to compare the responses of the natural world of tribal elders, who have engaged in considerable hunting and gathering activities during their lifetimes, with those of their grandchildren who have grown up fully exposed to television, prepackaged foods, and other trappings of modern life. With genetic lineages held constant but environmental conditions during early child development changed, we might begin to see how environmental influences affect the expression of biophilia.

To test such intergenerational differences in the real world is not as simple as we originally expected. Nevertheless, we believe that certain inferences from the ethnobiological knowledge, values, and behavior of different generations of the same cultures can provide some answers. For this approach to be useful, however, we must clearly define the conditions which must have prevailed during the course of human history for biophilia to have emerged through gene/culture coevolution. In particular, we hypothesize that the following conditions must have strongly influenced our hunting-gathering ancestors in the 2 million years since the emergence of language among them, which ushered in story as a way of distilling adaptive information:

Culture

1. The physical presence of impressive or salient species representing certain sets of creatures that obviously affected human survival over tens if not hundreds of thousands of years. For canids to have become prominent in our mythology or even in our consciousness, for example, populations of wolves, foxes, coyotes, or dingoes must have remained viable and widely dispersed over most of the period of hominid evolution. Conversely, it is unlikely that the megafauna which were extirpated early in the emergence of humankind remain universally present in our racial memory.

2. Frequent or at least significant human contact with these creatures such that people could recognize them, name them, and retain images of their characteristic traits. In other words, wild canids must have ranged within earshot and reach of human habitations. In addition, a critical mass of humans must have had direct experience of their presence, their tracks, their droppings, or their pelts.

3. Continuing oral transmission of stories about these species. When these stories were imbued with cultural values, they may have offered selective advantages to those who subscribed to them. Coyote myths, for example, may have reminded certain cultures to keep their resources safeguarded from these low-density, widely roaming omnivores which in some ways are competitors with human populations. Moreover, parents may have instructed their children to be distrustful of "coyote-like" behavior in other humans.

If we are to entertain biophilia as a working hypothesis to explain human reponses to nature, we must demonstrate that not only formally trained Western scientists and self-professed "nature-lovers" truly exhibit an affinity for a variety of creatures. People across the full spectrum of cultures must do so as well, especially those which have retained at least some of the attributes of our preliterate hunter-gatherer ancestors.

One useful measure of this affinity is the size and complexity of a culture's biosystematic lexicon (Brown 1985), for people must at least precisely name a variety of creatures to be able to encode cultural knowledge and values about them. We know that indigenous cultures vary greatly in the number of names they maintain for plants and animals within their ranges, and

The Extinction of Experience

a portion of this variance is due to the relative richness of species within a culture's home range. But the size of a lexicon of plant and animal names also depends on how much time people spend interacting with other species in their subsistence activities, as well as in their aesthetic and spiritual pursuits.

Such lexicons range from over 2,700 labeled categories of biota for Ifugao farmer-gatherers in the tropics of the Philippines to less than 200 for certain boreal hunter-gatherers and temperate-zone urban street gangs (Brown 1985). Brown admits that his own lexicon of names for familiar plants and animals is several times smaller than that of an Ifugao; we suspect that only a small percentage of Western-trained scientists can readily name more than 500 creatures by sight. The largest taxonomies tend to occur among small-scale agriculturalists who not only "overclassify" domesticated plant and animal species but also manage wildlands and field edges to increase the abundance of useful wild plants and animals (Berlin 1992; Anderson and Nabhan 1991). The lexicons of small-scale farmer-forager societies tend to contain a greater number of binomial labels for plant taxa than do those of hunter-gatherer societies, suggesting that farmer-foragers give greater attention to the morphological and behavioral details of local species, whether wild or cultivated. The depth of folk taxonomic knowledge in a cultural community is not, however, static; like biodiversity, it can be affected by social and environmental forces which change through time.

To understand how these social and environmental forces have influenced the expression of biophilia over the last few decades, we first sought to gain a sense of the cultural context of biophilia as it functions in a set of traditional societies native to a desert region. Even in an arid region of low biological diversity, the habitats in the imaginations of native dwellers are far from deserted. Yet generational differences within a single desert culture's responses to animals may be as pronounced as the differences between cultures of different subsistence strategies. In the three generations it has taken for some Indian families to move from a hunter-gatherer existence to dependence on centralized welfare systems in cities, perhaps we can see concomitant changes in their biological knowledge which mirror general differences between hunter-gatherer societies and urban societies. Therefore, we have sought to contrast the knowledge of desert biota among traditional elders with those now found among the youngest generation of O'odham and Yaqui Indians. We hope that such comparisons may allow us to evaluate the consequences of losing the three requisite conditions for the expression of biophilia:

1. The loss of biodiversity via habitat degradation or overexploitation of species
2. The extinction of experience—of hands-on, visceral contact with other forms of life
3. The demise of oral traditions of plant and animal stories

Our findings may help explain why the genes for biophilia now have fewer environmental triggers to stimulate their full expression among contemporary cultures compared to those in the past.

The Loss of Biodiversity

The Sonoran Desert of the U. S. /Mexico borderlands is perhaps the richest arid region in the world in terms of plant growth forms or adaptive strategies and species richness per unit area (Shreve and Wiggins 1964). Much of its wildland vegetation has been cleared or converted over the last century, however, and perhaps as much as 60 percent of the region's surface area is now dominated by nonnative species. Most of the hundred plant species at risk in the region have been threatened by recent anthropogenic landscape changes, including overgrazing by introduced livestock, aquifer depletion, and urbanization (Nabhan et al. 1991). At least another hundred vertebrate species have been extirpated or drastically reduced in numbers—mostly as a result of the drying up of formerly perennial streams and the associated loss of riparian corridors (Tellman and Shaw 1991; Rea 1983; Davis 1982). Historic environmental changes have triggered severe reductions in the abundance and distribution of several charismatic species such as desert bighorn, Sonoran pronghorn, desert tortoises, and Gila monsters. Today, such creatures are clearly less available for contact with Native Americans and recent immigrants alike.

The Extinction of Experience

Preliterate Native American societies within and beyond the desert borderlands have frequently been accused of being the primary agents of environmental degradation and species extirpation in the Americas (Redford 1985; Diamond 1986) or, alternatively, romanticized as the only ecologically sustainable societies that have ever existed (Orr 1992; Sale 1990). Usually, too little detail has been provided on the dynamics of these societies' interactions with local biota to accept or reject either notion. Between these two ideological ruts lies a vast terrain in which we might glimpse how indigenous cultures themselves have viewed and valued biological diversity.

The few empirical studies to assess the influences of contemporary indigenous cultures on habitat heterogeneity and biotic diversity suggest that factors other than hunting exert the strongest influences on faunal diversity and abundance (Nabhan et al. 1983; Vickers 1988). Animal numbers have been kept above extinction levels by indigenous practices for managing habitats and by switching to other resources when harvesting efficiencies begin to decline. Such practices have allowed numerous Native American cultures to draw upon nearly the same set of resources for hundreds of years. Active conservation management rather than passive acceptance of declining resources characterizes the behavior of numerous cultures from the Amazonian rain forests to the North American deserts (Anderson and Nabhan 1991).

In the context of biodiversity in the U. S./Mexico desert borderlands, we will address the indicators of biophilia found among two resident Uto-Aztecan cultures: the O'odham (Papago-Pima) and the Yaqui (Cahitan). Although these peoples have generally been considered desert farmers, certain members of their populations historically lived by hunting, gathering, fishing, and trading with relatives in permanent agricultural villages nearby. In fact, some nonagricultural bands of the O'odham traded songs and stories about wild animals and plants to their farming relatives in exchange for crops; the oral literature of these people remains rich in references to other creatures.

As Southwest Indian storyteller Leslie Silko (1987:87) has described for other cultures in the binational Southwest, these cultures perceived the world and themselves within that world as part of an ancient continuous story composed of innumerable bundles of stories. . . . Thus stories about the Creation and Emergence of human beings and animals into this World continue to be retold each year for four days and four nights during the winter solstice. . . . Accounts of the appearance of the first Europeans . . . were no more or less important than stories about the biggest mule deer ever taken or adulterous couples surprised in cornfields and chicken coops. . . . Traditionally, everyone, from the youngest child to the oldest person, was expected to listen and be able to recall or tell a portion, if only a small detail, from a narrative account or story. Thus the remembering and retelling were a communal process. Even if a key figure, an elder who knew much more than the others, were to die unexpectedly, the system would remain intact.

If we look at plants and animals mentioned in the transcribed and published corpus of O'odham and Yaqui stories, ceremonial orations, and songs (Russell 1975; Saxton and Saxton 1973; Giddings 1978; Rea 1983; Painter 1986), it is remarkable how many taxonomic classes are well represented. Rather than focusing their stories only on large fuzzy or scary animals, the O'odham refer to at least 26 taxa of plants, 16 taxa of invertebrates, 13 taxa of reptiles, amphibians, and fish, 28 taxa of birds, and 20 taxa of mammals. Only 7 large native mammals, 3 large domestic mammals, 8 large birds, and 7 economically important domesticated plants are represented among the 103 folk taxa featured in O'odham oral literature.

These numbers suggest that O'odham mythology is not merely focused on creatures that are conspicuous or have economic importance but extends to a wide range of local and extralocal biota. Some, as O'odham educator Danny Lopez (pers. comm.) recently told us, "are mentioned in stories so that our children understand that these animals are associated with the rains, and therefore need to be protected." In explaining O'odham taboos against eating "rain animals" such as mud turtles, herons, eagles, frogs, and horned lizards, Rea (1981:75) observes that "rain is an essential factor in the lives of desert agriculturalists, particularly dry farmers. Various plants and animals are incorporated into complex rain symbolism. For the Papago, eagle down feathers represent rain clouds; saguaro wine, the summer rain. The songs [referring to various plants and animals] during the summer wine feast help 'pull down the clouds.'"

Perhaps the clearest examples of the psychological power held by certain biota are the more than thirty-five plants and animals deemed by the O'odham to be too dangerous to abuse (Table 7.1). A psychosomatic "staying sickness" is cast upon O'odham who do not heed the taboos against disrupting these species (Bahr et al. 1973). Yet if one attempts to evaluate which organisms are taboo with respect to their direct utilitarian importance or their vulnerability to overexploitation, no single ecological determinant for inclusion emerges (Rea 1981). Some of these creatures may be indirect indicators of habitat quality; others appear to be food for thought alone.

Yaqui oral literature mentions no less than 29 folk taxa of plants, 13 taxa of invertebrates, 14 taxa of reptiles, amphibians, and fish, 18 taxa of birds, and 27 taxa of mammals. Of these 108 folk taxa, 18 large mammals and 6 large birds are included, as well as 7 domesticated plants. In their discussion of Yaqui deer songs, Evers and Molina (1987) discuss the centrality of two images in the Yaqui cultural landscape: the *huya ania* or "wilderness world" and the *sea ania* or "flower world." In Yaqui stories and ceremonies still ritually enacted by thousands of participants every year, the dozens of wild creatures that inhabit the *huya ania* and *sea ania* are communally remembered and paid respect by Yaqui communities on both sides of the U. S. / Mexico border. Reverence toward deer and other sacred creatures from these legendary worlds is publicly demonstrated so frequently in older Yaqui communities that it remains a vital component of contemporary life, not simply a vestige of the past. Again, many more plants and animals are distinguished by name than are used as foods, medicines, or ceremonial paraphernalia.

Citing numerous examples of ethnotaxonomies which extend far beyond the naming of useful species, Berlin (1992) has rejected Diamond's (1972) earlier contention that ethnozoological knowledge is fundamentally utilitarian in orientation. We agree with Berlin's insight that "the striking . . . structure and content of systems of biological classification in traditional societies from many distinct parts of the world are most plausibly ac-

TABLE 7.1. *Biota Associated with Psychosomatic Sicknesses and Taboos by Traditional Desert O'odham Communities*

Biota	Psycho-somatic Sickness	Cultural Food Taboo
Jimsonweed (*Datura* spp.)	X	
Peyote (*Lophophora williamsii*)	X	
Grasshopper (Acrididae)		X
Butterfly/moth (Lepidoptera)	X	X
Bee (Apoideae)	X	
Fly (Diptera)	X	
Mud turtle (*Kinosternon* sp.)	X	X
Gila monster (*Heleoderma suspectum*)	X	X
Desert iguana (*Dipsosaurus dorsalis*)		X
Chuckwalla (*Sauromalus obesus*)		X
Spiny lizard (*Sceloperus magister*)		X
Horned lizard (*Phrynosoma* sp.)	X	
Rattlesnake (*Crotalus* sp.)	X	
Quail (Odontophorinae)	X	X
Duck/goose (Anseriformes)		X
Heron/egret (Ardeidae)		X
Turkey (*Meleagris gallopavo*)		X
Roadrunner (*Geococcyx californianus*)		X
Great horned owl (*Bubo virginianus*)		X
Raven (*Corvus* spp.)		X
Vulture (Cathartidae)	X	X
Eagle/hawk (Accipitridae)	X	
Squirrel (*Spermophalus* sp.; *Ammospermophilus* sp.)		X
Pocket gopher (*Thomomys* spp.)	X	X
Mice (*Perognathus* spp.; *Peromyscus* spp.)		X
Kangaroo rat (*Dipodomys* spp.)		X
Hare (*Lepus* spp.)	X	X
Porcupine (*Erethizon dorsatum*)		X
Cat (*Felis cattus*)	X	X
Coyote (*Canis latrans*)	X	X
Dog (*Canis familiaris*)	X	X
Fox (*Vulpes* spp.; *Urocyon* spp.)		X
Bear (*Ursus* spp.)	X	X

counted for on the basis of human beings' inescapable and largely unconscious appreciation of the inherent structure of biological reality" (1992:3).

We wish to make it clear that O'odham and Yaqui knowledge of other
life-forms does not extend to all the species in their environments. Of the
more than 250 Western scientific taxa of plants known at desert oases inhabited by O'odham farmers and gatherers for centuries, only about 87 folk
taxa are named, representing a corresponding 96 Linnaean species (Felger
et al. 1992; Nabhan et al. 1983). Between 150 and 200 folk taxa of higher
plants are recognized by desert O'odham hunter-gatherers and farmers,
representing 250 or so Linnaean taxa, though there may be as many as 500
species within foraging distance from a village (Nabhan 1983). Similarly, of
more than 240 birds known in historic and recent times in the Middle Gila
River habitats where Pima Indians have lived for centuries, 75 O'odham
folk taxa have been recorded, corresponding to 85 to 90 Linnaean species
(Rea 1983).

What, then, is the relationship between local indigenous cultures and
the sensitive species most likely to be lost from the biological diversity of
the Sonoran Desert? Of more than thirty plant species now considered at
risk in the Organ Pipe Cactus National Monument vicinity, in the heart of
O'odham country, eighteen were historically used but only eight are used
today (Nabhan and Hodgson 1993). Two of these plants have been locally
depleted at their northern margins, but traditional uses have hardly affected
them elsewhere. Overall, other factors such as modern agricultural clearing
and overgrazing better account for the rarity of these two species over most
of their distributional ranges. Likewise, of the twenty-nine species of birds
that have been locally extirpated on the Middle Gila River during this century, the loss of twenty-five has been associated with riparian habit degradation caused by the non-Indian neighbors of the Pima. No losses seem to
be due to Pima hunting or agricultural practices (Rea 1983). The loss of
eleven of these species was recorded in Pima oral history. Of the twenty-
five other bird species which have conspicuously declined during this century, oral accounts by Pima elders comment upon population reductions in
nine of them. O'odham elder George Webb (1959) says that in contrast to
the verdure and diversity on the Gila River floodplain that he knew as a
child, he ended his days next to a "river that is an empty bed of sand. . . .

Where everything used to be green, there were acres of dust, miles of dust, and the Pima Indians were suddenly desperately poor." Davis (1982) and Rea (1983) have compiled numerous historic accounts of other animals that no longer occur in O'odham and Yaqui Indian country as a result of recent changes in river flow and groundwater levels, induced by non-Indians.

While these statistics and anecdotes do not fully answer how much the O'odham and Yaqui historically affected desert biodiversity, other studies suggest both positive and negative local effects—the negative ones largely created through overgrazing by introduced livestock (Felger et al. 1992; Nabhan et al. 1983). But how has the loss of biodiversity affected these cultures? It is clear from our review of O'odham and Yaqui ethnobiological knowledge that these cultures have had strong interest in a variety of life-forms and that their oral histories have been sensitive indicators of the loss of biological diversity locally. What we do not know with any confidence— if it is possible to know at all—is how long these regionally extirpated species will persist in the oral history, the songs, and the dreams of the O'odham and Yaqui, now that they have been removed from contact with the plants and animals in the flesh.

The Extinction of Experience

As Peter Steinhart (1989:816) has so eloquently observed: "Animals are far more fundamental to our thinking than we supposed. They are not just a part of the fabric of thought: they are a part of the loom." What happens when the loom is dismantled?

Plant and animal species need not be globally extirpated before declines in population sizes and ranges begin to have deleterious consequences for the human species. As insect ecologist Robert Michael Pyle (1992:65) has noted, local extirpations cause "the extinction of experience"—the loss of direct, personal contact with wildlife. Any conditions which reduce such intimate experience, Pyle claims, create a cycle of disaffection, apathy, and irresponsibility toward natural habitats.

In an effort to gauge the degree to which today's younger generation has lost direct experience with the natural world, we surveyed activity and attitudes of O'odham, Yaqui, Anglo, and Hispanic children. To what extent does a loss of direct nature experience correlate with reduced affinity for nature? In short, is the extinction of experience eroding biophilia just as relentlessly as is the extinction of species?

We interviewed fifty-two children living within a 25-mile radius of two national parks: Organ Pipe Cactus National Monument, on the U.S./ Mexico border south of Ajo, Arizona, and Saguaro National Monument, just west of Tucson. While the survey was neither large nor from a randomized sampling, respondents did represent a cross-section of urban and rural desert communities. If anything, the survey was biased toward children who live in small communities with considerable exposure to wildlands and farmlands.

We used the survey of Kellert and Westervelt (1983) as a point of departure, but we modified and expanded questions to address new issues and incorporate local geography and biology. We conducted the interviews face-to-face, reading the questions to the children either in Spanish or English depending on their dominant language. None of the indigenous children spoke O'odham or Yaqui as their primary language, but most did hear these languages spoken at home by relatives.

While most of the children claimed to have had some direct interaction with wildlife—either through hunting, plant gathering, or playful capture of small animals—the vast majority appear to be gaining most of their experience with other creatures vicariously. Some 35 percent of the O'odham, 60 percent of the Yaqui, 61 percent of the Anglo, and 77 percent of the Hispanic children responded that they had seen more wild animals on television or in the movies than in the wild. No doubt these figures would be even higher among a completely urbanized group with even less access to wild or open space.

Even in our sample, a clear majority of the children in each population group had never in their lives spent more than half an hour alone in a wild place (58 percent of O'odham, 100 percent of Yaqui, 53 percent of Anglos, 61 percent of Hispanics). These trends suggest that the personal, uninhibited, and spontaneous interaction with nature which solitude allows is seldom taking place today. A large portion of these same children, moreover, said they had never collected natural treasures, such as feathers, bones, insects, or rocks, from their desert surroundings (35 percent of O'odham, 60 percent of Yaqui, 46 percent of Anglos, 44 percent of Hispanics).

Now that the global electronic media dominate their knowledge of nature, these children are losing the kind of local awareness that television documentaries cannot supply. Basic facts that anyone living in the Sonoran Desert a century ago would know without even thinking are now known only by a limited segment of the population. When asked which plant smells the strongest when it rains, 23 percent of O'odham, 40 percent of Yaqui, 38 percent of Anglo, and 44 percent of Hispanic children responded that it was the prickly pear cactus or that they did not know, instead of correctly choosing the fragrant creosote bush. Similarly 23, 20, 15, and 16 percent, respectively, did not know that desert birds sing more in the early morning than around noon. Further, 17 percent of the O'odham, 20 percent of the Yaqui, 0 percent of the Anglo, and a startling 55 percent of the Hispanic children did not know it is possible to eat the fruit of the prickly pear cactus. Ironically, this fruit has been a major food source in the U. S. / Mexico borderlands for more than 8,000 years, continues to be sold fresh in markets, and even provides flavor to a popular variety of popsicle in markets near where many of these children live.

The potential problems of television, especially among indigenous populations, have received considerable attention by medical researchers, educators, and sociologists. Responding to the recent introduction of television to the Dene Nation in Canada's Northwest Territories, Dene educator Ernie Lennie (from Mander 1991:111) commented: "The type of learning we get in school and also on TV is the type of learning where we just sit and absorb. But in family life . . . children learn directly from their parents. Learning has to come from doing." Both television and certain formal education approaches run counter to the ways of indigenous education—for example, apprenticeships with elders—that have been most successful for previous generations.

Children's very ability to perceive the environment may be diminished by the replacement of multisensory experience in richly textured landscapes with the two-dimensional world of books or the audiovisual world of TV, videos, and movies. In Alaska, visual myopia set in within a generation's time after schools first introduced books and audiovisual media to Inuit children (Williams and Nesse 1991). Inuit children may have a genetic predisposition to myopia, but this condition was not expressed until books and screens began to dominate during critical stages of eyesight development when diverse stimuli are required (Wallman et al. 1987). When growing up more fully exposed to the rich visual stimuli in wildlands, Inuit hunter-gatherers never attained the current rate of 50 percent myopia among their current population. Shepard (1982) suspects that mental and emotional pathologies, too, might develop in individuals who have not had sufficient contact with the wild during the course of their early development.

Finally, replacing direct experience with television-mediated education discourages children from making their own observations and forming their own opinions of the natural world; instead they inherit those of the TV program's writers and editors. It is this quality of television that may explain the surprising similarity of responses among our four different groups—groups that in the past might have reflected significantly different knowledge and values. As technology critic Jerry Mander (1991:97) explains: "By its ability to implant identical images into the minds of millions of people, TV can homogenize perspectives, knowledge, tastes, and desires to make them resemble the tastes and interests of the people who transmit the imagery."

Even when television professes to impart an ecological message, it cannot supplant intimate experience. As Robert Michael Pyle (1992:66) writes, "A banana slug face-to-face means much more than a Komodo dragon seen on television. With rhinos mating in the living room, who will care about the creatures next door?" Peter Steinhart's (1991:11) warning reinforces this point: "Celluloid is not a substitute for experience. . . . If wildlife films ever become our only access to wild animals, we may be the less intelligent, perceptive, and imaginative for it." Not only does the exotic, unreal world presented on the television screen threaten to distract children from less spectacular local wildlife, but it fails to provide the personal stimulus awakened when you yourself choose the images, sounds, smells, and ideas that direct your experience of nature.

The Demise of Oral Tradition

As Italo Calvino observes in the chapter's epigraph, "new knowledge . . . does not compensate for the knowledge spread only by direct oral transmission, which, once lost, cannot be regained or transmitted." There is no doubt that linguistic diversity and its associated reservoirs of folk scientific knowledge have become as endangered this century as has biological diversity itself. If much of a people's knowledge about the natural world is encoded in their indigenous language, the same knowledge cannot easily be imparted in another foreign language which has not developed a specific vocabulary to describe local conditions, biota, and land management practices. Indigenous scientific knowledge is inevitably lost when cultures are faced with the demise in native languages recently reviewed by Dauenhauer and Dauenhauer (1992:119), who remind us that half or more of the approximately 200 Indian languages still spoken north of Mexico have already atrophied or are at risk: "If a Native American language dies, there is no place on earth one can travel to learn it. The public statements that some school administrators continue to make in opposition to teaching [oral] native languages would not be tolerated if made about some endangered species of bird or snail."

Stories of flora and fauna that have been told in native languages for dozens of generations are now being lost as linguistic acculturation proceeds. In our survey of desert-dwelling children, we observed that TV and classroom learning were not only usurping the time and status formerly given to direct experience; they also usurped time once dedicated to storytelling and personal instruction by elders.

Most children felt they were learning more stories about plants and animals from books than from grandparents or parents (58 percent of O'odham, 100 percent of Yaqui, 46 percent of Anglo, and 44 percent of Hispanic kids). And once again the presence of television was clear. When we asked Alvron, a young Hispanic boy, whether he had heard more animal stories from books or from family, he quickly answered, "Neither. Discovery Channel." A comparable percentage of the children—47 percent of O'odham, 100 percent of Yaqui, 61 percent of Anglo, and 55 percent of Hispanic kids—claimed that most of their learning about plants and animals was coming from school, not family. When we posed the question to Misty, a long-time patron of the Arizona-Sonora Desert Museum's summer nature programs, she, like Alvron, dismissed the instruction of either source. Television, she said, had been her best teacher.

How does storytelling differ from other sources of knowledge? Dene educator Barbara Smith explains it to Mander (1991:112) in this way: "Legends are tools that help people grow in certain ways. A lot of what matters is the power and the feeling of the experience. . . . But when you put something in a museum, or even on TV, you can see it all right, but you're really looking only at the shell." Mander then goes on to point out that storytelling is a stimulus for children's imagination, requiring their mental contributions to the storyteller's words, not just their passive absorption of information.

Yaqui educator Felipe Molina (pers. comm.) recalls that when he was growing up, the Yaqui legends were never "uplifted" or given credence in school; instead, the Western scientific mode of analysis dominated all learning about nature. "We might learn about plants in science," he explained, "how to name their parts or how they grow. But we never went the next step, which was to talk about how to *care* for them." Yaqui legends, he said, cultivated this kind of ethic.

Apart from this loss of legends, a rough census of language knowledge among O'odham and Yaqui children revealed a significant loss of animal and plant names, which encode considerable cultural information about nature not found in American English. For the twelve O'odham and Yaqui children who were shown pictures of familiar plants and animals and asked how many they could name in their indigenous language, they averaged only 4.6 names out of the 17 native species shown. In contrast, their grandparents averaged 15.1 names for the 17 species. Virtually every O'odham or Yaqui living at the turn of last century would have named all seventeen of them. Children from both a new Yaqui community in the midst of farmland and the poorer O'odham communities in Mexico seemed to reflect the greatest intergenerational loss of native language and its encoded biological knowledge.

It is perhaps worth noting that the Yaqui community in which our in-

terviews took place represents an extreme of cultural and natural disloca-
tion. Marana is a recent settlement of young families several hundred kilometers distant from the Yaqui aboriginal homeland, so the children living
there have less than traditional contact with grandparents and reduced exposure to Yaqui ceremonies, folk customs, and language. This may explain
why none of the kids felt they had learned more stories from grandparents
and parents and why none of them knew the Yaqui names for the animals
and plants on our list. Further, the community is surrounded on all sides by
farmland, which might explain why none of the children felt they had ever
been alone in a wild place for any length of time.

Most of these children and most of the entire group surveyed felt that
they were learning more about plants and animals in school than their
grandparents ever learned (58 percent of O'odham, 60 percent of Yaqui,
and 61 percent of Hispanic kids). Only 38 percent of the Anglo group
shared this belief—perhaps because there is not much difference between
the kind of learning the Anglo kids are currently getting and that which
their grandparents received. That being the case, their grandparents would
only be wiser and more experienced because of their age. But for the other
groups, it appears likely that the older generation's orally transmitted, indigenous science is not being given the same status as Western science formally transmitted through schools.

We suspect that information derived from even the most casual observations written down by Western scientists is more likely to be treated as
"fact" than is similar information from oral traditions. For example, Grant
and Grant (1983) "discovered" within a few nights of observation that
sphingid moths behave as though intoxicated after imbibing the floral nectar of jimsonweed (*Datura*). Similar information has, however, been in the
oral traditions of the O'odham for at least a century (Russell 1975:x: "I
drank the jimsonweed flowers / and the drink made me stagger . . . / the
drunk moths / the drunk moths / they too drop / fluttering their wings").
Despite the astute observation of these and other folk scientists, we suggest
that their progeny may not be considering that stories or songs embody
much information of scientific merit. Although our evidence to support
this statement is only anecdotal and conjectural, we think it deserves consideration for more detailed study.

The Extinction of Experience

Toward a Cross-Cultural Concept

The term *biophilia* may have first been used in the present context in the *New York Times Book Review* in 1979. Its originator, E. O. Wilson, was not only an accomplished practitioner of Western science himself; he claimed (1984:x) that "modern biology has produced a genuinely new way of looking at the world that is incidentally congenial to the inner direction of biophilia." Thus, while biophilia is hypothesized as an innate tendency of *all* humans to focus on life and lifelike process, "this instinct has the opportunity to be 'aligned with reason'" by the empirical insights of logical positivists who are informed by modern evolutionary ecology.

We assume that biophilia is a phenotypic expression of gene/environment interaction capable of being produced in most if not all human societies. One might infer from Wilson, however, that Western-trained scientists somehow have a greater capacity to express their innate biophilia than do nonscientists—or, for that matter, than do hominid hunter-gatherers over the last hundreds of thousands of years. We doubt whether Wilson intended to load his concept with a Western scientific bias. To do justice to this notion, biophilia may still require elaboration in less ethnocentric terms or at least definition by criteria other than those peculiar to Western scientific competency. As David Orr (1992) recently observed, all societies presumed to have been ecologically sustainable for centuries—from the tropics of the Amazon to the deserts of the American Southwest—have been nonliterate, guided only by their orally transmitted folk science of natural resource management. If we are to assume that non-Western societies may indeed have cultural practices, myths, and tenets that encode an ethic compatible with our genes for biophilia, then we must use more universal criteria to evaluate them. At the same time, we must recognize intergenerational differences within each ethnic population, rather than assuming that each culture is internally homogeneous with regard to its views of the natural world.

Other surveys regarding various ethnic groups' knowledge of nature and biases toward or against it have tended to treat information derived from Western scientific sources as fact, but observational or culturally encoded information as merely anecdotal. Such biases can easily be avoided with some forethought. Among the possible traits we propose for further discussion as cross-cultural indicators of biophilia, here are four:

1. Personal identification with the behavior and adaptive strategies of other creatures
2. Positive valuing of wildness in the lexicon of a culture and its land management
3. Continued oral traditions of storytelling about biota
4. Interest in heterogeneity and diversity of any sort, but especially in environments and life-forms unmanipulated by human handiwork

These same universal indicators of biophilia should be used by environmental education programs. As Arturo Gomez-Pompa and Andrea Kraus (1992:272) have recently observed of environmental educators versed in Western science: "We assume that our perceptions of environmental problems and solutions are the correct ones, based as they are on Western rational thought and analysis. [But] the perspectives of the rural populations are missing in our concept of conservation. Many environmental education programs are strongly biased by elitist urban perceptions of the environment and issues of the urban world. This approach is incomplete and . . . neglects the perceptions and experience of the rural population, the people most closely linked to the land, who have a firsthand understanding of their surrounding natural environment as a teacher and provider."

While formal environmental education programs cannot supplant spontaneous hands-on experience of nature or the richness of intergenerational storytelling, they can incorporate certain universal stimuli for biophilia into their format. Such programs should contain the following elements: direct exposure to plants and animals; emphasis not merely on domesticated species but especially on wild ones; incorporation of traditional knowledge and lore into the curriculum, preferably through guest teaching by community elders.

As Cherokee folklorist Rayna Green (1981:212) has argued: "Every group has a 'science' based on observation, experiment and tradition whether the 'science' is called 'belief' or 'hypothesis'; whether 'science' is imparted traditionally through 'shamans' or 'books,' . . . whether knowledge is organized into a 'systematic body' or diffused throughout the range of cultural expression and behavior." The richness of each culture's traditions of biological sciences has much to do with its expression of biophilia. Until orally transmitted indigenous scientific traditions are treated with as much respect as Western science, the Euro-American monoculture will continue to drive into extinction a diversity of human adaptive responses to local environments and their biota, and the biota may be negatively affected as well.

REFERENCES

Anderson, Kat, and Gary Paul Nabhan. "Gardeners in Eden." *Wilderness* 35(194) (1991):27–30.

Bahr, Donald M. *Pima and Papago Ritual Oratory*. San Francisco: Indian Historian Press, 1975.

Bahr, Donald M., Juan Gregorio, David I. Lopez, and Albert Alvarez. *Piman Shamanism and Staying Sickness (Ka:cim Mumkidag)*. Tucson: University of Arizona Press, 1973.

Berlin, Brent. *Ethnobiological Classification: Principles of Categorization of Plants and Animals in Traditional Societies*. Princeton: Princeton University Press, 1992.

Brown, Cecil H. "Mode of Subsistence and Folk Biological Taxonomy." *Current Anthropology* 26(1) (1985):43–64.

Calvino, Italo. *Mr. Palomar*. Translated by William Weaver. New York: Harcourt Brace Jovanovich, 1983.

Curtin, L. S. M. *By the Prophet of the Earth*. Santa Fe: San Vicente Foundation, 1949.

Dauenhauer, Richard, and Nora Marks Dauenhauer. "Native Language Survival." In special issue on extinction. *Left Bank* 2 (1992):115–122.

Davis, Goode P. Jr. *Man and Wildlife in Arizona: The American Exploration Period, 1824–1865*. Phoenix: Arizona Game & Fish, 1982.

Diamond, Jared M. *Avifauna of the Eastern New Guinea Highlands*. Cambridge: Nuttal Ornithological Laboratory, 1972.

———. "The Environmentalist Myth." *Nature* 324 (1986):19–20.

Evers, Larry, and Felipe S. Molina. *Yaqui Deer Songs: Maso Bwikam*. Tucson: University of Arizona Press, 1987.

Felger, Richard S., Peter L. Warren, L. Susan Anderson, and Gary Nabhan. "Vascular Plants of a Desert Oasis: Flora and Ethnobotany of Quitobaquito, Organ Pipe Cactus National Monument, Arizona." *Proceedings of the San Diego Society of Natural History* 8 (1992):1–39.

Giddings, Ruth Warner. *Yaqui Myths and Legends*. Tucson: University of Arizona Press, 1978.

Gomez-Pompa, Arturo, and Andrea Kraus. "Taming the Wilderness Myth." *Bioscience* 42 (1992):271–279.

Grant, Vern, and Karen Grant. "Behavior of Hawkmoths on Flowers of *Datura metaloides*." *Botanical Gazette* 144(2) (1983):280–284.

Green, Rayna. "Culturally-Based Science: The Potential for Traditional People, Science, and Folklore." In M. Newall, ed., *Folklore in the Twentieth Century*. London: Rowman & Littlefield, 1981.

Kellert, Stephen, and Miriam Westervelt. *Children's Attitudes, Knowledge and Behaviors Toward Animals*. Washington: U. S. Department of the Interior, Fish and Wildlife Service, 1983.

Mander, Jerry. *In the Absence of the Sacred*. San Francisco: Sierra Club Books, 1991.

Nabhan, Gary Paul. "Papago Fields: Arid Lands Ethnobotany and Agricultural Ecology." Doctoral dissertation, University of Arizona, 1983.

———. *Enduring Seeds: Native American Agriculture and Wild Plant Conservation*. San Francisco: North Point Press, 1989.

Nabhan, Gary Paul, Amadeo M. Rea, Karen L. Reichhardt, Eric Mellink B., and Charles Hutchinson. "Papago Influences on Habitat and Biotic Diversity: Quitovac Oasis Ethnoecology." *Journal of Ethnobiology* 2(2) (1983):124–143.

Nabhan, Gary Paul, Donna House, Humberto Suzan A., Wendy Hodgson, Luis Hernandez S., and Guadalupe Malda. "Conservation of Rare Plants by Traditional Cultures of the U. S./Mexico Borderlands." In Margery L. Oldfield and Janis B. Alcorn, eds., *Biodiversity: Culture, Conservation, and Ecodevelopment*. Boulder: Westview Press, 1991.

Nabhan, Gary Paul, and Wendy Hodgson. *Pilot Research Project to Prepare an Ethnobotanical and Rare Plant Database System for Biosphere Reserves*. Technical Report NPS/WRUA/NRTR-92-48. Tucson: U. S. National Park Service, in press, 1993.

Orr, David. *Ecological Literacy*. Albany: State University of New York Press, 1992.

Painter, Muriel Thayer. *With Good Heart—Yaqui Beliefs and Ceremonies in Pascua Village*. Tucson: University of Arizona Press, 1986.

Pyle, Robert Michael. "Intimate Relations and the Extinction of Experience." In special issue on extinction. *Left Bank* 2 (1992):61–69.

The Extinction of Experience

Rea, Amadeo. "Resource Utilization and Food Taboos of Sonoran Desert Peoples." *Journal of Ethnobiology* 11 (1981):69–83.

———. *Once a River*. Tucson: University of Arizona Press, 1983.

Redford, Kent. "The Ecologically Noble Savage." *Orion Nature Quarterly* 9(3) (1985):24–29.

Russell, Frank. *The Pima Indians*. Tucson: University of Arizona Press, 1975.

Sale, Kirkpatrick. *The Conquest of Paradise*. New York: Knopf, 1990.

Saxton, Dean, and Lucille Saxton. *O'othham Hoho'ok A'agitha: Legends and Lore of the Papago and Pima Indians*. Tucson: University of Arizona Press, 1973.

Shepard, Paul. *Nature and Madness*. San Francisco: Sierra Club Books, 1982.

Shreve, Forrest, and Ira Wiggins. *Vegetation and Flora of the Sonoran Desert*. Palo Alto: Stanford University Press, 1964.

Silko, Leslie Marmon. "Landscape, History, and the Pueblo Imagination." In Daniel Halpern, ed., *On Nature*. San Francisco: North Point Press, 1987.

Steinhart, Peter. "Dreaming Elands." In Robert Finch and John Elder, eds., *The Norton Book of Nature Writing*. New York: W. W. Norton, 1989.

———. "Electronic Intimacies." *Audubon* (1991):10–12.

Tellman, Barbara J., and William W. Shaw. "The Natural Setting." In Robert Varady, ed., *Preserving Arizona's Environmental Heritage*. Tucson: University of Arizona Press, 1991.

Vickers, William T. "Game Depletion Hypothesis of Amazonian Adaptation: Data from a Native Community." *Science* 239 (March 1988):1521–1522.

Wallman, Josh, Michael D. Gottlieb, Vidya Rajaram, and Lisa A. Fugate-Wentzek. "Local Retinal Regions Control Local Eye Growth and Myopia." *Science* 237 (1987):73–77.

Webb, George. *A Pima Remembers*. Tucson: University of Arizona Press, 1959.

Williams, George C., and Randolph M. Nesse. "The Dawn of Darwinian Medicine." *Quarterly Review of Biology* 66 (1991):1–21.

Wilson, Edward O. *Biophilia*. Cambridge: Harvard University Press, 1984.

CHAPTER 8

New Guineans and Their

Natural World

Jared Diamond

I SHALL DISCUSS the relationships of people to wild animal and plant species in New Guinea and other southwest Pacific islands. My discussion is
based on having spent parts of the last thirty years on these islands, studying birds and living with native peoples.

There are several reasons why New Guinea (together with other Pacific
islands) seems a good place for gathering empirical data relevant to testing
the biophilia hypothesis. First, all humans were wholly dependent on
stone technology until about 5000 B.C., and only since then has one human
group after another begun using metals. New Guineans were among the
last to switch to metal: all New Guineans depended on stone technology
until European colonization began in the nineteenth century. Of the
groups of New Guineans with whom I have worked, most remained un-
contacted by the outside world and still used stone tools until the 1950s, and

some continued in that way up to the 1980s (Souter 1963; Connolly and Anderson 1987). Thus New Guineans furnish some of our best surviving models of the human conditions that have prevailed for millennia.

Moreover, while it may at first seem that New Guinea peoples offer a narrow data base, constituting less than one-thousandth of the world's human population, they actually represent a large fraction of surviving human diversity—whether diversity is measured genetically, linguistically, culturally, economically, politically, or ecologically. To appreciate how this condition arose, recall that New Guinea has been occupied by people for at least 40,000 years and was reached by multiple colonization waves whose descendants still occupy the Pacific today (White and O'Connell 1982). Waves of the earlier settlers contributed to the modern peoples of the New Guinea highlands and certain islands of the Bismarck and Solomon archipelagoes. Subsequent waves of Austronesian peoples contributed heavily to the modern coastal peoples of northern New Guinea and most of the Bismarck and Solomon islands. About 3,600 years ago the ancestors of the modern Polynesians arrived (Bellwood 1987). In addition to this human diversity arising from multiple waves of colonization, much further diversity has arisen *in situ* as a result of New Guinea's rugged topography, which has permitted the human population of each valley to develop its own characteristic genes, language, culture, and diseases.

For example, about 1,000 of the modern world's 5,000 languages are confined to New Guinea (Ruhlen 1987). These are not mere dialects but mutually unintelligible languages, many of them far more different from each other than English is from Hungarian. While New Guinea thus harbors about 20 percent of the modern world's languages, it actually harbors more than 20 percent of the world's linguistic diversity, because most of the 4,000 languages spoken elsewhere in the world fall into seventeen families such as Indo-European or Finno-Ugric. Each of these families is derived from a protolanguage or group of languages spoken within the last 10,000 years. Thus almost all languages outside New Guinea trace back to one of only about seventeen recent ancestral languages. This situation arose because other languages were eradicated as a result of expansions of a few dominant peoples—especially farmers, herders, and peoples organized into centralized and militarily potent states. However, New Guinea languages appear to represent dozens of independent languages or language stocks, with no proven relationship to each other nor to any other language stock in the world (Foley 1986). Hence New Guinea may actually harbor most of the linguistic diversity that has survived into the modern world.

Because of New Guinea's rugged topography and many waves of colonization, it also harbors great cultural diversity. For example, child-rearing systems range from extreme permissiveness (even letting babies extend an arm into the fire) to extreme repression resulting not infrequently in child suicide. Sexual customs are equally variable, ranging from institutionalized heterosexual promiscuity to institutionalized child/adult homosexuality.

New Guinea peoples are also diverse economically. A few are still hunter-gatherers. Most New Guinea peoples and Pacific islanders, however, do some farming supplemented by much hunting and gathering. The farming itself is variable—ranging from casual swidden agriculture, overshadowed by hunting and gathering, to intensive irrigation agriculture at population densities approaching those of Holland. Coastal peoples consume much fish and shellfish, while inland peoples consume none. There are three species of domestic animals (pig, chicken, and dog), but they make only modest or infrequent contributions to people's protein and caloric needs.

As for political organization, traditional New Guinea had no permanent unit larger than the village, but parts of Polynesia had more complex organization, including two archipelagoes verging on state organization.

With respect to habitat diversity, most of New Guinea is covered with closed forest. Southern New Guinea has three areas of savanna woodland similar to that of northern Australia. Other natural habitats include the seacoast, rivers, and alpine grassland, while large areas of the highlands are clothed in anthropogenic grassland. The closed forest itself is very varied, ranging from lowland tropical rain forest and swamp forest to montane oak forest and montane beech forest. Some of the largest areas of intact tropical forest in the world are to be found in New Guinea. For example, I have frequently had the experience of flying for 200 miles by small aircraft over the New Guinea lowland forest and seeing no signs of human occupation except for one or two huts of nomads. I should add that I am puzzled, in other

New Guineans and Their Natural World

discussions of the biophilia hypothesis, by what seems to me an exaggerated focus on savanna habitats as a postulated influence on innate human responses. Humans spread out of Africa's savannas at least 1 million years ago. We have had plenty of time since then—tens of thousands of generations—to replace any original innate responses to savanna with innate responses to the new habitats encountered. The earliest attested behaviorally modern *Homo sapiens* were the ancestors of aboriginal Australians, who between 40,000 and 30,000 years ago occupied Australia's entire span of habitats—from desert, the seacoast, tropical rain forest, dry forest, mulga, Mediterranean scrub, and heath to cold temperate rain forest and periglacial grassland.

My own fieldwork in New Guinea consists of studies of animal evolution, ecology, and behavior, mainly of birds (Diamond 1972, 1973, 1984). Almost all of my hours in New Guinea are spent in the company of New Guineans, either because I am living with them in forest camps or living in their villages and walking accompanied by them when I go out bird-watching. My time with New Guineans is spent in almost incessant conversation, much of it consisting of comparing our respective lifestyles—how much money did you pay for your wife, how many children do you have?—or else discussions of each species of local animal and plant in turn and in great detail. I also have the opportunity to watch what New Guineans do and say to each other. Many of these conversations, too, turn out to involve social relationships, daily life, and local animals and plants.

The main languages in which I converse with New Guineans are Indonesian (Bahasa Indonesia) in the western half of New Guinea, now Irian Jaya province of Indonesia, and Neo-Melanesian (New Guinea Pidgin English) in the eastern half of the island, now part of the independent nation of Papua New Guinea. Today, most New Guineans are fluent in one of these two languages, depending on the half of the island on which they live. I have also learned something of one indigenous language, Foré. In each area that I visit, I always learn the names of local animal and plant species in the local native language, since these are almost exclusively the names by which people refer to local species (Diamond 1966, 1984, 1989a, 1989b; Diamond and LeCroy 1979); Indonesian and Neo-Melanesian have only a few, mostly higher-level, names for animals and plants. To date, I have thus learned local species names in the languages of seventy-five New Guinean and Pacific island peoples, approximately equally divided between Papuan and Austronesian languages plus two Polynesian groups.

I should give some feeling for the lifestyles and environments that native peoples experience in New Guinea, since these lifestyles and environments are so different from ours. People in Western industrialized societies today receive most of their input from other people indirectly and noninteractively via media such as television, books, radio, newspapers, magazines, and lectures. Our physical environment is mostly human-created or heavily modified. Our experience of the natural world, if any, is confined to infrequent occasions when we go bird-watching, hiking, fishing, or off on fieldwork. Our major concerns are constructs of our industrialized society—especially our cash economy, professions, and politics.

In contrast, for traditional New Guineans and other Pacific islanders, all input from people is direct and interactive via conversation. As a result, New Guineans spend far more time talking with each other than we do. Much of the physical environment experienced by New Guineans is the natural world: they spend much of their time actually inside closed forest, and their houses are so close to the forest that they can hear fifty bird species while still lying in bed. Most of their concerns involve either immediate social relationships or else subsistence, much of which is drawn from nature. Thus, for New Guineans and other traditional Pacific islanders, the natural world forms a far larger part of their experience than it does for us.

Knowledge of the Natural World

This involvement with the natural world is reflected in local names for plant and animal species. Most land vertebrates—birds, mammals, reptiles, and amphibia—are named to the species level. Just to give a sense of the local-language names themselves, here is a sample: *i, wo, kwok, screw, hihi, piss-piss, yor-bichul-bichul, wai-squirty-squirty, kuntry-kuntry-knai, dinigahawaso-geri, i-brochit-cauley*, and *go-go-harigi*. I found that the Foré people apply 110 different Foré names to the 120 different bird species occurring regularly in the Foré area (Diamond 1966). Of these 110 names, 93 correspond one-to-one to individual bird species recognized by Western taxonomists. Eight

Foré names apply separately to the male or female of four sexually di-morphic bird of paradise or bowerbird species. Finally, there are nine lumped Foré names, each applying to two or more related species recog-nized by Western taxonomists, although in each case the Foré are aware that their name covers several different species.

The most detailed account of the natural lore possessed by a New Guinea people was assembled for New Guinea's Kalam people by a remark-able collaboration between a Kalam man, Ian Saem Majnep, and a New Zealand anthropologist, Ralph Bulmer (Majnep and Bulmer 1977). The Kalam distinguish by name more than 1,400 wild animal and plant species. Majnep and Bulmer's book is essential reading for anyone interested in the biophilia hypothesis.

Along with this detailed naming system come detailed knowledge of habits and acute abilities of field identification. For example, one of the banes of ornithologists in New Guinea is the warbler genus *Sericornis*, con-sisting of two dozen drab, very similar populations about whose relation-ships and grouping into species taxonomists have been arguing for a cen-tury. In the area occupied by the Foré people of New Guinea's eastern highlands occur two of these warblers, *Sericornis nouhuysi* and *Sericornis perspicillatus*, which differ slightly in size and in depth of olive and orange wash on the face. When I first arrived in the Foré area and began catching these species in mist nets, I could not distinguish them as I held them alive in the hand. It took me several weeks of work in the reference collection of the American Museum of Natural History, comparing and measuring my stuffed study skins, before I was able to separate most specimens. Even to-day, authorities who have published taxonomic revisions of *Sericornis*, such as Ernst Mayr, Erwin Stresemann, Mary LeCroy, and myself, still disagree over the question to which species of *Sericornis* some specimens belong.

It was therefore humbling to discover that the Foré themselves had sep-arate names for the two *Sericornis* species in their area—referring to *Seri-cornis nouhuysi* as the "*mabisena*" and to *Sericornis perspicillatus* as the "*pas-agekiyabi*" (Diamond 1972). To make matters more embarrassing, the Foré distinguished the *mabisena* and *pasagekiyabi* in the field without binocu-lars, in silhouette, when the birds were over 10 meters away. I eventually re-alized that they did so not by relying on the minute differences in plumage, which Western observers waste their time trying to distinguish with binoculars, but on differences in behavior and song. (*Sericornis nouhuysi* feeds closer to the main trunk of trees, stays closer to the ground, and has a trilling song, whereas *Sericornis perspicillatus* spends more time gleaning from leaves near the tips of branches, feeds higher toward the canopy, and has a song consisting of an ascending series of dry notes.)

While on the island of Kulambangra in the Solomon Archipelago, I had two free days at the end of my fieldwork, and so I spent that time transcribing the knowledge of local birds possessed by one Kulambangra villager, Teu Zinghite. For every one of Kulambangra's eighty resident bird species, Teu dictated to me an account consisting of its name in the Kulambangra language, its song, preferred habitat, abundance, size of the group in which it usually foraged, diet, nest construction, clutch size, breeding season, seasonal altitudinal movements, and frequency and group size for overwater dispersal.

Many of us Westerners remember in detail certain people whom we met only once, several decades ago. New Guineans have such detailed recollections not only for people but also for certain bird and animal species that they encountered only once. As an example, in 1974 I was surveying birds at the village of Sasamonga on the Solomon island of Choiseul. As usual upon landing, I hired one of the older men in the village to guide me through the forest in search of birds, and we began talking about all the bird species he knew, one after the other. After we had talked about each species that we encountered plus other species that he said occurred regularly but we had not yet encountered, my guide started talking about rare species. He mentioned that there was one species, called the *kurulilúa*, which he had seen only once in his life, long ago in his childhood. As he recounted it, a group of white men had arrived at Sasamonga in a yacht whose engine didn't work, in order to collect birds and other animals. On the beach was the carcass of a dead porpoise that had washed up and that the white men boiled down in order to carry back its skeleton. The white men also shot the only specimen of the *kurulilúa* that my guide ever saw, which he proceeded to describe to me. From his description it was obviously a large ground pigeon, evidently the yellow-legged ground pigeon *Columba pallidiceps*.

New Guineans and Their Natural World

From my guide's account, I guessed that the white men belonged to the famous Whitney South Seas Expedition of the American Museum of Natural History, which combed most Pacific islands collecting birds during the 1920s and 1930s. When I returned to New York, I pulled out the field diaries of the Whitney Expedition stored in the museum's bird department. There I read that Hannibal Hamlin and several other Whitney collectors visited Sasamonga in 1929, at a time when their yacht's engine had broken down. They found on the beach a porpoise carcass, which they cleaned and boiled in order to prepare the skeleton. It was also at Sasamonga, Hamlin noted in his diary, that he collected the only specimen of the rare yellow-legged ground pigeon that the expedition was to encounter on Choiseul. Thus after forty-five years my guide still remembered and described accurately to me a rare bird species that he had seen on that one occasion and never saw again.

Uses of Knowledge of Nature

Why do New Guineans and other Pacific islanders devote so much stored memory to the names and habits of so many plant and animal species? What use do they make of all this knowledge, and why do they involve themselves so intimately in the natural world?

One obvious reason is the direct economic utility of many of the species named. Species that are consumed as major sources of dietary protein include birds, mammals, fish, crustacea, mollusks, snakes, lizards, frogs, and roots and fruits and seeds and pith and leaves of many plants, plus certain insects and spiders and worms. Traditional clothing is made from grass or the bark of certain trees. Containers are made of bamboo. Canoe hulls are hollowed out from certain tree species; outriggers and lashings are made of other tree species. Houses are made of local woods, and different tree species are used for different structural parts of the house. Tools and implements are made from naturally available materials as well. Thus, the natural world was the traditional source of all material objects without exception—a condition that is difficult for us today to imagine. Of course, people have to know the distribution and habits of each needed species in order to be able to locate and harvest it.

While these are some of the direct economic uses of wild animals and plants, other species are valued for arbitrary reasons and used as decorations, status symbols, or almost as cash. In the highlands of New Guinea, for example, cowry shells traded up from the coastal lowlands were traditionally worn as decorations and used virtually as large-denomination banknotes. New Guineans are famous for decorating themselves on ceremonial occasions with plumes of birds of paradise, certain species of parrots and bowerbirds, the New Guinea harpy eagle, the long-tailed buzzard, and pieces of fur of certain mammals. Most valuable are the feathers of Pesquet's parrot, *Psittrichas fulgidus*, whose value considerably exceeds the purchase price of a wife in New Guinea. Bird of paradise and other plumes are not only worn as status decorations but are also essential portions of the bride's purchase price. Moreover, many low-status objects are used for minor daily decoration—such as inserting the wingbone of a bat through one's nose or perforating one's earlobe and then using a piece of dried snakeskin as an earring.

Still, these uses of animals for direct economic value or decoration or status symbols fail to explain why New Guineans also distinguish so many species of apparently no utility. Why do they have names for all those species of dull little warblers, rats, and frogs? First, New Guineans actually eat not only the tree kangaroos and imperial pigeons about which they like to boast, but also any small bird, rats, snakes, spiders, and other unheroic prey. Second, all these other species, even if not actually utilized, furnish part of the background noise and visual stimuli and signs that must be distinguished in the forest in order to locate species that are truly of value. For example, the calls of ribbontail birds of paradise (genus *Astrapia*) are very similar to the calls of certain frogs, which must be learned in order not to waste time chasing after frogs when one actually wants *Astrapia* plumes. Third, unutilized species serve as ecological indicators of the likely presence of valuable species occupying the same habitat. For instance, a small dull honeyeater, *Ptiloprora guisei*, occurs in the same habitat as the King of Saxony bird of paradise *Pteridophora alberti*, whose long occipital plumes are among the most prized decorations of New Guinea highlanders. Hence the calls of this honeyeater are a good indicator of suitable *Pteridophora* habitat.

Finally, animal and plant species form the major part of the salient environment experienced by New Guineans, apart from the human environment itself. Since traditional New Guineans did not have Bibles and books about cowboys and Ninja Turtles, their songs, stories, art, and myths draw heavily instead on species of the natural world. I can illustrate this point with one typical New Guinea story and one typical myth.

The story concerns a common chickenlike bird known to ornithologists as the scrubfowl *Megapodius freycinet*, which weighs up to two pounds but whose meat begins to rot very quickly after the bird is killed. New Guineans told me the following story about what you must do if you want to eat a scrubfowl before its meat begins to stink too much. During the day you walk through the forest looking for white excrement on low branches, an indication of where a scrubfowl habitually roosts for the night. Then you return to the forest at night, carrying a container of water and a bow and arrow. You sit down immediately below the branch where the scrubfowl is now roosting, light a fire, and heat the container of water. When the water finally boils, the moment has at last arrived to shoot the scrubfowl with an arrow, so that it plummets directly from its perch into the boiling water. Only in that way can you be assured of cooking a scrubfowl so soon after death that it will not begin to rot. (Naturally, this story was recounted to me without knowledge of the English name "scrubfowl" or the Latin name *Megapodius freycinet*, but with the local native name that was applied to this species and that I had learned.)

As an example of a New Guinea myth, here is the origin myth recounted to me by several peoples in the Lakes Plains of Irian Jaya. The Lakes Plains are a mountain-ringed basin, about 200 miles by 50 miles in extent, occupied by peoples speaking several dozen mutually unintelligible languages. Lakes Plains peoples recount the following origin myth to explain why so many drastically different languages should be spoken within this one basin. In the beginning, all people of the Lakes Plains spoke the same language and lived in the forest near a big ironwood tree (referred to by its native name, of course). One man suffered from the common condition of filariasis or elephantiasis, a condition caused by a parasitic worm that blocks the lymph ducts and causes soft tissues such as limbs, breasts, or testes to swell enormously. This unfortunate man had enormously swollen

Culture

testes. Since their size interfered with his walking or sitting, he spent his time sitting on a branch of the ironwood tree, where his testes rested comfortably upon the ground. Ground-dwelling marsupials called bandicoots in English, a common source of game meat and also referred to by their native name, were regularly attracted out of curiosity to the testes and would start to nibble at them. While thus distracted, the bandicoots could easily be killed by the other people.

Thus, everybody enjoyed a comfortable life of easy hunting—until two brothers killed their brother-in-law, the brother of the wife of one of the two men. The murdered man's relatives chased the murderers and all of their relatives and friends, who climbed the ironwood tree for safety. Hanging from the ironwood tree were many lianas, referred to in the story by their local native name. The pursuers pulled on the lianas in their efforts to bring the crown of the ironwood tree closer to the ground, where they could seize or shoot their quarry. As they pulled harder and harder on the lianas and the crown bent lower, the murderers and their friends realized that they were doomed unless they did something drastic. In desperation, they cut through the lianas. The bent tree snapped back—hurling all the poeple out of its crown with such force that they became scattered in all directions, so far apart that they were never able to find each other again. In their isolation they proceeded to develop the diverse languages now spoken in the Lakes Plains.

I have illustrated how native plant and animal species are named and observed for many purposes, both narrowly economic and otherwise. Is there any naming and accumulation of knowledge motivated by no reason at all other than interest and natural affinity to the living world—simply because species are there? This seems not to be the case. Let me cite two examples.

First, individual stars and constellations are named by Polynesians because they still are used as navigational aids in canoe voyaging. Since I have long been a stargazer and almost became an astronomer, I assumed that inland peoples of New Guinea would similarly name the stars so clearly visible in their night skies and in fact was looking forward to learning their stories about the constellations. To my disappointment, the first highland people with whom I worked, the Foré, told me that they referred to all stars by the single term *nori* and did not distinguish individual stars. Ever since then, I have found one inland group after another that failed to distinguish individual stars. From their point of view, naming of stars would be useless, since no one but an inexperienced Westerner would be foolish enough to try to navigate by the compass or stars over New Guinea's rugged inland terrain.

The second example concerns butterflies. I expected New Guineans to have separate names for species of butterflies, which are as conspicuous and easily identified by sight as bird species. Again to my surprise and disappointment, the Foré told me that they used the single term *poporiya* for all butterflies and did not distinguish species. Again from their point of view, this knowledge would be useless, no matter how easily one can recognize individual butterfly species. New Guineans had no use for butterflies—at least until Western collectors began paying impressive sums of money for New Guinea's giant species of birdwing butterflies.

Responses to Animals

So far, I have discussed New Guineans' knowledge and discrimination of their natural world. Do they exhibit any positive emotional responses to animals as living creatures—responses such as love, reverence, fondness, concern, or sympathy? New Guineans certainly are capable of positive responses to at least one species of domestic animal, the pig, which serves as a major status symbol and with which they live on intimate terms. Young pigs often sleep in the same hut with their human owners, and New Guinea women sometimes nurse a piglet at one breast while nursing their own infant at the other. I have seen New Guineans become upset when their pig dies and get angry when they think somebody else is responsible for the death of their pig.

It is rare, however, to see corresponding signs of New Guineans recognizing individual wild animals as living creatures to which one can form a bond. Wild animals are only rarely kept as pets in New Guinean or Melanesian villages. The only examples I have encountered are a couple of pet hornbills, several cases of imperial pigeons, and one case of a young tree kangaroo. (I do not count as pets the possums that one sometimes sees trussed up in villages, being kept alive after capture until their intended sacrifice for food.) This infrequency of village pets is surprising to Westerners, because so many New Guinea wild mammal and bird species tame well and make cute, responsive pets much beloved by Western expatriates living in New Guinea. New Guineans' disinterest in pets can certainly not be generalized to other Stone Age peoples of the modern world: I saw numerous pets in a single village of recently contacted Ishcanahua Amerindians whom I visited on a tributary of the Peruvian Amazon.

It is not only that New Guineans fail to cultivate friendly responses from wild animals. They also seem not to take account of the fact that these are living creatures capable of feeling pain. For example, when a wild animal is captured in the forest early in the day and is to be transported alive for the rest of the day so that it can be killed and eaten fresh in the village that evening, the animal's legs may be broken to prevent it from escaping. One painful experience stands out vividly in my mind. On one occasion it turned out that the man whom I had hired as my guide for the day in the Aru Islands was a bird trapper retrieving wild cockatoos that he had snared. The cockatoos were not intended for consumption but for sale to a trader who would eventually sell them for illegal export to foreign fanciers of cockatoos as pets. In this case the man immobilized the cockatoos for transport during the day by the barbarous method of bending their wings behind their backs and then tying several of their primary feathers together in a knot. Since the man was armed with a bush knife and I was alone with him in a remote area of forest, I was helpless to interfere.

As another example of indifference to needless suffering of wild animals, I recall an occasion on my first New Guinea expedition when I heard squealing and shrieking ahead of me as I returned to my campsite. At the camp I found several New Guineans holding a large fruit bat of a species whose long and slender wingbones are used as nose decorations. The men wanted to cut these bones out of the bat's wing, but they did not bother to kill the bat before doing so. Instead, the two men spread out the bat's wings, another tied up the bat's mouth with a vine to prevent it from biting, and the other man then proceeded to dissect and scrape through the joints and muscles of the live bat so as to extract the bones.

In another case, I found men intentionally inflicting pain on captured live bats for no other reason than amusement at the reactions of the tor-

New Guineans and Their Natural World

tured animals. The men had tied twenty-six small *Syconycteris* blossom bats to strings. They lowered one bat after another until it touched the red-hot embers of a fire, causing the bat to writhe and squeal in pain. The men raised the bat, lowered it again for another touch to the red-hot embers, repeated this process until it was dead, and then went on to the next bat, finding the whole proceedings funny.

This treatment of animals is not so different from ways in which New Guinea highlanders treated other people, namely, captured warriors of enemy tribes. An early account from the highlands describes a game in which an unarmed prisoner was placed in the center of a wide circle of his captors armed with axes. One captor after another in turn would enter the circle with his axe and take a swing at the captive, who tried to dodge. Eventually, one of the axemen succeeded in landing a blow on the leg of the captive, knocking him down screaming. The man's leg was then chopped off with the axe, then his other limbs in turn.

Biophobia

If there is thus little evidence of love or other tender feelings for individual wild animals, is there any evidence of fear of animals, such as snakes and spiders, to which people in industrialized societies are often described as possessing an innate fear (biophobia)? If there is any single place in the world where we might expect to find an innate fear of snakes among native peoples, it would be New Guinea, where one-third or more of the snake species are poisonous and certain nonpoisonous constrictor snakes are sufficiently big to be dangerous. One of the few well-attested examples of a large snake actually killing and eating a human involves a reticulated python consuming a fourteen-year-old boy on an Indonesian island.

In one case I did observe fear of snakes on the part of New Guineans: two men returned from a morning's pursuit of birds in the forest to report that they had encountered a very large python, that they were afraid of it, and that they did not want to return on that trail again. This can hardly be considered an example of irrational innate fear; instead, it is a completely appropriate learned response to a dangerous animal. New Guineans certainly possess no generalized fear of snakes and spiders as many Westerners do. They are well aware which species are poisonous and which are not. When a New Guinean and I encounter a snake in the forest together, the New Guinean simply explains to me matter-of-factly whether that particular snake is dangerous. Nonpoisonous snakes are routinely captured by children and women, as well as by men, for eating. Children capture large spiders, singe off the legs and hairs, and eat the bodies. Asked whether they have a generalized fear of snakes, New Guineans laugh in scorn and say that that is a reaction for ignorant white men too stupid to distinguish poisonous from nonpoisonous snakes.

Just as New Guineans distinguish dangerous from nondangerous snakes and spiders, they similarly distinguish dangerous or poisonous plants, mammals, fish, and mollusks from harmless or nonpoisonous species. Most Westerners inexperienced as woodsmen never learn reliably to distinguish poisonous from nonpoisonous snakes, but they do recognize that some plants (such as poison ivy) are poisonous while others (such as dandelions) are not. Similarly, even urban Westerners know that, among mammals, lions and bears are dangerous while rabbits and deer are not.

From identification of burnt bones in hearths, we know that snakes have been regularly consumed by hunter-gatherer people at least as long ago as the Upper Paleolithic. From ethnographic evidence, we know that modern hunter-gatherers as well consume snakes. I can confirm from personal experience that snake meat has a good taste and texture (like a cross between chicken and rabbit). All this suggests that, throughout most of human history, hunter-gatherer peoples living in areas with snakes learned routinely and readily to distinguish dangerous from harmless species just as they distinguished dangerous from harmless species of other animals and plants.

What about the supposedly widespread, innate fear of snakes reported in cross-cultural studies—of people as seemingly diverse as Americans, Europeans, Japanese, white Australians, and Argentineans? These studies actually refer to only a tiny slice of human cultural diversity—a slice composed of modern industrialized metal-using peoples living in centralized political states. Such people have good reason to fear snakes. Distinguishing poisonous from nonpoisonous snakes can be difficult, especially since certain harmless species have evolved into excellent mimics of poisonous ones. (How many American readers of this book actually remember the mnemonic verse by which the red-yellow-and-black banding patterns of our harmless milk and king snakes and our deadly poisonous coral snake can be distinguished?)

Foraging peoples who eat snakes have to learn these distinctions. However, snakes do not make a major contribution to the diet of any foraging people. Over the past 10,000 years, as domesticated animals and plants have increasingly replaced wild foods in the diets of most peoples, snakes must have been among the first wild food items to be dropped from the diet: too much danger, too much specialized knowledge required, too little payoff. In nonforaging societies it doesn't make sense for people to waste time learning to distinguish snake species; it's better to learn as babies from their parents' frightened responses a generalized fear of snakes, to be passed on in turn to their children. The same reasoning applies to spider species: worth distinguishing and selectively eating, for New Guinea children; not worth distinguishing, and only worth generically learning to fear, for children in nonforaging societies. Probably it was not until recent millennia, as snakes and spiders lost their traditional value as minor food items, that most human societies developed a learned generalized biophobia of snakes and spiders.

These thoughts were on my mind as I watched my three-year-old son Max encounter his first snake, a recently killed garter snake. It was love at first sight. Max picked up the snake and refused to be parted from it for several days. When we finally insisted on burying it, Max announced that he would find another snake. Accompanied by his biologist father who prides himself on his skills as an experienced field observer, and who has the advantage of being able to scan the ground from a height double Max's, Max led me off on a two-mile hike. After an hour, a thrilled Max was the first one to spot a snake, at which he exhibited no more fear than would a New Guinean. When Max insisted on bringing the snake home as his pet, he and his twin brother Joshua held it all day (even while eating dinner or watching television). They argued incessantly over whose turn it was to hold the snake, and whether the other boy had been holding it for too long. Family peace was restored only when we bought another snake at a pet store so that each twin could have his own personal private pet snake.

The Ancient Tragedy of the Commons

Do New Guineas exhibit conservationist attitudes toward wild animals and plants? Do they avoid overharvesting the species on which they depend? Something of the sort can be said for species to which individual rights are owned by individual people. For example, trees that birds of paradise use for their traditional lek displays are owned by individual villagers, who hold the sole right to shoot birds of paradise from that tree. Similarly, individual trees of species preferred for hollowing out as canoe hulls are recognized as being the property of the person who discovers the tree and who will eventually fell it when he needs to make a new canoe.

However, there is little conservation of species viewed as community property, which suffer the tragedy of the commons. Anthropologists sometimes report stories of taboos supposedly observed by men before they will hunt cassowaries—for example, sexual abstinence and other rituals practiced for a month beforehand. In reality, the notion that New Guineas would hunt specifically for cassowaries is absurd, since it is difficult to predict when a cassowary or any other specific game animal is going to be found. Instead, New Guineas after big prey go out hunting with dogs, which run down anything worthwhile they encounter, whether it is a cassowary, wild pig, or wallaby. The hunters then follow the barking of the dogs and kill the cornered animal—including a cassowary if that is what the animal turns out to be.

Only recently have zoologists begun to realize how heavy an impact New Guineas have had upon their native biota, even in the days when they were equipped only with stone tools. For example, among New Guinea's phalangers (a group of arboreal possumlike marsupials) the largest species is *Spilocuscus rufoniger*, known only from a few widely scattered localities in the New Guinea lowlands (Flannery 1990). It is very easy to kill: when detected in a tree at night it does not flee but just curls up into a ball, making it easy for a hunter to climb and retrieve it. Similarly, New Guinea's largest native mammal, the tree kangaroo, *Dendrolagus scottae*, is now known only from a single mountain in the Torricelli Range (Flannery and Seri 1990). This large phalanger and large tree kangaroo were prime targets of hunters.

No specimen of a tree kangaroo of any species is known to zoologists from the entire Central Range of Irian Jaya, even though zoologists were the first Westerners to contact the Stone Age society of the Baliem Valley in the heart of the Central Range in 1938. These prime targets must have been hunted out over most of Irian Jaya's Central Range before 1938, although they still survive on certain outlying ranges of both Irian Jaya and Papua New Guinea and on parts of Papua New Guinea's Central Range. Paleontologists have discovered subfossil bones of two species of large wallabies that disappeared around 3,000 years ago and are not known anywhere as living species (Flannery 1990). What were formerly the largest of all native New Guinea mammals, the rhinoceros-like marsupials known as diprotodonts, disappeared tens of thousands of years ago, some time after New Guinea was first reached by humans.

European explorers have always found it difficult to view native peoples of other parts of the world as fully human. European attitudes have oscillated between regarding natives on the one hand as subhuman animals deserving of extermination without compunction and, on the other hand, as environmentally minded paragons of conservation living in a Golden Age of harmony with nature, in which living things were revered, harvested only as needed, and carefully monitored to avoid depletion of breeding stocks. In reality, modern and prehistoric peoples throughout the world are human: neither animals, nor paragons, but human. Like other humans throughout the world, New Guineans kill those animals that their technology permits them to kill. The more susceptible species become depleted or exterminated, leaving less susceptible species which people continue to hunt without being able to exterminate them. When technology improves, as it did in New Guinea with the arrival of the bow and arrow, or with the arrival of dogs a few thousand years ago, or with the arrival of shotguns within the present generation, some species that have been able to survive previous hunting technology become susceptible and disappear. A quick first wave of exterminations, followed by a slower trickle, has similarly marked the arrival of humans in all other areas of the world explored paleontologically: the islands of the Mediterranean, Pacific islands, Madagascar, Australia, and the Americas (Diamond 1992).

Culture

Learned Knowledge and Attitudes

New Guineans' knowledge of local plants and animals is certainly a learned product of their experience. At least in large part, their attitudes toward their natural environment are also a learned product of experience. Let me cite two examples.

First, Indonesians and New Guineans from forested areas are at ease in the forest and live contentedly inside it for weeks at a time. Indonesians from deforested areas are afraid of the forest and reluctant to enter it. When I was carrying out a national park survey of the Fakfak Mountains of Indonesian New Guinea in 1981 and 1983, a wonderful Indonesian forester named Lefan explained to me how he and other members of his department felt about the forest. Lefan and some of his close colleagues had grown up at home in the forest, enjoyed working in it on timber surveys, and were accustomed to spending long periods in it dozens of kilometers from the coast. Lefan's boss, however, was an Indonesian who came from a deforested area and preferred to stay at his desk. When I asked Lefan whether his boss entered the forest at all, Lefan replied (in Indonesian): "*Ya, masuk ke hutan. Masuk seratus meter, dan kembali.*" ("Sure, he goes into the forest. He goes in 100 meters, then he comes out again.")

My second example concerns young New Guineans whose parents used Stone Age technology and exploited the forest, but who themselves go to school and eventually move to a town in search of work. Naturally, these young New Guineans have no opportunity to acquire detailed knowledge of the forest. Perhaps surprisingly, they exhibit little interest in it either. The national parks and zoos that the governments of Papua New Guinea and Indonesia have established near urban areas, in hopes of stimulating interest by urbanized peoples in their country's natural heritage, elicit disappointingly little interest, except by people using the open spaces for picnicking and Sunday outings. Their fear of the forest, as well as their negligible interest in the natural heritage with which their ancestors lived so intimately for tens of thousands of years, are a big problem for the governments of Indonesia and Papua New Guinea today in their effort to develop indigenous support for conservation.

New Guineans and Their Natural World

A Question of Definitions

In short, New Guineans and other Pacific islanders still practicing traditional lifestyles possess a deep and detailed knowledge of wild plant and animal species, a dependence on their natural environment for their economy, and a use of wild species for their decorations and status symbols and myths—to a degree, indeed, that is difficult for those of us reared in urbanized Western societies to grasp. If biophilia is defined as human affinity—regardless of whether learned or innate—for other species, New Guineans serve as textbook examples.

If, however, biophilia is specifically defined as an innate affinity, it is not presently clear to me what might constitute evidence for such a genetic basis. Yes, New Guineans with traditional lifestyles grow up to depend on other species, live surrounded by them, and learn a great deal about them. Most of us Westerners instead grow up dependent on human artifacts, live surrounded by them, and learn a great deal about them. To a striking degree, the human brain is a generalized data-processing organ capable of absorbing enormous bodies of information that did not exist throughout our evolutionary history: chess gambits, stock prices, baseball records, organic chemical syntheses, and so on. Each of these bodies of information becomes the mental world of different individuals in our society. To assess the biophilia hypothesis, we shall have to identify and evaluate evidence that acquisition of knowledge about the natural world might have an innate basis lacking in the acquisition of knowledge about our worlds of human artifacts.

REFERENCES

Bellwood, P. 1987. *The Polynesians*. Rev. ed. London: Thames & Hudson.

Connolly, B., and R. Anderson. 1987. *First Contact*. New York: Viking Penguin.

Diamond, J. M. 1966. "Zoological Classification System of a Primitive People." *Science* 151:1102–1104.

———. 1972. *The Avifauna of the Eastern Highlands of New Guinea*. Monograph 12. Boston: Nuttall Ornithological Club.

———. 1973. "Distributional Ecology of New Guinea Birds." *Science* 179:759–769.

———. 1984. "The Avifaunas of Rennell and Bellona Islands." *Natural History of Rennell Island, British Solomon Islands* 8:127–168.

———. 1989a. "This-Fellow Frog, Name Belong-Him Dakwo." *Natural History* 98(4):16–23.

———. 1989b. "The Ethnobiologist's Dilemma." *Natural History* 98(6):26–30.

———. 1992. *The Third Chimpanzee*. New York: HarperCollins.

Diamond, J. M. and M. LeCroy. 1979. "Birds of Karkar and Bagabag Islands, New Guinea." *Bulletin of the American Museum of Natural History* 164:469–531.

Flannery, T. 1990. *Mammals of New Guinea*. Carina, Australia: Robert Brown.

Flannery, T., and L. Seri. 1990. "*Dendrolagus scottae* n. sp. (Marsupialia: Macropodidae): A New Tree-Kangaroo from Papua New Guinea." *Records of the Australian Museum* 42:237–245.

Foley, W. A. 1986. *The Papuan Languages of New Guinea*. Cambridge, England: Cambridge University Press.

Majnep, I. S., and R. Bulmer. 1977. *Birds of My Kalam Country*. Auckland, N.Z.: Auckland University Press.

Ruhlen, M. 1987. *A Guide to the World's Languages*. Stanford: Stanford University Press.

Souter, G. 1963. *New Guinea: The Last Unknown*. London: Angus & Robertson.

White, J. P., and J. F. O'Connell. 1982. *A Prehistory of Australia, New Guinea, and Sahul*. Sydney: Academic Press.

Part Four

SYMBOLISM

CHAPTER 9

On Animal Friends

Paul Shepard

During nearly all the history of our species man has lived in association with large, often terrifying, but always exciting animals. Models of the survivors, toy elephants, giraffes and pandas, are an integral part of contemporary childhood. If all these animals became extinct, as is quite possible, are we sure that some irreparable harm to our psychological development would not be done?

G. E. HUTCHINSON[1]

BEHIND ALL DISCUSSION of the relation of humans to other animals is the final and irresolvable enigma of our identity—personal, social, and as a species. This essay begins with the "savage" mode of self-identification by reference to others and then arcs forward through the end of wildness to modern narcissism as the failure of such a reference.

Among the charities that impinge on our daily lives are those for saving, rescuing, rehabilitating, and protecting animals. We are exhorted to treat them as fellow beings, even as friends, to extend our horizons to include them within our circle of familial and social responsibility, and to be gratified in this "humane" caring for all things great and small. Is this partnership what is meant by the "innate tendency to affiliate" that defines biophilia? "Innate" implies evolutionary roots and species-wide characteristics. It evokes the past in the strict sense of our ecology and the Pleistocene perception of animals by our ancestors who lived in small, foraging groups.

Classical scholars of the past often characterized the "barbarian" mind as capable only of fuzzy distinctions between the self and its environment—the muddled thinking of a tribal horde "at one" with nature. It appears, however, that primal peoples (variously termed primitive, savage, tribal, subsistence, hunting-gathering, Paleolithic, indigenous, and ethnic) elaborate the distinctions between culture and nature—beginning with that which is human and that which is not—and then pursue ever-finer refinements. This line between culture and nature, defined by Claude Lévi-Strauss as totemic, creates "a homology between two systems of differences, one of which occurs in nature and the other in culture."[2] He refers to animal species on the one hand and a taxonomy of human groups on the other, and it is evident that the "system of differences" includes ecological and ethological characteristics among the nonhumans and a parallel track of social behavior among the humans.

Contact across this polarized field is fraught with power and danger. Among primal peoples, contact with the living nonhuman world is hedged with circumspection, caution, and sometimes ceremonial formality. Invisible realities play through this concern—animal spirits, sacred beings, and ancestral presences. Since they have no domestic animals, the physical presence of living creatures in close contact is unusual, even though the use of animals as food and skins is extensive.[3] One does not abuse their remains or consort familiarly with such very different "peoples" with impunity, for they are sentient, alive or dead, and can influence human well-being. It is inappropriate to behave in a companionable way with a deer or a wolf, even if the deer is prominent in a material or spiritual sense and the wolf a fellow-hunter, both major figures in the mythology. As a consequence of this division, reaching across the boundary between culture and nature, the exercise of utility obligates humans to canonical acts of depressurization resembling diplomatic obligations. The hunt and other encounters with animals in daily life are framed in aversion, circumspection, convention, protocol, thanksgiving, and acts of contrition or apology, but not as ordinary gregarious conviviality, thoughtless exploitation, or the dominance relationships of slavery. The killing, skinning, cooking, and other uses of an animal are circumscribed by customary, rhetorical practices at once efficacious and mitigating. Reticence marks the contract between human

Symbolism

and the other. Not only is the deer seen in this way but all species, members of a polythetic cosmos, a community with its own implicit, intergroup formalities, as if the universe were a vast social drama.

This does not mean that the Lévi-Strauss line between the humans and the others is simply an edge, the crossing of which requires traditional courtesy. It is also the zone of translation between the distinct domains of nature and culture. The ecological side is perceived as a coded reference for the rationalization of human society. It is as though all human societies were endlessly clarifying their internal distinctions. Endowed with the furor and tumult of primate group life, humans relentlessly repair and rephrase the social contract, often with reference to some external model. The naming of clans after animal species is an example. A consequence of this reference is, Lévi-Strauss notes, that culture is not regarded as an improvisation but is perceived from within as "natural." At the same time this allusion bridges the artificial distinction between nature and culture, mending the conceptual damage done by such dualism. If a binary approach to the world makes metaphor possible, it is healed in due course by myths of a shared ancestry of humans and animals characteristic of totemic societies and the perspicacious idea that nature is a language and guide to human life.

The wild animals composing nature are an array of species whose differences and interactions are observed and translated by keen, lifelong attention to the nuances of natural history. Out of this observation emerges the notion of a comity to which human society is analogous. Lévi-Strauss describes wild species as the concrete model of categorical forms, from which societies give names to their clans or other subgroups, each having its eponymous or totemic animal. In short, people justify group relationships by such a poetic reference, giving them expression in narration, art, fetes, culinary life, and intergroup protocol—all out of a kind of logic, not of emulation but of parallels.

If this nature/culture correlation is as widespread among tribal peoples as Lévi-Strauss implies, it is probably extremely old. Some prehistoric art can be fitted to its general framework, suggesting that the system of parallel worlds might very well be basic to the evolution of human consciousness. A possible stage in the evolution of ecological patterns as poetic models of a human social matrix may be what Paleolithic cave art was about—that is, as the middle period between the earliest symbolic reference to animals and the full flowering of cynegetic thought at the end of the Pleistocene. Bertram Lewin argues that cave art may represent the deliberate effort to collectively internalize key images in a group of observers, perhaps initiates. The darkness of the cave is experienced as a shared reference to the darkness of the cerebrum where in imagination we glimpse figures, as it were, in a brief light—a mode of bringing a fauna into the head.[4]

Regional differences in such practices may have tended to isolate human groups from one another, leading to cultural differences such as the emergence of linguistic dialects and the islanding and evolution of races, giving cohesion and identity that bind members to the group or set groups apart. If this nature/culture line characterizes the evolutionary origin of human cognition, a root of biophilia, it is in us still, the tug of attention to animals as the curved mirror of ourselves—not as stuff or friends but as resplendent, diverse beings, signs that integrity and beauty are inherent in the givenness of the world.

Minding of other animals is far older even than the human arts of the Paleolithic, however, and is widely shared. We are not its sole heirs. Harry Jerison has described the Cenozoic evolution of mammalian brains, notably the savanna predator/prey system in which the attention structure was focused on other species in the mutual honing of pursuit and escape strategies and the reading of signs, resulting in progressive, reciprocal enlargement of the brains of hunters and hunted. The advent of protohumans into this savanna game as part-time carnivores and occasional prey corresponded, he believes, with the emergence of speech and the cognitive need to determine what constitutes "objects" or "categories" in the perceptual field to which words could be attached and which thereby become real. In short, as we became open-country hunters and hunted, we entered an ongoing system of brain-making, using our advanced primate vocal and visual apparatus instead of an olfactory system. This process centered visual imagery on the simultaneous and tireless scrutiny of other animals and the emergence of self-consciousness.[5]

The metaphysical role of animals in primal societies which shaped and defined our species is not just a vague "reverence" for animals but a body of procedure and narrative which acknowledges kinship and the necessity of killing, a sinew of sentience and spiritual power linking death and love. The Pleistocene offers us no compassion for animals in the warm idiom of the teddy bear. It constitutes instead the genesis of self-consciousness as assimilation, an endless scrutiny that is both instrument and synonym of becoming—the eco-psychology of predator and prey. Eating, in-taking, is the culmination of the holy hunt, a sacred meal in which not only energy but qualities are internalized. This was the new integration of an anthropoid mind with savanna thinking. And it continues to remind us of the reality and significance of wildness in our individual becoming and our extraordinary origin in the game.

As its strategies became conscious, our savanna-based, hominid, omnivorous ancestor assembled the self cognitively by reference to an extensive fauna swallowed bit by bit. As E. H. Lenneberg recognized thirty years ago, semantic structure is native while vocabulary is acquired. But both depend on a given world. Category making relies on external prototypes.[6] Inherent grammatical structure articulates with a perceived array of species. Human development—as a result of 2 million years of participation in the savanna game—endows each of us with an epigenetic ontogeny, personal development combining inherent predisposition and the right experience, in which speech and cognition are keyed to the natural world.

This is why awareness of animal taxonomy is so prominent in childhood. Species recognition has been described by Eleanor Rosch and others as a two-step process in which the whole animal is first identified in a major category, such as "bird," and then defined to kind, such as "sparrow," by specific cues.[7] Such cues are compelling because of the early attention to body parts or, one might say, the "butchering perception." Everybody is familiar with the infant's interest in external anatomy, its shifting attention between the other and the self in an association of tactile and visual play—the touch and say "eye," touch and say "nose," touch and say "ear" with a caregiver. It is antecedent to pointing. The reciprocity between infant and mother in these rituals of naming and touching toes, nose, fingers, and belly is perhaps the procedural model for a lifelong process of naming based on distinct, detachable characteristics. For these reasons the principal nouns of interest to small children are body parts and animals.

On Animal Friends

Life is centered on the enigmas of "I," "we," "you," and "they." The pronouns are conceptually slippery. As a group or clan member, the self could be identified as part of a social species in the way Lévi-Strauss suggests. But we know the self from an inner life as well, which is more obscure and more personal. One's own "body percept" is built up incrementally, first outer and then inner. Our external anatomy has corresponding parts among animals, while our inner anatomy also matches the body parts of nonhumans. This physical correspondence is a kind of map of feelings. The visceral terrain is composed of felt events. In their organs and behavior—and thereby in their names and images—animals provide the concrete reference for sorrow, pain, moods, tempers, dispositions, and all those vivid, chromatic tides that wash through our being, which we know as our own life. In this way animals fill the void to which our self-consciousness would be subject without prismatic representations of qualities or bits of experience. This discovery of the external existence of animals who correspond to our inner reality began not as an invention but in the natural history and evolution of mind. Both character-filled story and anatomical taxonomy depend on the close association of speech with the early naming of body parts and actions, followed in childhood play by the mimicry of animals. From external morphology to the identity of internal organs (mythically animated in some people as having lived independently like other animals) to the less tangible parts of a self there is the progressive construction of the world as pronouns.

The various steps by which we discover the inner animals have both phylogenetic and ontogenetic aspects. Verbalizing the names of animals may have come early in the evolution of speech and is a stage event in infancy and childhood in which the creatures are swallowed by attention—as though humans had not been convinced in their nascent carnivory that the animals eaten were not still somehow alive inside. The verbing of these nouns—the vast body of verbs, gerunds, and infinitives: to cow, fish, duck, quail, clam up, skunk, weasel, outfox, hound, dog, goose, horse around, lark about, hawk, worm your way, bug, ram, pig out, hog, grouse, fawn, buffalo—constitutes the lexicon of an inner structuring dependent on a system of referent species whose actions are a speaking of individual experience as a fauna. Having recognizable external parts resembling my own, animals are an invitation to think about my own visceral landscape, of our corresponding organs and behavior, so that the fish seems to embody all random seeking and searching in our lives, as though created to present us with that particular representation of ourselves.

In the human developmental calendar the animal is first a noun and then is verbed not only in speech but in play. Stylized and conventional movements (prefiguring ritual in adult life) are essential to play—in which we find the "predication of the animal on the inchoate pronoun" in games such as Sharks, Wolf and Sheep, or Fox and Goose, as the participants define themselves briefly and successively as different animals.[8] The enactment of the verbed animal captures the evanescent feelings that go with "wolfing" one's food or "chickening out" of a dangerous situation.

Such games are like enacted stories. And among these stories is a genre with happy endings in which the normal concerns of childhood, personified and distanced from the self, are defeated in the end, often with the help of animal figures from deep in our organic substrate. The personified feelings or states that the protagonists in fairy tales embody in listeners include different animals representing slightly disguised (that is, perceptible to the unconscious) aspects of the self as a kind of promise. The hearers of stories that begin "once upon a time" and have a happy end ruminate on the tales as they foreshadow intrinsic solutions to childhood worries and register a deep correspondence based on the double life of the protagonists—for example, the dove as one's spiritual side, the wolf as aggression, or the frog as the inevitability of metamorphosis and growth.[9]

In yet another mode, adults can visualize and summon the animal avatars of their own diverse physiological and psychological interiors as aspects of experience and personality. This mode has emerged recently as a distinct therapy. First with a guide and then independently, one meditates on an encounter and imaginary conversation with a series of inner animals who emerge from their places in the vital (*chakra*) centers or from the bodily organs to speak of their needs and fears.[10] The human capacity to generate these images, under circumstances less focused than hypnosis, testifies to a facility too easily characterized as archetypal and too casually dismissed as imagination. The train of fictive interaction between these animals and their human host is astonishing evidence of the fundamentally zoomor-

On Animal Friends

phic perception of one's inner life—a domain immensely troubling and destructive since the reign of "axial man" sought to banish its numinous presence and mechanistic biology reduced our concept of the organs to that of a vegetable-like stasis. The interior fauna reveal a realm of dynamic animal mediators who both embody and represent that which is otherwise obscure to the conscious self. They confirm the self as alive within—a community of being congruent with the outer, living world and dependent on it in the cognitive development of the individual.

These three modes—story, animal games, and visual imagery—are the heritage of 2 million years or more of the assimilation of animals in diet and mind. It should be added that such self-realization is still an enigma. The awakening of the mediators requires devoted attention, for animals are more complicated and interesting than we think. There is a wildness about our own feelings that they embody—a wildness which resists final capture, a strangeness which is itself an aspect of being, a hidden side to our otherness which is like the creatures of an ecosystem who remain underground.

The foregoing may be represented by the schema of Figure 9.1. Beginning with Lévi-Strauss's simple diagram, I at the top of the figure, I have elaborated this binary culture/nature system in II and further in III. Part 1 at the bottom shows Strauss's totemic patronymics derived from species A, B, C and so on. Parts 2 and 3 at the bottom show that emotions and other features of the external world which need "organizing" may also be perceived as analogies to the animal taxonomic system.

If the development of the person's sense of his own structure depends on the beauty, strangeness, and diversity of a wild fauna, assimilated ceremonially as food and perceptually as the plural assembly of the self, what can the collapse of the Lévi-Strauss line mean as animals are domesticated? Very few species cross the borderline to live in the human domain. They are like refugees from a ruined nation or guerrillas in support of a failing ministry. Once across, captive and bred, domestic animals become numerous, docile, and flaccid, their brains diminished, their anatomy and physiology subject to dysfunction, and their ethology abbreviated. At the same time the remaining wild fauna recedes from human sight and table. The anthropological literature bulges with human attempts to make do cognitively

Symbolism

I. In *The Savage Mind:*

"Totemic Institutions"

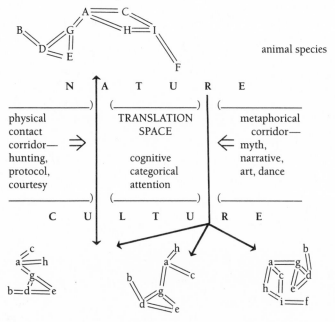

NATURE   species 1 = species 2 = species 3 = . . . species *n*

CULTURE   group 1 = group 2 = group 3 = . . . group *n*

II. As seen under the hand lens:

species, the
A = B = C = D = E = F = G = H = I   paradigmatic categories
a = b = c = d = e = f = g = h = i   human social groups

III. As seen under the microscope:

animal species

N A T U R E

physical
contact
corridor—   TRANSLATION   metaphorical
hunting,   SPACE   corridor—
protocol,   myth,
courtesy   cognitive   narrative,
                 categorical   art, dance
                 attention

C U L T U R E

1. Totemic society
   (clans, sex, rank, age,
   kinship)

2. Internal structure
   (organs, feelings,
   emotions, other
   intangible elements
   of self)

3. Other analogies
   (space, zodiac, tools,
   terrain, deities)

FIGURE 9.1. *The Lévi-Strauss or Nature/Culture Line.* Part **I** reproduces the nature/culture diagram exactly as shown in Lévi-Strauss's *The Savage Mind.* Note that the analogy is between the signs (=) designating relationships. It represents in an elemental way the basic use of an external model for cultural analogy. Part **II** emphasizes the one-way flow of this connection—entirely as a mental function of humans—and suggests that there are many species. In part **III** the concept is elaborated. The species relationships are shown in constellated rather than linear form. The boundary between nature and culture is punctuated by two types of connection: the one on the left involves the physical use of animals; that on the right indicates metaphorical use that breaks into three categories. Number 1 at the bottom shows a pattern of clan relationships representing a portion of the animal ecological pattern as given paradigmatic application to human group membership. Number 2 is another pattern, also "taken from nature," indicating internal psychological structures that parallel animal interactions. Number 3 suggests a number of other analogical uses of the ecological system of concrete images as the means of giving coherence to an otherwise inchoate structure.

with cows, pigs, horses, dogs, and chickens, their impoverished zoology and hybrid extravagance.[11]

Domestic animals (and plants) are the products of the first genetic engineering. They have served for millennia as material substance and workers, though celebrated in principle as sacred gifts and epiphanies. In recent times their role has evolved from barnyard utility to companionship. As pets they contributed to human well-being long before anyone asked why. Now there is statistical evidence from therapeutic programs proving domestic animals to be anodynes for human suffering ranging from penal incarceration to Down's syndrome. No one who looks at the evidence can doubt that animals in hand improve the quality of modern human life, whether measured in terms of longevity or recovery. Pet keepers can take consolation in knowing that their animals are proven health additives. The efficacy of this dumb-beast panacea has an impact comparable to that of antibiotics. Animal-facilitated therapy cuts a swath through despair, loneliness, genetic impairment, the terminal blues, pain, the black holes of autism and schizophrenia, the chutes of age, boredom, and immobility. As medicine nips the usual killers in the bud and people live longer with their geriatric syndromes and other civilized maladies, the potential for animal companion tonic increases. Apart from helping the sick and disabled, pet therapy has created jobs for health care specialists, pet mortuary and bereavement experts, grooms, keepers, handlers, new categories of veterinary specialties, breeders, middlemen, the makers of pet food and clothing, designers and builders of special facilities, along with a vast corporate, academic, and medical contingent, and acquires an easy marriage to the animal shelter industry. There is both altruism and financial profit. And the alliance of pet keeping and corporate medicine does not end there. It generates moral judgment, as well, and facilitates animal rightists, animal ethics theorists, antihunters, vegetarians, and the rising tide of fringe animal lovers right down to the zany keepers of legions of cats and dogs in their bedrooms who might otherwise kill themselves or burn down city hall.

From this heterogeneous multitude there emerges a great cloud of righteous socializing with the entire animal kingdom—projecting onto wild nature unlimited fraternization with its baggage of care, compassion, and kindness and its overbearing social injunctions, ideal personal standards, and other traits and rules that shape human society: in short, the invasion of the ecological world in a spirit of human arrogance.

What can be said of the existential status of the institutionalized domestic animals as well as the family dog and cat? Clearly they are happy friends of humankind—just see how they wag their tails and purr. Indeed, they *chose* this life, they prefer it, and they are far better off (as all slave keepers have said). In fact it is part of nature's grand plan. I offer two recent examples of this thinking from different ends of the popular/scientific spectrum: a journalist writing of the "covenant of the wild," arguing that domestication is just recent evolution, and the editor of *Science* saying, "Over evolutionary time the friendliest of wolves (and possibly the most intelligent) learned that wagging their tails and delivering slippers was an easier way to earn a living than hunting caribou in the wilds. . . . An original wolf might say to the dog, 'You have lost your freedom. Your obsequiousness is humiliating to the family of Canidae.' The dog could reply, 'I am much less warlike, far more altruistic, and besides, it's a wonderful standard of living.' Whether society prefers to have wolves or dogs remains to be seen."[12]

The editor of *Science* is not just being down-home with his tail-wagging and slippers bromides. Even school kids know that wolves did not *decide* to become dogs. The part about warlike and freedom and standard of living exploits our bias in favor of dogs. It is a lie. As for subservience, the worst thing about the editorial is its breezy participation in the larger techno-

On Animal Friends

philic fiction that everything is getting better and better through the control of plants and animals—an attempt to coopt the whole momentum of evolution in the advocacy of genetic engineering.

To understand what the collapse of the Lévi-Strauss line means, we must ask: who are these animals who, domesticated, came to be the model images for all animals? Two biologists, Konrad Lorenz and Helen Spurway, were not so foolish as to confuse evolution and domestication. She, a geneticist, referred to domestic animals as "goofies" because of their addled genetics and the resulting phenotypic mélange.[13] Lorenz made the comparison clear: the domestic had been shorn of what was subtle, complex, and unique in the wild ancestor, including the intelligence and independence meant by "wildness."[14] He loved dogs, but he knew them to be gross outlines of the wolf. With diminished brains and congenital defects, these abducted and enslaved forms are the mindless drabs of the sheep flock, the udder-dragging, hypertrophied cow, the psychopathic racehorse, and the infantilized dog who will age into a blasé touch-me-bear, padding through the hospice wards until he has a breakdown and bites the next hand.

If domestication breaks up wild genotypes, which are continually honed in a kind of DNA harmonic with the environment, it is hardly appropriate to say that the animal is "adapted" to human care. This bottleneck breeding with its catastrophic breakdown of genetic equilibrium, releasing broods of monsters, is no "evolutionary adaptation." From the standpoint of their relationship to people it is hard to know what to call this protoplasmic farrago of dismantled and reassembled life. All of those living animals drawn onto this side of the nature/culture line enter the human social system—as healers, friends, siblings, offspring, competitors, entertainers, caricatures, sexual mates, bodyguards, protégés, saviors, litigationists, princes, and fairy godmothers, not to mention their participation in the social categories of the homeless, the unemployed, the sick, the deserving, and so on.

In a radio call-in program recently I referred to animal companions as "slaves." The telephones buzzed with angry callers who insisted that their dog was a comfortable, privileged family member. Their reaction was like that of certain pre–Civil War southern plantation owners who could point with defiant compassion to their grateful, singing cottonpickers. The truth is that pets are subject to their owner's will exactly as slaves. Yet the term *slaves* may not be suitable, since human slaves can be freed by political and social action. The goofies, congenitally damaged, cannot. If freed they die in the street or become feral liabilities. We could simply quit breeding them. Their relationship to us is not symbiotic, either, or mutual or parasitic. None of these biological terms is suitable to describe organic disintegration in a special vassalage among creatures whose heartwarming compliance and truly therapeutic presence mask the sink of their biological deformity and the urgency of our need for other life.

Less than kindly euphemisms for "companion animals" come to mind—crutches in a crippled society, candy bars, substitutes for necessary and nurturant others of the earth, not simply simulations but overrefined, bereft of truly curative potency, peons in the miasma of domesticated ecosystems. The corporate takeover of the pet is merely a recent step in institutionalizing, rationing, and marketing these ameliorations for something essential and missing in human health.

My concern here is not the destiny of these lumpish, hand-licker-biters among humans who are desperate for the sight of nonhuman creatures because they touch some deep archetypal need. Nor is it the monumental logistics of the 40 million dogs and 40 million cats in American homes, not to mention the rest of the world, and all the animals in the world's zoos. My focus is the effect of the replacement of domestic for wild animals in our psychological development, especially in the formative processes by which we mature. The colossal upsurge of the pet as an industrialized healer brings the issue of our inner life before us, along with the planet's diminishing wild abundance and diversity. The agro-urban world replaced a way of life centered on the elegant courtesies of totemism and the brain-making hunt, its roots deep in the Pleistocene and deep in the human heart. When animals as domestics came literally into our households across that Lévi-Strauss line, they filled the lowest ranks of our society. There was the end of respect for the other on its own terms.

The projection of the domestic milieu has infected our perception of all animals. The breakdown of the Lévi-Strauss line results in the impaired ability of the ordinary person to make critical distinctions between dena-

On Animal Friends

tured goofies and wild animals. The idea of responsibility for the animal kingdom as a whole is clearly neobiblical, especially "caretaking" and all its benevolent expressions. These are three: the Noah Syndrome, which puts us in charge (as God's steward) of *all* the animals; the hagiographic model of Saintly Hermit before whom the beasts, recognizing human holiness, gladly enter into cringing servitude; and the Peaceable Kingdom, the prototype for our perception and regulation of nature as if it were a nursery school playground.

To recreate the peaceable kingdom on earth requires our invasive acts of protecting the weak from the strong, the imposition of interspecies ethics, the infusion of intention such as kindness or mercy, the projection of the domestic world onto nature. Our Noachian authority requires that we choose which individuals shall breed to survive the catastrophes we have brought upon the wild world, followed by their incarceration in the zoo cells of the modern ark. Our saintly status of righteousness and dualistic thought justify these extensions of our power over the wild others and the expectation that they will submit to our moral eminence. All three concepts take wild animals one step closer to becoming slaves along with their domesticated cousins. Wild animals are not our friends. They need protection from those who, with good intentions, would harm their biology to save them by extending our obligations of care to them, along with its trail of ethics. Paradoxically, the most characteristic feature of modern "animal rights" is its withdrawal. It smells of the armchair philosopher who, as insulated from nature as possible, advocates "letting them be" as a moral stance. It seems to contradict the notion of benign care. But it is merely the final hubris, the sanitized isolation from those whose gaming made us what we are.

Across the fence of the Lévi-Strauss line, tribal peoples see the ecosystem as a code and mnemonic device for the organization of society. With the collapse of that line, the human social system becomes the standard for managing the wild world and reducing the others to individuals. Lévi-Strauss has described intermediate cultural practices between the totemic and caste cultures, for example, one in which the members of clans—such as raccoon and bear—are expected to behave like raccoons and bears. Although Lévi-Strauss does not define domestication explicitly as a breakdown in the totemic, metaphoric system, it is clear from his study that in "caste" (or domestic) society metonymy permeates and disintegrates an older structure the way mycelia of a fungus leave the outward form of its host empty of its original function. Domestic culture replaces totemic or analogous thought with physical conjunction and leads to literal-mindedness. As Lévi-Strauss shows, clans identify themselves by an eponymous animal. Intermediate societies, becoming domesticated, cease to ponder the animal as heuristic and begin to imitate it. On the face of it, behaving like a bear or raccoon may not be so bad. But when the pig and the dog have become the animals of reference instead of the bear and raccoon, the animal as model of human degradation cannot be far off and the "beast" in Mary Midgley's sense is born.[15] Thus does civilization slander savagery by transforming its elegant meditations into gross actions. Indeed, the antinomy of attitudes toward dogs throughout the world displays the contrast of the beauty of the wild canid with its domestic shadow. By human standards the dog is incestuous, shameless in its excretory habits, and evil as a latent killer of sheep and among humans as a rabid brute, yet it is admired for its helping habits as herder, hunter, and protector. The dog's modern incarnation as individual personality has not reduced its ambiguity. In any issue of *The New Yorker* magazine you will find dog/man cartoons that reveal the confusion and angst incident to the fuzzy boundaries of identity among an educated elite who expiate their stress as humor.

From this metonymic stew of the animal as friend and object emerges the paradox that primal peoples kept their distance from animals—except for their in-taking as food and prototypes—and could therefore love them as sacred beings and respect them as other "peoples" while we, with the animals in our laps and our mechanized slaughterhouses, are less sure who they are and therefore who we are. The surprising consequence is that "nature" is more distanced, not less. Lévi-Strauss diagrams this domesticated situation by enclosing the animal species and the human group in a box to indicate that the relationship is one between terms rather than an analogy between two sets of relationships. Difference, he says, has overwhelmed the connectedness otherwise drawn from a poetic translation of concrete, ecological relationships. Lévi-Strauss's language has beleaguered generations of students—the "paradigmatic" and the "syntagmatic," the "metaphor" and the "metonym." But we can understand these terms as "figurative" and "literal" and recognize the breakdown of the distinction. If we in the pursuit of progress are destroying species, it may not be due only to the effects of industrial, multinational, corporate greed but to our unconscious resentment against the animals themselves, whose analogical and perceptual roles—anticipated by our human nature—are no longer given credence. The animal, refusing eye contact with us in the zoo, seems to convey a final insolence and abandonment in which we, mistaking who has done what, feel ourselves to be forgotten. This situation is schematized in Figure 9.2, showing the incorporation of a few species (species B becomes B', C becomes C', and so on) as they are shunted by way of a syntagmatic domestic ecology into the human social order (a', b', and so on), as objects and surrogates. Some cognitive and metaphorical categorical use of the species system continues, such as athletic teams, but generally the civilized world uses other, less vital, less definitive, and less heuristic systems for defining human groups or other nebulous complexes.

The puncture of the Lévi-Strauss line by literalness heightens our sense of anguish about nature and its destruction, but the final loss is our own. Identity is the issue—our difference from and likeness to the nonhumans. According to Julia Kristeva, the issue in civilized society is characterized by infatuation with the self. The Greek myth of Narcissus was an attempt "to tackle a problem that the ancient world had not solved—otherness."[16] The "murky, swampy, invisible drama must have summed up the anguish of a drifting mankind, deprived of stable markers."[17] Narcissus, a hunter, having rejected the love of Echo, pauses to drink while hunting and falls in love with his reflection in a pool. Ultimately he "discovers in sorrow the alienation that is the constituent of his own image"; he commits suicide and his substance is transformed into a flower. That is, he regresses into a vegetative state in which the questions of "psychic space" do not arise. His anguish reflects an immense loss in which Narcissus "no longer has the thinking *nous* of the ancient world that would have enabled him to approach the other as plurality, as a multiplicity of objects or parts."[18] Kristeva's insights are worth exploring, although she does not seem to realize that the story is not

I. In *The Savage Mind:*

NATURE	species 1	species 2	species 3	species *n*
CULTURE	group 1	group 2	group 3	group *n*

"Endogamous and Endo-Functional"

II. The concept enlarged:

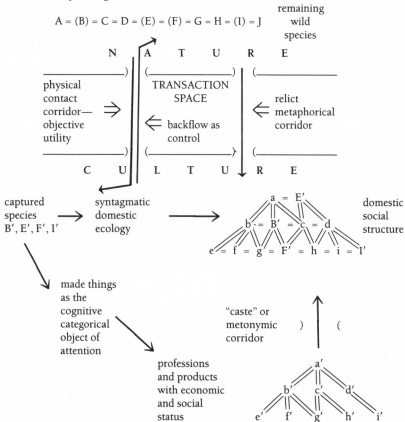

A = (B) = C = D = (E) = (F) = G = H = (I) = J   remaining wild species

N A T U R E

physical contact corridor— objective utility

TRANSACTION SPACE

backflow as control

relict metaphorical corridor

C U L T U R E

captured species B′, E′, F′, I′

syntagmatic domestic ecology

a = E′
b = B′ = c = d
e = f = g = F′ = h = i = I′

domestic social structure

made things as the cognitive categorical object of attention

"caste" or metonymic corridor

professions and products with economic and social status

a′
b′ c′ d′
e′ f′ g′ h′ i′

FIGURE 9.2. *Domestic Deformation of the Lévi-Strauss Line.* This diagram repeats the schema of Figure 9.1 with changes brought about by domestication—which Lévi-Strauss labels "caste." Part **I** is Lévi-Strauss's diagram. He emphasizes the locking of species 1 to human group 1 by boxing them together. No longer does the relationship between species signify as in totemic thought. In part **II**, the concept is enlarged. Species that have been domesticated are shown in parentheses, although their wild ancestors may remain part of nature. At the left the physical use continues, although it is no longer characterized by formal protocol. The reverse flow along this same corridor, from bottom to top, signifies the projection of human social ideas upon the species system, imposing new rules on the animal ecology. Some residual, metaphorical uses are represented by the arrow at the right. The captured forms have been brought across at the left and domesticated, becoming, at bottom left, part of the system of production—the "things" that have become the concrete models of categorical thought. These are associated with the social class system that emerges from the value and power of the professions, taking a pyramidal form at the bottom and, finally, a domestic social structure. The wild animals are no longer a point of reference, having been replaced by the various products of society. The domestic animals are members of the society, ranked among people rather than studied as model peoples.

actually about totemic culture but a myth, projected back upon hunters, about despair by urban societies.

What was that multiplicity of which Kristeva speaks? Narcissus's inner vacuum, the empty psychic space, is a result of the disarray and the paradox of the solitude of cosmopolitan life, the doubt, disintegration, and estrangement of the soul in the urban stewpot, the deracination and cacophony of the city, a dilemma remaining with us still: a "falling apart . . . for which the present-day equivalent would be advanced mass mediazation."[19] Seeking origins, she refers unfortunately only to the archaic world of Homeric Greece and the fading of the goddess, and in doing so falls short in chronological explanation.

Kristeva is interested in the recovery of the feminine (and, by implication, the goddess), but neither the goddesses nor gods of archaic history were themselves the primal others. She notes that the cultures of the ancient world (early civilizations) had not solved the problem. The reason was, in fact, that they had created it. The epiphany of the goddess may have indeed represented a lost sensibility more organically tuned than that of the modern world. But what it replaced was a pantheon of wild others, the many-specied consort of preagricultural humanity, part of the genesis of imagination itself, a company in whom both the condition and fulfillment of life were incarnate. The nature of the others was a matter of consuming interest, not a Pleistocene "problem." It was only after the defeat of that numinous, nonhuman presence of animals as a meditation on the nature of the self, in the era of the "ancient" world of cities, goddesses, and gods, that

Symbolism

Narcissus rises and falls. All the humanized deities were insufficient substitutes for a zoological theriophany. As the subject and object of its own meaning, the human figure produced disillusion and inner crisis, the dead end of remaking the gods in human form.

Subsequently there have been three expressions of the relationships of human and other within the major Western tradition: Gnostics, who reduced the others to personifications of evil after the Fall; Christians, preoccupied with the godlike self and the difficulty of integrating it with the singular other as Kristeva says; and the Neoplatonic Cartesians for whom the others were reduced to simple, finite blobs.[20] These solutions to the "problem" of the dynamic connection of the self and the other have since occupied much of philosophy, narcissism having become a major trait of Western subjectivity.

Plotinus, Kristeva notes, rehabilitated the Narcissus myth for Christianity by calling upon "dignity inherited from ascetic solitude" and identifying the mirrored other as God. The object was to love one's self, since one is created in the image of God, as the first step toward loving God. Bernard of Clairvaux further argued that Narcissus and his image were the corporeal passion and the spiritual passion, respectively, two aspects of the self, keeping "the flesh from a spirituality that would thus become too ethereal, without forgetting, nevertheless, the presence of a spirit in a flesh."[21] Not that Bernard saw any other saving grace in the organic world or the body: "Be ashamed, my soul, for having exchanged divine resemblance against that of beasts, for wallowing in the mud, having come from heaven."[22] Narcissus, according to this thesis, was merely the failure of true self-love as a fusion with God. If the love of God was inseparable from a love of the human figure, God's image in the self should be One. It had nothing to do with nature—that scorned, plural realm of the beasts. This "privilege of self-love," which began as an Aristotelian idea, was formulated by Plotinus and Christianized by Bernard. The church made a parable of narcissistic despair as a failure of the spiritual imagination to accomplish the necessary unity with God as One. Man, "seized by an unnameable solitude, was called upon to withdraw into himself and discover himself as a psychic being. . . . Henceforth, there is an inside, an internal life, to be contrasted with the outside."[23] "The loving soul must therefore give up its otherness

On Animal Friends

and give in to the sameness of a single light where it loses itself as other . . . as nonbeing." And so, "otherness disappears when we merge with the One."[24] The natural others—those diverse beings who previously had been thought to be coded clues to the nature of the inner life—disappear or become irrelevant or destructive.

As the church tailored the myth to represent a failed meditation on the unity of human and God, all beings except the human became spiritually irrelevant, leaving the external world as evil (in Persian and Gnostic thought) or available as neutral material for the pragmatic followers of Francis Bacon and René Descartes to do with as they would. The heroic notion of *ego cogito* leads to "the conquest of the outside . . . the outside of nature, to be subjugated by science."[25] Neither the Christians nor modern theorists have recognized that Narcissus begins, not with the loss of the anthropomorphic deities, but with the loss of diverse, therioform others.

The loss of the wild others leaves nothing but our own image to explain ourselves by—hence empty psychic space. Portraiture, for instance, arising late in history, is narcissism magnified: the reassertion of the human primate's obsession with the visage. Portraiture presents us with the double bind of a creature that gained in evolution the means of knowing the self as a polythetic being and lost it in repudiation of the biodiversity upon which it depended. The magnification of the face, be it a Rembrandt painting or the latest tabloid photo of a pop celebrity, must, we insist, reveal the person. The art of caricature arises along with portraiture as an aggressive and cynical mockery of this frantic pursuit in which individuals are assaulted by deriding their appearance in terms of bestial qualities that have been made to seem ridiculous and banal after ten thousand years of life with domesticated animals. As William Gregory points out, the closer you look at the face the more animalistic it becomes because of the shared bony structure among the vertebrates.[26]

Mirrors fail us. They reflect that image whose significance is the problem, reminding us of the discomforts of puberty. Adolescent interest in personal appearance is part of a normal ontogeny. In the tribal world the impulse is redirected toward initiation in clan membership—into sacred stories that include composite animals as the instrument of relating disjunct ideas and the ceremonial use of animal masks, giving access to the multiplicity of the self and to cosmological diversity. In a world of mirrors, however, the reflection degenerates into an assessment of our vertebrate skull or the scars of life. Narcissus represents the story of the failed attempt in the quest to know the self by substituting the human image for the mosaic of a wild fauna. It was not, as the Greek myth would have it, the mistake of the hunter and totemism at all. That was a false attribution of civilized problems onto the "savage," a projection by the suffering, civilized mind.[27]

The psychiatric concept of countertransference throws light on Narcissus and the lost role of the others. An analyst may project elements of his own psyche onto the patient and then read them as if they were the patient's. By contrast, when animals are the objects of rumination about the self they reveal shared elements in the context of otherness. There is no countertransference—no ricochets like the passive surface of the pool (or the ogled looking-glass eye of a French poodle) from which Narcissus got back, to his grief, only his frustrated simulation.

No wonder the chimpanzee is such a problem. In our Pleistocene psyches we know that its image is not trivial. In spite of our post-Christian, Neoplatonic, and Cartesian self-esteem, we cannot quite dismiss the primates. Their appearance infects and subverts conviction. They seem to combine the mirrored reflection and the intransigent persistence of the other. A colleague, living with a family of Pennsylvania Hutterites, a branch of fundamentalist Christian Amish, went with them to the zoo, where they refused to look at the monkeys. They knew the trap similitude set for their dogma when they saw it: they felt the numinous power of the monkey mocking their belief that the true god looks like a man. They could repress turtles and birds, but the assault of such likeness was too much, for it carried the full weight of archetypal recognition. This chimpanzee confusion is only the most incisive instance in which we are unable to look away and yet unwilling to accept that in all animals we are seeing aspects of ourselves through a compelling, uncanny lens.

As for the Baconian / Cartesian notion of animal as mindless, to which animal protectionists often refer as the root of the abuse of animals, it was that objectivity after all which led to the laboratory experiments that in turn triggered the humane movement. Perceiving the animal as a machine does not make it hateful—it took the medieval Christians to do that—but merely reduces it to a utility. The medieval abuse of cats, owls, and toads was the result of their being loaded with demonic powers by evangelists who allowed the pagans to keep their spiritual animals in these degraded forms.

The foregoing argument raises two important questions. It may be asked first whether I have not confused the use of imaginary animals as important mental furniture with real forms. Nothing has been said here about going outdoors, wading through swamps and thickets, or watching grasshoppers. The animals in the psychogenesis of an internal self are those of the mind's eye, not the field. One may wonder whether simulations and virtual realities might not be sufficient. After all, what children actually experience is mostly pictures, toys, films, stories, and other representations. Even the crucial ceremonies of theriology among tribal peoples are centered on storied, drawn, sculpted, and costumed representations. If the occupants of future starships are to spend years away from a wild biota and scatter human colonies on moons without wild species, would not some good holographic animation be more to the point and easier to take along? What has all this development of the inner self to do with actual wild animals?

The answer involves the ecology of maturity. The developmental process is not an end in itself. Cognitive taxonomy and artifacts are indeed the tools in the perceptual work by which the whole person is achieved. But the effect of a healthy identity and maturity is realized in attitudes toward the environment, a sense of gratitude more than mastery, participation in a rich community of organisms, a true biophilia or polytheism. The images—animal guides and mediators—are the representations of an outer world that made our own being possible and toward which our maturity has its end: the preservation of the world. The obligations of having evolved in natural communities constitute a kind of phylogenetic felicity in which we acknowledge that the fish, amphibian, mammal, and primate are still alive within us and therefore have a double existence. They are present as bits of DNA, affirming kinship, and also in the world around us as independent others. The concept of biodiversity as a social value grows from

Symbolism

an inner world and creates respect for a mature ecology, that is, "climax" ecosystems with their diverse inhabitants.

The second question concerns the apparent adoption of a dualistic mode of thought in the nature/culture division. Binary distinctions are necessary for the development of the cognitive skills of categorization. But a holistic or truly ecological perspective should transcend and supplant such schemes of oppositions. Dichotomy, however, characterizes much of Western thought as a full-blown ideology, lending power to modern social and ecological imperialism. The wisdom of the nature/culture line can only be defended as a preliminary step in a process of recognizing, clarifying, and naming plurality and ambiguity—to be succeeded in more mature reflection by a structure of overriding interrelationships. At the end of *The Savage Mind*, Lévi-Strauss observes that the division is merely methodological.[28]

Kinship is the transcendent issue of maturity because of the necessary equilibrium of likeness and difference. Immaturity perceives a world of unresolved ambiguity and contradiction, as though a plurality of powers (and the multiplicity of the self) were defects. What the mature self (and Paleolithic culture) understands is that ambiguity is an intrinsic characteristic, not a deformity resulting from a Fall or a dialectical problem. Such misinterpretation marks an estrangement from others, no longer balanced by the sense of generic relationship. Domestic therapies cannot succeed any more than shock can "heal" infantile trauma or a warm puppy can restore the victim of Alzheimer's disease.[29] In his brilliant book *Masks, Transformation and Paradox*, A. David Napier describes the primal function of the mask as affirming ambiguity.[30] It is a marvelous, mitigating device of perceptual and philosophical meaning—not in the reduction of diversity but in its acceptance and in the principle of transformation. The basic mask is that of an animal, worn or carried in a ceremony or dance. Its message is that we are each two or more things at once—human and animal—an insight welcomed by our intuition. The mask and its dance remind us that transformation is closer to the heart of life than stasis or an abstract essence, that flesh and appearance mean more to our identity than ideology, that incarnation, not ideas or heaven, is what life and death are all about. The animal

On Animal Friends

mask is a premier sign of our connection with animals as the framework of our humanity because it affirms the continuity arising in transience between forms.

Animals and their representations constitute essential elements in human mental life: cognition and psychogenesis, individuation, personal and social identity, surrogate and symbolic figuration, and the conscious and unconscious iconic repertoire by which emotion and other internal states are integrated, coded, and communicated. Animals connote fields of action and power—the objects of attention acquired during the evolution of human ecology as the neurophysiological structure of knowledge and speech. The characteristic mode of most of these processes is metaphorical. They came into existence in connection with a wild fauna directly experienced, typified in art, and translated in their metaphorical implications.

The substitution of a limited number of genetically deformed and phenotypically confusing species for the wild fauna may, through impaired perception, degrade the human capacity for self-knowledge. The loss of metaphorical distance between ourselves and wild animals and the incorporation of domestic animals as slaves in human society alter ourselves and our cosmos. Without distance and difference, the others remain monsters of a terrifying jungle or, dissolved in our own unconscious minds, monsters of a chaotic and undifferentiated self.

ACKNOWLEDGMENT

Thanks to Flo Krall for her critical reading and suggestions.

NOTES

1. G. E. Hutchinson, "The Uses of Beetles," in *The Enchanted Voyage* (New Haven: Yale University Press, 1962), p. 74.
2. Claude Lévi-Strauss, *The Savage Mind* (Chicago: University of Chicago Press, 1966), p. 115.
3. This is not to deny that many tribal peoples keep captive wild animals. The distinction is between them, however, and animals bred and genetically altered in captivity.
4. Bertram Lewin, *The Image and the Past* (New York: IUP, 1968).

5. Harry Jerison, *Brain Size and the Evolution of Mind* (New York: American Museum of Natural History, 1991).

6. E. H. Lenneberg, *Biological Foundations of Language* (New York: Wiley, 1967).

7. Eleanor Rosch, "Principles of Categorization," in Eleanor Rosch and Barbara B. Lloyd, eds., *Cognition and Categorization* (New York: Wiley, 1978).

8. James Fernandez, "Persuasions and Performances: The Beast in Every Body and the Metaphors of Everyman," *Daedalus* 101(1) (1972).

9. Only since the nineteenth century have we recognized the fairy tale as a distinct genre. Perhaps it crystallized out of the larger narrative domain in which it had been embedded, possibly because of the needs of the time. The idea that animals in fairy tales are one's own organic substrate comes from Bruno Bettelheim, *The Uses of Enchantment* (New York: Knopf, 1976).

10. Eligio Steven Gallegos, *Animals of the Four Windows* (Santa Fe: Moon Bear Press, 1991).

11. One thinks, for example, of the famous studies of the Nuer (most notably E. E. Evans-Pritchard, *The Nuer: A Description of the Modes of Livelihood and Political Institutions of a Nilotic People*, Oxford, 1940), who create a conceptual cosmos from the cow. In general we tend to admire any culture that is traditional and intact and coherent. But the cow, in its earthly overabundance, devastates the environment, precipitating or confirming a rigid philosophy of duality. Among the Nuer the cow is the One. But as in all monotheisms, the cosmos as cow cannot be realized because the One inevitably evokes the Other—that which is not the cow. It is popular in academia to tell freshmen that polytheism does not exist because most peoples believe in a creator. But in real life the contrary is true: monotheism does not exist because the One inevitably implies the Other—the devil, the nonbelievers, and so forth. Perhaps the yearning for this impossible concept of unity is at the root of the modern hatred of nature—as our actual experience, to the contrary as Lévi-Strauss says, is of "the ultimate discontinuity of reality."

12. Daniel E. Koshland, Jr., editorial, *Science* 244, 16 June 1989, p. 1233.

13. H. Spurway, "The Causes of Domestication: An Attempt to Integrate Some Ideas of Konrad Lorenz and Evolutionary Theory," *Journal of Genetics* 53 (1955):325.

14. Konrad Lorenz, *Studies in Animal and Human Behavior*, vol. 2 (Cambridge: Harvard University Press, 1971).

15. In totemic society much time is spent watching wild animals, whose relationships include the normal violence of predation. Flo Krall has observed

that this observation sublimates their own violent impulses. This is quite different from the intraspecies violence on TV which, as a literal reference, is an incitement and model. Having nothing but the mild domestic animals to watch, our murderous species may keep its aggression repressed only so long before people start killing each other. In any case, as I have noted in *The Tender Carnivore and the Sacred Game*, subsistence peoples on small islands, without large mammals to hunt, regularize and institutionalize war.

16. Julia Kristeva, *Tales of Love* (New York: Columbia University Press, 1987), p. 119.

17. Ibid., p. 376.

18. Ibid., p. 212.

19. Ibid., p. 376.

20. Ibid., p. 120.

21. Ibid., p. 157.

22. Ibid., p. 159.

23. Ibid., p. 377.

24. Ibid., p. 120.

25. Ibid., p. 278.

26. William K. Gregory, *Our Face from Fish to Man* (New York: Capricorn, 1965).

27. This idea was suggested to me by Flo Krall (pers. comm.). See also Jane Flax, *Thinking Fragments* (Berkeley: University of California Press, 1990), and Harold Searles, *Counter-Transference and Related Subjects: Selected Papers* (New York: IUP, 1979).

28. Lévi-Strauss, *The Savage Mind*, p. 249.

29. Ambiguity marks the limits of taxonomy to deal with heterogeneity, especially in terms of growth, change, transformation, or metamorphosis. Inevitably there are creatures along the boundaries of definition. This is why all marginal forms are cognitively emphatic and all cultures give special attention to them and create composite, imaginary forms.

30. A. David Napier, *Masks, Transformation and Paradox* (Berkeley: University of California Press, 1986).

CHAPTER 10

The Sacred Bee, the Filthy Pig, and the Bat Out of Hell: Animal Symbolism as Cognitive Biophilia

Elizabeth Atwood Lawrence

THE HUMAN NEED for metaphorical expression finds its greatest fulfillment through reference to the animal kingdom. No other realm affords such vivid expression of symbolic concepts. The more vehement their feelings, the more surely do people articulate them in animal terms, demonstrating the strong propensity that may be described as cognitive biophilia. Indeed, it is remarkable to contemplate the paucity of other categories for conceptual frames of reference, so preeminent, widespread, and enduring is the habit of symbolizing in terms of animals. The famous dictum of Claude

Lévi-Strauss—that animals are "good to think" as well as good to eat (1963:89)—holds true not only for primitive cultures but for the most complex societies, and not only for the past but for the present.

Recently, when black motorist Rodney King was beaten by policemen in Los Angeles, transcripts of police conversations used in the officers' trial revealed "cops referring to blacks as 'gorillas in the mist' and calling King 'a lizard'" (Reynolds 1992:15A). During the 1992 political campaign, in describing the candidates, one columnist characterized George Bush as a dog, specifically a Corgi, "dashing around trying to please everyone," appearing "yappy," lacking true conscious awareness by being merely "focussed on the moment, the instant at hand," and having demonstrated "'dogged' determinism" in waging the Gulf War. Bill Clinton is seen as a cat, "fuzzy," with "feline smugness," a "weakness for pleasure," and possessing multiple lives as evidenced by "resurrecting himself at each new scandal." Jerry Brown is a ferret, a clever critter whose "existence is rather pointless." Paul Tsongas is a hamster, "cuddly and vaguely pathetic," who "runs really fast on that wheel, round and round," getting nowhere. Patrick Buchanan is "a vampire bat" who "usually comes out at night, on CNN, and drinks from the veins of helpless liberals." "Like the winged rodent . . . some of his ideas are rather batty" (Achenbach 1992:A11).

Whenever a human being confronts a living creature, whether in actuality or by reflection, the "real-life" animal is accompanied by an inseparable image of that animal's essence that is made up of, or influenced by, preexisting individual, cultural, or societal conditioning. Thus "nature," as represented by the actual biological and behavioral traits of a particular animal, becomes transformed into a cultural construct that may or may not reflect the empirical reality concerning that animal but generally involves much embellishment. Moreover, underlying the process of symbolizing a species is the extreme flexibility that characterizes the categorization of animals. That even the natural history itself is shaped to suit certain purposes is illustrated by the example of the capybara, a huge South American rodent. Observing the creatures' aquatic habits, the sixteenth-century church officials who accompanied the conquistadores into the new lands classified them as fish. Today, though science has long known capybaras are mammals, "for religious and dietary purposes" they retain their status as fish.

Therefore, their delicious flesh may be eaten in good conscience by Catholics during Lent. Venezuela ranchers have developed a profitable industry ($250,000/year) by letting these four-foot-long rodents who were once considered pests reproduce during most of the year and rounding them up for slaughter in February for consumption during the forty days of the pre-Easter ban on meat (Fur 1992:93).

Among the many animals that are endowed with rich symbolism in Western culture are three diverse species that illustrate especially well the power of metaphorical attribution in influencing and being influenced by human interactions with them. As forms of life that are heavily laden with meanings whose roots lie deep in the Judeo-Christian tradition, I have chosen the honeybee, the pig, and the bat—denizens of the sky, the earth, and the underworld—and have limited my consideration to their roles in that tradition. These animals have elicited a variety of religious and societal responses, ranging from worshipful admiration to ambivalence to revulsion, that are part of the human process of affiliating with the nonhuman realm.

The Sacred Bee

Honeybees have had some of the closest interactions with humankind experienced by any species of animal. Often referred to as "the smallest domesticated animal," this unique insect has been regarded from antiquity as especially endowed with wisdom and sensitivity and hence worthy of veneration. These perceptions developed out of appreciation for the valuable contributions that bees make to human life and the unparalleled gifts they bestow upon people. For thousands of years the honey they produced was the only known sweetening food. Because the delicious golden liquid was beneficial to health and could be transformed into a sparkling fermented drink, mead, it was called "giver of life." The bee who produces the precious substance and moreover exhibits such extraordinary social structure and behavioral traits came to be regarded as an exalted creature, ultimately sanctified and associated with life's central events—birth, marriage, and death—and with belief in resurrection (Ransome 1986:19). Bees were drawn into the human life cycle, were regarded with awe, and were perceived as possessing spiritual qualities and supernatural abilities. They became closely associated with Christianity and assumed powerful symbolic and allegorical roles within the religion.

Bees are believed to recognize, respond to, and cooperate with their caretakers (Shaler 1904:196). Out of the special rapport between bees and their human associates arose the widespread custom of "the telling of the bees," a practice that lasted into the early twentieth century and still survives in some areas. This tradition, once prevalent in Central Europe, the British Isles, and North America, mandates that beekeepers must inform their bees of important family events such as births, weddings, and deaths. To fail in these communications might mean that an infant would die or a marriage dissolve—or that the offended bees would flee the hive and die of grief and resentment (Travers 1989:81–82; Henderson 1879:309–310; Ransome 1986:218, 220–221, 271). In America, John Greenleaf Whittier wrote a poem describing "The Telling of the Bees," and Mark Twain's *Huckleberry Finn* includes a scene in which Jim explains the custom to Huck.

It was especially urgent to tell the bees about the death of their owner. For if they received no notification of it, they would disappear into the sky in order to seek their master there. This custom relates to the idea that bees are souls who can "fly up into the heavens from whence they had come." Another belief is that bees, as messengers to God, would carry the news to the spirit land of the impending arrival of their dead master (Ransome 1986: 172, 218–219).

In many areas, after the death had occurred, hives were draped with black crepe that might be lifted as the corpse passed by so that the deceased could bless them for the last time. Neglect of this procedure meant that the bees would desert the hives. It was inadvisable to buy the hives of a dead person, for the purchased bees would surely leave to follow their departed owner. Traditionally, a beekeeper should will his hives to a successor before his death. After inheriting them, the heir must make known to the creatures his status as their master, asking them to stay with him and assuring them of his continued good care. When told, the bees may begin to hum, indicating their consent. If the new master failed to tell the bees of his new position, they would follow their old master and die (Ransome 1986:161, 172–173, 221; Clausen 1962:117–118).

As further indication of the bond between bees and their keeper, portions of the funeral feast were provided to the hives. In France, the dead person was always referred to as a relative of the bees. Some of the clothes of the deceased were fastened to the hives in the hope that the bees would believe their keeper was still there and not attempt to follow him. In many areas, at the time of a marriage bees would be given a portion of the wedding cake. Pieces of the wedding dress or white linen were attached to the hives. For celebrations such as baptisms or for general rejoicing as in a good harvest, hives were decorated with brightly colored cloth (Ransome 1986:218, 220–221, 235).

The unusual intimacy between bees and their caretakers is revealed etymologically in the old German title for a bee master, *Bienenvater* or *Immenvater*, "bee father," for there is no comparable designation for those who tend other kinds of animals. In Germany and France, the singular identification of humankind with the bee is linguistically evident in the use of the same verb when a bee dies and when a person dies. All animals except bees *perish*, whereas bees and people *die*. Similarly, in parts of Germany bees are set apart because they are said to *eat* like people rather than *devour* like other animals. Interestingly, in some areas of Europe and in Palestine, bees, like persons, were customarily summoned by a hiss (*tst!*) (Ransome 1986:155, 169, 239; Wood 1880:686).

The well-being of beekeepers is interwoven with that of their bees, and the hive's prosperity is dependent upon the health of its master. The number of bees diminishes as the keeper grows old. People once made written contracts with bees assuring that they would love and care for them. Belief held that the insects would not remain with a quarrelsome family or one in which members deceived one another. Under the care of a miserly person, bees would refuse to work and perish. Bees would not live unless owners were "good and intelligent," moreover, and would pursue and sting people who shirked their duty. Swearing at bees causes them to die or to sting those who use profanity. In France and Germany, bees were said to sting a girl who had lost her virginity and an immoral man, but not idiots or people with good dispositions. In keeping with the spiritual nature of bees, they should never be bought with money but, rather, bartered or exchanged.

The Sacred Bee, the Filthy Pig, and the Bat Out of Hell

When inherited or received as a gift, hives flourish, whereas stolen bees cease producing honey and die (Clausen 1962:115, 118; Ransome 1986:169–170, 174, 227, 236, 239).

The biological and behavioral traits of bees, and the ways in which these characteristics have been perceived, explain the close bond between bees and people and fit the insects for their many roles in the Christian religion. Those who work intimately with bees find that the insects "hear music unknown to human ears" (Longgood 1985:8). Bees sing hymns, "love music and song," and are able to "talk and understand the speech of men" (Ransome 1986:175). A devoted bee master could "hear the things bees say to each other" (Stratton-Porter 1991:201). Among the Hebrews, bees were related to the idea of language. The Hebrew word for bee, *debvorah*, comes from *dabvar*, to speak (Fisher 1972:35, 38; Wood 1880:683). The etymology derives from the insects' powers of communication, which are suggested by the constant hum of their wings.

Possession of communicative abilities exceeding those of other animals has facilitated the involvement of bees in human affairs, especially in the spiritual realm. A vital factor in their sharing of human speech is the widely held notion that bees are the only creatures in the world that have come unchanged from paradise. If we remember that before the fall of Adam and the expulsion from Eden animals and people spoke a common language, the unique intercommunication between bees and people is a logical consequence of the bees' original state of grace. According to legend, when Adam and Eve were banished from the Garden of Eden, "the bees accompanied them—but with an all-important blessing instead of a curse" (Scott 1980:7).

Bees go beyond mere language to attain eloquence, a quality they symbolize. The insects' creation of honey, with its unparalleled sweetness, became a sign of special religious eloquence expressed by "honeyed words." By virtue of producing sweetness, bees are associated with the powers of orators and singers (Clausen 1962:127; Ransome 1986:155, 196, 240; Ferguson 1961:170; Charbonneau-Lassay 1991:324). St. John Chrysostom, called "Golden-Mouthed," St. Ambrose, and St. Bernard were all born with a swarm of bees hovering around their mouths to symbolize the sweetness of their preaching (Mercatante 1974:179). The capacity of honey to soothe

a sore throat makes it a basis for cough suppressors and sore throat reme-
dies. This medicinal usage continues into contemporary times and is an interesting example of convergence between belief and pragmatism.

Honey is a food of great potency in both the Old and New Testaments.
The Hebrews were promised a "land flowing with milk and honey" (Exo-
dus 3:8), as honey was symbolic of fertility and abundance (Cansdale
1970:245). John the Baptist lived in the wilderness on a diet of locusts and
wild honey (Matthew 3:4). Ancient people looked upon the making of
honey, whose derivation comes from the Germanic root meaning golden,
as miraculous, and even today the substance retains a quality of mystery regarding some of its constituents. It is said to be the only food that requires
no digestion, as it passes directly into the bloodstream (Norman 1990:4;
Longgood 1985:195). Since honey is produced from the nectar that bees collect from flowers, it is associated with the image of purity, fragrance, and
beauty. Flowers impart their own symbolic connotation of rebirth and immortality to the bees. Plants and bees are interconnected biologically
through their role in pollination. In poetic terms, "bees do the flowers'
courting for them" (Stratton-Porter 1991:198).

Honey was important in many religious rituals, especially those dealing
with birth and death. In the Christian sacrament of baptism, honey mixed
with milk to represent infancy signified a ritual second birth and stood for
a foretaste of heavenly sweetness and bliss, an indication that the baptized
person would reach the land of promise. Regarded as "the dew of heaven,"
honey was made by the "bees of the paradise of delight" from the "dew of
the roses of life that are in the Paradise of God." Thus eating the celestial
substance during baptism ensured immortality. Related to this symbolism
is the widespread use of honey as an offering to the dead (Ransome
1986:161, 279, 281–283).

Honey is considered "pure," for there is no way to synthesize, cheapen,
or improve it. One of its most striking characteristics is its stability, for it
does not spoil like other foods but remains antiseptic. A contemporary authority calls it "one of nature's most powerful germ killers," claiming that
"germs simply cannot survive in honey" (Parkhill 1980:36; Clausen
1962:121). Thus it has been used as a preservative and for embalming the
dead. Because of its antibacterial capacity, as well as beliefs about its origins

The Sacred Bee, the Filthy Pig, and the Bat Out of Hell

and powers, honey has been important in healing from ancient times through the present. Medicinal uses include therapy for wounds, especially those of the mouth and throat, treatment for ulcers, and application as a skin balm. Since antiquity, honey has found favor as an aid to digestion and has been esteemed for its invigorating and health-promoting capacities (Clausen 1962:121–123, 125). Even the sting of the bee is considered therapeutic, and today bee venom still retains its reputation as an arthritis remedy.

Because it is manufactured by a sacred and gifted insect, beeswax holds a special place in religious observances. Like honey, it lasts indefinitely. Great quantities are used by the Catholic church—indeed in some areas, due to the blessed nature of the bees who drew their origin from paradise, Mass cannot be celebrated without beeswax candles providing light for the altar. Wax candles symbolize the virginal body of Christ, since the bees who make it frequent the best and sweetest-smelling flowers. The wick signifies the soul and immortality of Christ and the light represents his divine person (Clausen 1962:127; Ransome 1986:148).

The role of honeybees in Christianity is not confined to association with honey and wax production but is intimately related to the insects' social structure and behavior. The organization of the hive represents what Maurice Maeterlinck called "the most orderly community in the universe" (1954:1). As a parallel with humankind, honeybees form societies with specialization of labor, permanent settlements, and the production of food. They are among the finest of animal architects, building their dwelling place by means of remarkable feats of engineering that are still incompletely understood, the hexagonal cells of their honeycombs "combining beauty, great strength, and efficiency" (Longgood 1985:64). As a noted scholar of domestication points out, human beings and bees are both "self-domesticating" species in which some members live at the expense of others, forming a unit in which all benefit from division of labor (Zeuner 1963:506).

Observers from the earliest times noted that in producing honey, bees work tirelessly to meet the needs of the hive, individuals laboring together for the common cause. The insects came to represent the Christian virtues of industry, order, purity, economy, courage, prudence, and cooperation.

Symbolism

The perceived asexual nature of the honeybee has been used from antiquity as a parable of moral behavior. The great poet Virgil, who anticipated Christianity, promulgated Aristotle's idea that bees do not copulate or engage in lovemaking, nor bear their young in travail, but rather "unmated / Gather their children in their mouths from leaves and fragrant herbs" (1982:130–131). An early Christian text praises bees "who produce posterity, rejoice in offspring, yet retain virginity," making the insects' purity a model for chastity (Ransome 1986:144).

Reflecting the assumptions of his times, Virgil celebrated the "king bee"—for the unique and indispensable member of the hive, the queen, was believed to be male until early in the seventeenth century when the truth was discovered that the most important bee was female. Before the spread of that knowledge and its acceptance, the alleged "king" ruled the hive, and the activities of bees were compared to the lives and duties of Christians, especially monks and clergymen. Hives suggested cloisters, and monasteries generally kept apiaries, the bees serving as models of industrious creatures of God who lived in pious unified communities. Worker bees, dedicated to the colony, were likened to clergy who took vows of chastity, poverty, and obedience. St. Jerome advised: "Make Hives for bees. . . . Watch the creatures and learn how to run a monastery and control a kingdom!" (Mercatante 1974:179). The hive was an "ideogram of a community life that is wisely ruled, peaceful, and fruitful under the governance of one single head." The church with the pope as leader was compared to the hive: "Within its shelter, the pontiff must preserve the sacred teaching which is symbolized by honey, and set forth the discipline which gives each one his place in the Christian society" (Charbonneau-Lassay 1991:325). Just as one king bee lived in virgin purity in an ordered commonwealth, it was argued there should be one "king pope." Because the king bee seldom used his sting, it followed that bishops should be mild and gentle. Honey symbolized Christ's gentleness and compassion, whereas the bee's stinger represented his role as judge of the world. A person stung by a bee could also be receiving a message from a soul in purgatory who was requesting prayers (Ransome 1986:145–146, 149; Ferguson 1961:12; Longgood 1985:177; Matthews 1986:21).

Christ was symbolized as a honeybee. His vivifying spirit in the church was likened to the role of the queen in the hive. Especially in his miraculous birth Jesus resembled the bees who brought forth their young through their mouths as he sprang, allegorically, from his father's mouth. The virtue of cleanliness was possessed in common, for bees live and reproduce through association with fragrant plants and avoid anything dirty. His immaculate conception found ideal expression in the alleged asexual reproduction of bees. The ancient belief that bees sprang from the carcasses of dead oxen also emphasizes the bees' chastity, for oxen, being castrated, possess no sexuality (Charbonneau-Lassay 1991:324–325; Ransome 1986:146–147, 153–154; Matthews 1986:21).

The honeybee symbolized Christ's incarnation, and its capacity for parthenogenesis represented the virgin birth. The bee stood for the virginity of Mary, who was apostrophized as the "Mother Bee." Just as the bee produces honey, the Blessed Virgin produced the Savior. Mary was envisioned as a beehive, for in her womb she had carried Jesus in all his sweetness (Matthews 1986:21; Ransome 1986:148; Travers 1989:81). Even after the scientific facts of their reproduction had been determined, bees continued to symbolize sexual purity, since the queen and the one drone who mates with her are the only individuals in the hive who experience sexual union. All other members remain chaste. Thus the beehive stands as an example of a community with restriction of sexual activity for the good of the society.

Honeybees are closely associated with the parents of Jesus. Mary, invoking the name of the Holy Trinity, once persuaded a swarm of bees to settle in a hive that Joseph had made, promising them health and prosperity. An old charm recited to bees reveals that Mary raised her hand to cause the "dear bees" to fly about and obtain honey for the Christ child. A Central European tradition dictates that a person who refuses to give honey to children sins against Mary and Joseph. Beehives in areas of Europe were sprinkled with holy water and smoked with incense on the eve of church festivals. The bee worships God not only by producing wax for altar lights and honey for human consumption but by humming anthems of praise on holy occasions (Ransome 1986:165–166, 169, 173, 217; Radford and Radford 1974:38).

Because of their association with Christ, honeybees were thought to have originated from drops of water from his hands, to have come out of his forehead, or to have sprung from the tears he shed on the cross. The number three, sacred to Christianity, is represented by the honeybee and its life cycle. Three main parts—head, thorax, and abdomen—make up the insect's anatomy. Parallels were drawn between the hatching of honeybee eggs in three days and the body of Christ remaining in the tomb for three days. The reappearance of bees from the dark hive after three months of winter darkness was a sign of resurrection and immortality. Tradition dictated that a newcomer to beekeeping should begin with three hives, the holy number. Since belief held that bees never slept, they represented Christian vigilance and zeal in acquiring virtue. The diligence with which they store up honey, producing enough food in summer to nourish the colony during the winter, became symbolic of the Christian working to lay up treasure by accumulating merits on earth in order to ensure life everlasting (Ransome 1986:238, 245, 247–248; Charbonneau-Lassay 1991:323, 329; Cooper 1978:19; Ferguson 1961:12; Leach 1949, 1:130).

The community of the hive in which the interests of the individual are subordinated to the common welfare provided inspiration for Christians to work tirelessly for the realization of Christ's kingdom in which personal interests must be sacrificed for social progress. Although, according to one authority, "no other species is known to lavish such attention on their newborn" (Longgood 1985:79–81), the effort is ultimately for the benefit of the hive. Through intricate mechanisms that assure the succession of generations, bee society gains an ongoing existence, a kind of collective immortality. It is only through identification with the larger whole that an individual becomes immune from personal extinction. Bees cannot live alone, apart from the colony, and by working together as a unit their corporate identity represents the essential oneness that "lies beneath the surface of life" (Chetwynd 1987:40).

For Christians, bees represent immortal souls. Migrating from their hives in swarms, they symbolize souls swarming from the divine unity. They carry messages between this world and the realm of the spirits (Cooper 1985:70). St. Bernard saw bees as "an image of souls which . . . lift themselves on the wings of contemplation" (Charbonneau-Lassay 1991:328). As noted by P. L. Travers, "The bee has at all times and places been the symbol of life—life as immortality." Languages reflect this concept:

"The Cornish 'beu,' the Irish 'beo,' and the Welsh 'byw' can all be translated as 'alive' or 'living.'" The "Greek 'bios' and the French 'abeille'" are also "akin to these" (1989:81).

Christianity incorporated old traditions that acknowledge the insight and knowledge of bees and attribute to them "the inspiration or wisdom which discerns the essence of life" (Chetwynd 1987:40). Modern scientific knowledge confirms the existence of the honeybees' remarkable abilities that have been intuited throughout the ages. Indeed today their powers are considered "the most astonishing achievement of nature," for bees exhibit not just instinct but true intelligence (Longgood 1985:203).

The research of Karl von Frisch, first published in the 1950s, proved that bees possess communicative abilities that make them unique among animals. Through studies that have stood the test of time and have been confirmed by other scholars, von Frisch showed that by means of "dances" honeybees communicate to their fellows precise information regarding the distance, direction, and desirability of available food sources and other important objects the dancer has discovered. The bees' "superb intelligence system" represents "the most complicated language of signs and symbols in the animal kingdom" (Sparks 1982:185, 188). Modern research has revealed that bees are even more remarkable than von Frisch imagined, for they have "formidable powers of learning" (Sparks 1982:188). Honeybee behavior indicates conscious thinking and intentional action. Cognitive ethologist Donald Griffin concludes that bees indeed fulfill the criteria for possessing mind, since they communicate about objects remote from the immediate situation where the communication occurs. Bees demonstrate an analogy to human linguistic exchanges, as they "speak" and "listen," with a system that is close to human speech in symbolization and flexibility (1981:6, 18, 41, 46, 49).

In his classic monograph on the honeybee written about half a century before von Frisch's studies, Maeterlinck extolled the intelligence of bees and wrote that "the discovery of a sign of true intellect outside ourselves procures us something of the emotion Robinson Crusoe felt when he saw the imprint of a human foot on the sandy beach of his island. We seem less solitary than we had believed." In our endeavor to understand the intellect of bees, he noted, "we are studying in them that which is most precious in our own substance"—our mental capacities (1954:76–77).

According to the New Testament, "In the beginning was the Word, and the Word was with God, and the Word was God" (John 1:1). Thus in Christian belief language is divine, and because of linguistic expertise humankind has generally considered itself different from and superior to the rest of creation. But the dance of the honeybees constitutes language and thus challenges human uniqueness (Barth 1991:283–284). It is noteworthy that the Bible compares God's word to honey as sustenance for man, indicating that his ordinances are sweeter even than honey and the honeycomb (Psalm 19:10). Taking into account the mental qualities of the honeybee, its complex social organization, its role in pollination, and its production of substances that are useful and pleasurable to humankind, the reverence inspired by this unique creature is not surprising. What Virgil said of bees— that they possess "a share / Of the divine mind and drink ethereal draughts" (1982:131)—expresses a truth that has become more evident with the passing centuries.

Bees were once regarded as friends and protectors of a house or town. In Germany they were called "the Birds of God," or "Mary's birds," and were "in communication with the Spirit" (Baring-Gould n.d.:14). A prevalent belief likens the covenant between humankind and God to that between people and bees (Stratton-Porter 1991:254). A practice once existed in America whereby a Catholic priest would place a morsel of the consecrated wafer from the Mass into the hive for the bees (Ransome 1986:217). From earliest times, people everywhere have regarded the honeybee as God's gift (Charbonneau-Lassay 1991:319), and this idea persists into the present day. In Yugoslavia, belief holds that bees are divinely given. If some calamity befalls a hive, that is an act of God which the beekeeper must accept. Bees are said to choose their host, and custom still dictates that only a person of physical and spiritual purity dressed in clean clothes can approach the hive; no one who drinks, swears, shouts, or steals can retain the bees (Domacinovic and Tadic 1991).

Although the symbolism surrounding bees that was so prevalent in Western culture during the Middle Ages gradually declined and began to

The Sacred Bee, the Filthy Pig, and the Bat Out of Hell

disappear early in the present century, the old aura persists. In a contemporary American bee journal, for example, bees are called "little angels of agriculture" for their role in pollinating crops (Graham 1992:259), retaining religious terminology that preserves the idea of a supernatural role. Gene Stratton-Porter's classic, *The Keeper of the Bees*, written in 1925 and reissued in 1991, perpetuates the image of the honeybee as a paragon of morality and a teacher of humankind. Against a backdrop of bees as perfect beings who feed on a specially planted garden of blue flowers—the "perfect color"—parallel human values are extolled. Clean outdoor living in accord with nature, good health, cleanliness, purity, restriction of sex to marriage, reverence for women as mothers, and strictly delineated gender roles are essential to the good life. Belief in God is central. Only God, as the "Master Mind," could have designed the beehive. Strong sympathy exists between bees and their keeper, who is able to "magic" them to control their behavior, and they sting only the villain (1991:196, 211, 410–411).

Use of the bee for Christian moral instruction continues into the present. A 1992 Sunday School resources catalog features Honeybee Bible Lessons and sells the items needed to start a Honeybees Bible Club, including lesson plans, teaching outlines, and T-shirts with a bee logo (Miley 1992:4–5). A birthday card for Sunday School children featured in a religious store depicts busy bees "all abuzz," carrying presents to the recipient, who would "do a lot of fun things" on the special day. A contemporary children's sermon asks youngsters: "Isn't God wonderful, the way in which he makes the bees so that they know so much, and do so much?" The bee, who "has a wonderful brain," is not to be feared, for he "will not bother you if you don't bother him." The minister celebrates bees for their "many trades," for the fact that "each one sticks to his own trade," as well as for their praiseworthy "busy-ness," cleanliness, and unselfishness. Children are sternly cautioned not to be like the "boys and men of the bee cities," the drones, "lazy good-for-nothings" who are "contemptible"—being "untidy, greedy, and selfish." The fact that the sentry bees and workers ultimately take revenge on the drones by killing them or turning them out to starve is given didactic force. Christian adulation for bees is strongly emphasized as the lesson ends: "Jesus came to show us that the loveliest work anyone can do is to work like the worker bees: to help one another, and to make others happy. After He had risen from the tomb on Easter, he met His pupils on the lake shore, and asked them if they had anything to eat. They gave Him some cooked fish and a piece of honeycomb! I am sure he said, 'Thank you, God, for food so sweet!'" (MacLennan 1966:273–275).

The Filthy Pig

As von Frisch said of the honeybee, "each species is a magic well—the more that you draw from it, the more there is to draw" (Russell 1991:11). The clear waters from which Western culture has drawn the pure image of the bee are decidedly muddied in the case of the pig. Expressions such as making a bee-line, holding a husking bee or spelling bee, and being busy as a bee all derive from desirable qualities emphasizing the bees' capability of flying directly to sources of nectar, their spirit of cooperation, intelligence, and social solidarity, and their industrious behavior. Contrasting these metaphors with epithets such as filthy pig, fat pig, and descriptions like pigheaded, pigging out, and living in a pigsty reveals much about differences in symbolic perceptions of the two species.

Unlike the celestial bee, the pig is earthbound. Rather than visiting fragrant flowers, he is observed rooting in the dirt and wallowing in the mud or even in his own excrement. Although pigs have their defenders and admirers, these stand out as exceptions to the general view in the Judeo-Christian tradition and Western society according to which the porcine animal is vile and repulsive. As pig advocate William Hedgepeth notes, "there is no other earthly life-form toward which men so eagerly apply so many overgeneralized negative judgments" (1978:198). And according to Franco Bonera's appreciative study of the species: "No animal on earth is more unjustly treated than the pig. Abused, mocked, insulted, vilified, exploited—and in the end, slaughtered" (1990:6). Thus, at the opposite pole from the tiny, elegant, nonedible bee who is God's elect and is perceived as innocent, intelligent, industrious, altruistic, and socially cooperative; the huge, corpulent, dirty, foul-smelling, clumsy, stubborn, stupid, lustful, selfish, and slothful hog, whose succulent flesh is eagerly consumed in so many forms, is cursed and reviled by both God and man, the object of scorn and taboo.

Religious and secular aspects of animal taboos are discussed in anthropologist Edmund Leach's celebrated essay about animal categories and verbal abuse. Leach sheds light on the matter of animal terms that constitute insults and express obscenities. Human beings are equated with animals, he points out, because certain species are the focus of attitudes that relate to edibility and are linked to taboos and rules concerning the killing and eating of those animals. Whatever is taboo holds special interest and is the cause of anxiety. Taboo animals can be sacred and powerful, but they are also regarded as untouchable and filthy. Humankind's shame in killing animals of substantial size causes their carcasses to be called by different names: dead cows become beef; dead pigs become pork, bacon, or ham. "You swine!" is a potent epithet in Western society due to the status of the pig. Leach points out that

> some animals seem to carry an unfair load of abuse. Admittedly the pig is a general scavenger but so, by nature, is the dog and it is hardly rational that we should label the first "filthy" while making a household pet of the second. I suspect that we feel a rather special guilt about our pigs. After all, sheep provide wool, cows provide milk, chickens provide eggs, but we rear pigs for the sole purpose of killing and eating them, and this is rather a shameful thing, a shame which quickly attaches to the pig itself. Besides which, under English rural conditions, the pig in his backyard pigsty was, until very recently, much more nearly a member of the household than any of the other edible animals. Pigs, like dogs, were fed from the leftovers of their human masters' kitchens. To kill and eat such a commensal associate is sacrilege indeed! [1975:28–29, 47–48, 50–51]

The Old Testament taboo against the consumption of pig flesh and even the touching of pigs is one of the most fascinating and provocative of human prohibitions. As delineated in the Book of Leviticus, God revealed to the people of Israel through Moses and Aaron the rules according to which certain animals must not be eaten. "Every animal that has divided hoofs but is not cleft-footed or does not chew the cud is unclean for you; everyone who touches one of them shall be unclean" (11:26). "The pig, for even though it has divided hoofs and is cleft-footed, it does not chew the cud; it is unclean for you. Of their flesh you shall not eat, and their carcasses you shall not touch; they are unclean for you" (11:7–8).

Symbolism

How is the biblical unclean status of the pig to be interpreted in the light of the human propensity to symbolize through animals—to exhibit cognitive biophilia with a negative as well as a positive orientation? Many explanations have been put forward. Some postulate that the Jews simply turned away from pigs because the animals were used in the rites of the hated Egyptians who had held them in bondage, or because the animals were sacred to certain heathen deities. A commonly cited reason for the Hebrew condemnation of pigs is hygiene—the belief that the animals were vectors of trichinosis (not scientifically proved until the nineteenth century) or other diseases such as leprosy. Biological rationalizations of the banning of pigs are rooted in the fact that the animals are unsuited to the Jews' hot, arid environment and could not survive the rigors of their nomadic life. According to the cultural materialist viewpoint championed by Marvin Harris, the omnivorous pig is "a direct competitor of man" for food and is not a "practical source of milk." The Hebrews were culturally adapted as nomadic pastoralists, and pigs are "notoriously difficult to herd over long distances." In sum, then, "the divine prohibition against pork constituted a sound ecological strategy," for "pig farming was a threat to the integrity of the basic cultural and natural ecosystems of the Middle East" (1974:40–42). Other reasons cited for the taboo arise from the notion that the porcine habit of feeding on garbage disgusted the fastidious Jews and the idea that the animals became moral lessons for the avoidance of vices like gluttony, sensuality, and sloth, possibly acquired by eating pork.

Political, aesthetic, hygienic, or biological causes, and factors relating to cultural materialism, however valid as surface phenomena, still do not satisfactorily address the taboo surrounding the pig. For a more valid explanation, it is necessary to take into account the powers and imperatives inherent in humankind to create a symbolic universe—to make sense out of existence through the imposition of a cognitive order. In essence, the rationalist motivations just described do not speak to the same level of human experience that is vital to the domain of religious belief (the level that constitutes "what the bee knows"—images from the wellsprings of a deeper reality mediated by symbols). As the editors of the *New Oxford Annotated Bible* point out, the difference between clean and unclean animals is not based on sanitary or hygienic considerations in the modern sense of "cleanliness is next to godliness." Rather, unclean means "ritually impure, and therefore the opposite of holy" (Metzger and Murphy 1991:137).

In exploring what she calls "the abominations of Leviticus," anthropologist Mary Douglas elucidates the question of why certain animals are not holy. As she so aptly points out: "Even if some of Moses' dietary rules were hygienically beneficial it is a pity to treat him as an enlightened public health administrator rather than as a spiritual leader" (1976:29). The concern is holiness, and holiness and impurity are at opposite poles. Old Testament injunctions, Douglas asserts, do not deal with hygiene, aesthetics, morals, and instinctive revulsion. Each is prefaced by the command to be holy, and the contrariness between holiness and abomination makes sense of the restrictions. For the Israelites, blessing was the source of all good things, and God's work through the blessing was essentially to create order by which human affairs prospered. Infringement of God's rules was the source of all dangers (1976:49–50).

The key to understanding Mosaic dietary laws is this: like other taboos involving impurity and defilement, they are symbolic systems. Symbols of pollution like the pig, Douglas argues, represent anomalies—elements that do not conform to their class or category and hence signify disorder. Pollution behavior is "the reaction which condemns any object or idea likely to confuse or contradict cherished classifications." Anomalous entities evoke anxiety and thus are avoided or stigmatized with taboos (1976:34, 36). Holiness requires that individuals conform to the class to which they belong, and this means keeping distinct the categories of creation. "Hybrids and other confusions are abominated" and ambiguous species are unclean. Holiness involves order, unity, and integrity; the basis for clean and unclean meats is a metaphor of holiness. Their herds of cattle, sheep, and goats constituted the livelihood of the Israelites and these forms of livestock were regarded as clean animals. As is the case with most pastoralists, "cloven-hoofed, cud-chewing ungulates" became "the model of the proper kind of food." It is noteworthy that the failure to conform to the two criteria is the only reason given in the Old Testament for avoiding the pig; its dirty scavenging habits are not mentioned. Because those species are unclean that are imperfect members of their class, or whose class confounds the general scheme of the world, the pig's lack of cud-chewing determines its taboo status. For the Israelites, the dietary laws were signs that "inspired meditation on the oneness, purity and completeness of God. By rules of avoidance holiness was given a physical expression in every encounter with the animal kingdom and at every meal" (1976:53–55, 57, 73).

Noah, the Old Testament patriarch chosen by God to preserve each animal species, is said to have utilized the pig "to clean up the filth that had accumulated in the hold of the Ark" (Lavine and Scuro 1981:9). Throughout the Bible, pigs are denigrated. "Like a gold ring in a pig's snout is a beautiful woman without good sense" (Proverbs 11:22) encapsulates a clear moral lesson through oppositions. The New Testament perpetuates the hog's bad reputation. The oft-repeated admonition "Do not throw your pearls before swine, or they will trample them under foot and turn and maul you" (Matthew 7:6) is an image embedded in Western culture. The story of Christ casting demons out of people and into a herd of swine is repeated three times in the Gospels (Matthew 8:30–32; Mark 5:11–16; Luke 8:32–34). Peter warns that a person who accepts Christianity and then turns back to a sinful life is like the sow that is washed only to go back and wallow in the mud (2 Peter 2:20–22). And the didactic force of the celebrated Prodigal Son's ultimate degradation focuses on his detestable occupation as a swineherd and his willingness to eat the pigs' food (Luke 15:15–16). Conspicuously absent from Nativity scenes is an adoring pig among the traditional oxen, sheep, asses, and camels at the manger.

In Christian religious contexts during the Middle Ages, pigs, signifying earthly vice, were vehicles for malice. To ridicule the pope, Henry VIII depicted him as a hog wearing a tiara. As an attack on Judaism, churches in eastern Germany, Basel, and Salzburg once used carvings showing Jews sucking milk from sows, the animals they scorned. The Church of Rome represented its fight against Judaism as a duel in which the enemy, the Synagogue, is astride a pig. Ecclesiastical decorations and carvings often made use of swine to signify sin, foolishness, lust, greed, and other vices. The sow symbolized sinners and unclean persons, heretics, wanton behavior, and foul thoughts. In illustrating the seven deadly sins, the vices of anger, unbridled passion, and greed were personified in one church by figures mounted on pigs. Pigs not only illustrated moral lessons but satirized certain ecclesiastical officials or prevalent practices. In numerous instances, pigs are shown playing musical instruments, and in one case the piglets who listen to their mother's music are said to symbolize "worldlings who lose spiritual opportunities through preoccupation with the needs of the flesh." In Norman sculpture, the forces of evil opposing Christianity are represented by the boar of the 80th Psalm (8–13) who ravages the vine of Israel (or Tree of Life). A symbolic carving at Rouen cathedral shows a woman emptying a basket of marguerite blossoms before her pigs. The meaning derives from the fact that the Latin for pearl is *margarita*, and thus she is flouting the injunction from Proverbs not to cast pearls before swine. Some porcine figures may represent the swine into which Christ cast the demons. Possessed by evil spirits, the pig became the devil incarnate—indeed, proof of this status was found in the animal's forefeet. Between its front toes are small holes appearing as though burned into the skin (actually scent glands). Traditional belief held that the devil entered through these openings, leaving his clawprints that were henceforth called the devil's marks. In this regard, the pig is associated with another animal often held to be in league with the devil: the rattlesnake. Impervious to the rattler's bite, pigs can even eat the hated reptile, for the venom runs out through the holes in their forefeet (Hedgepeth 1978:191–192; Bonera 1990:43; van Loon 1980:27; Ash 1986:42, 75; Sillar and Meyler 1961:16, 18, 20–26; Lavine and Scuro 1981:25).

Pigs occur in the lore of saints. Satan once appeared to St. Anthony the Abbot in the guise of a pig in an attempt to seduce him with the pleasures of sin and persuade him to renounce his faith. Recognizing the "close links between this creature and eternal damnation," who symbolized vices and temptations, Anthony repulsed the devil. Once Satan had been exorcised, the pig became a docile pet who remained with the saint. The pair are often depicted as an allegorical representation of the victory of good, as portrayed by Anthony, over evil in the form of the pig. Anthony became the patron saint of swineherds (Bonera 1990:62–65; Bowman and Vardey 1981:55; Ash 1986:40).

Influential scholars of natural history, who might have been expected to redeem the pig from such ignominy, instead reinforced negative religious and popular perceptions, as exemplified by classic Roman and eighteenth-century British and European sources. In A.D. 77, Pliny the Elder in his *Natural History* called the pig "the stupidest of animals." In 1788, celebrated French naturalist Georges-Louis Leclerc, Comte de Buffon, stated that "of all quadrupeds, the pig seems to be the ugliest animal; its imperfections of form appear to influence its nature; all its habits are clumsy, all its tastes are filthy; all its feelings amount to no more than violent lust and brutal greed which make it devour indiscriminately anything it happens to find." British artist/naturalist Thomas Bewick (1753–1828) wrote of the species that it "is of all other domestic quadrupeds, the most filthy and impure. Its form is clumsy and disgusting and its appetite gluttonous and excessive" (Bonera 1990:9–10; Sillar and Meyler 1961:1).

Though their defenders point out that pigs are clean by nature and only become "filthy hogs" through the intervention of humanly imposed conditions of domestication and husbandry, porcine biology does predispose the animals to a disdained image. Unlike other farm species, pigs root in the earth. Their ancestors were forest animals, living in the shade, adapted to a relatively cool climate. Since they have few sweat glands (contradicting the common expression "sweat like a hog"), in warm climates they need a constant supply of water for cooling their bodies. Thus pigs seek relief from excessive heat by wallowing in the mud in order to form a protective layer against the sun. Pigs have voracious appetites and generally push aggressively at the feeding trough. They may shove food around with their snouts, reportedly to release the aroma, a trait that offends human sensibilities. Because they are omnivorous, pigs can subsist on garbage, an attribute that leads to low status and makes their habitation notoriously foul smelling. Hogs' grunts are less pleasing to human ears than neighs, moos, and baas. Their fat bodies make them repulsive and their sparse hair deprives them of the aesthetically pleasing sleek coat of other familiar animals. They have small eyes, a feature distasteful to humans, and their facial expressions often appear dull and drowsy.

The stereotype of porcine stupidity persists in spite of contrary evidence of a high level of intelligence compared to other animals. Far from having dull senses, pigs used to obtain truffles can detect the presence of the delicious and exorbitantly expensive fungi growing 10 inches deep from 20 feet away. Pigs have been useful as hunters—one British porker named

"Slut" gained renown for pointing and retrieving game. Owners of pet pigs, who have become more numerous since the recent popularization of the Vietnamese pot-bellied variety, praise the intelligence, affection, and loyalty of their companions and find them not only smart but also obedient, protective, and adaptable to family life. In 1984, a pig named "Priscilla" showed her devotion by rescuing a child from drowning and was honored with the American Humane Association's award for animal heroes. "Sapient and learned" pigs once served well in circuses, performing intricate feats and earning a solid reputation for trainability. (See Ash 1986:9, 25; Britt 1978:398; Towne and Wentworth 1950:20; Atwood 1982:52, 58; Warshaw 1986:82–83; Coudert 1992:150–154; Pet Pigs 1986:4; Jay 1987:9–23, 25–27.)

Presently, though, pigs in performance have lost their "learned" status and now take a different role in modern circuses and rodeo novelty acts. As demonstrated by Paul Bouissac in his semiotic study of circus, clowns often "address very sensitive areas in our cultural system" (1990:195). A frequently enacted circus scenario involves a clown in grotesque female attire who enters the arena pushing a baby carriage. Suddenly loud cries are heard and the "mother" asks a nearby person to fetch some milk. Circus hands bring a huge bottle (or tank) full of milk. The rubber pipe attached to the bottle is put into the carriage, presumably into the baby's mouth. The bottle is rapidly emptied, and cries are again emitted. In response, the clown picks up the "baby," which turns out to be a small piglet. As the animal is handed over to another person, it urinates abundantly, and is carried off still dripping. The clown is then expelled from the ring in mock disgrace accompanied by much laughter. In this act, the piglet's behavior pertaining to ingestion, elimination, and vocalization is similar to an infant's, stressing likenesses between raising a child and a pig. But society takes great pains to separate humans from pigs, creating demarcations that are blurred by the clown. The severity of the confusion is enhanced by the fact that the pig is our most edible animal, raised solely for slaughter. Absurdity results from equating a human infant, a cherished being, with a pig—the object of disgust in Western culture (Bouissac 1990:200–202).

As Bouissac points out, "Many characteristics both anatomical and ecological concur to situate the pig in the immediate proximity of humans: the pigmentation of its skin; the unusually light body hair which makes it appear almost naked; the relative expressivity of its face which lends itself so well to anthropomorphization; the fact that in traditional subsistence farming it is kept closer to human habitations than the other domestic animals and is fed in a human way with leftovers"; and "the rumor that nothing tastes more like pork than human flesh." Thus "the closely intertwined lives of pigs and men lead to the necessity of strongly disjuncting them from each other on the cognitive and symbolic levels." Accusations of behaving like a pig or being a pig signify exclusion from humankind. Yet, paradoxically, " a pig sleeps in every man's heart" (1990:201).

Rodeo clown acts reflect the same themes. During my fieldwork on human / animal interactions in rodeos, I often saw a classic clown and pig act in which the clown shouts to the announcer: "There's a mad pig loose on the fairgrounds. It's a thousand-pound Russian boar hog with tusks two feet long. It's got rabies and swine flu!" Then the "dangerous animal" appears—a baby pig named Mary who follows the clown as he feeds her with a nursing bottle. The announcer comments: "Yes, Mary is quite a ham, and there are two hams with her" (two additional clowns in the arena). Thus the frightful monster is only an innocent baby; but the message is clear that the pig is, after all, meat. The interchangeability of pigs and people and the use of pigs for ridicule are demonstrated by rodeo acts such as the one in which a performer comes to the stage holding a piglet he feeds with a nursing bottle. He reveals to the audience: "I had a blind date last night, and this [the piglet] was it." He follows that statement by announcing, "I've traveled with enough pigs." Then, turning to the suckling animal, he tells it, "Come on, little pig, grow up!" (Lawrence 1982:199–200).

The symbolic status of the pig may be related to some remarkable similarities between the animal and human beings. In contrast to the very different bees who frequent a remote, mysterious realm of sky, flowers, and hives, pigs, before factory farming, were near neighbors, backyard dwellers, sharing food and space. Familiarity may indeed breed contempt. Ironically, pigs have a physiological likeness to humankind that is surpassed only by primates. Glandular and chemical derivatives from pigs are commonly administered to alleviate pain and extend human life. There are anatomical similarities in the digestive system, teeth, blood, and skin of the two spe-

The Sacred Bee, the Filthy Pig, and the Bat Out of Hell

cies. Pigskin is highly beneficial in the treatment of burns. Pigs are used to test the effects of radiation. Porcine heart valves are frequently implanted in cardiac patients. Both pigs and humans develop peptic ulcers as a result of tension and confinement. Pigs, like people, will consume intoxicating liquids and hence are used as experimental animals for alcoholism studies (Britt 1978:406, 411–413; Hedgepeth 1978:172–179).

The curious relatedness of pigs and people is illustrated by the old belief that if one of your pigs dies suddenly, someone in your family will become ill, as well as the tradition of transferring mumps from a sick child to a pig by rubbing the patient's head on the pig's back (Opie and Tatem 1989:306). Probably the physical closeness between pigs and people fosters a sense of identification, causing the animals to become vehicles for human feelings. Raising hogs has always been a profitable venture, but the animals have rarely received the gratitude they deserve for elevating their owners' standard of living. Humans may feel a deep guilt for being so mercenary about pigs. Hogs are generally only commodities without respectability or identity, and they are harvested without a qualm. Seldom is another species served so grotesquely—its head on a platter with an apple in its mouth. Whereas other food animals may also provide milk, fiber, or traction, pig husbandry involves only meat. People love pigs only when dead: "The hog is never good but when he is in the dish" (Ash 1986:47). As repositories for our own fears of ourselves and the animal within us, pigs bear the brunt of our self-reproach. In transferring human frailties to them through verbal abuse, they become scapegoats. Undoubtedly, we project onto hogs our own perceived vileness; hence they may literally "die for our sins." Pigs not only embody our shame and uncertainties but may represent our fear of dying. Hogs' bodies do not go back to earth when they die. As William Hedgepeth points out, they are totally liquidated after slaughter: they have no graves other than in our own bodies (1978:262). Lynn White, in his celebrated essay about the religious roots of our ecological crisis, suggests that butchering pigs is symbolic of human mastery over nature (1973:24).

Literature reflects the symbolic connection between pigs and people. In William Golding's *Lord of the Flies*, for example, "Piggy," the fat, clumsy boy who is scorned by his peers, is a sacrificial victim, just as are the wild pigs that the children hunt and kill. No reader of Thomas Hardy's *Jude the*

Obscure can ever forget Jude's reaction to his task of slaughtering a pig—a fellow mortal he had fed with his own hands. Branded as a tender-hearted fool, Jude's empathy places him outside of society, an outcast. Going from the sublime to the ridiculous, popular culture features "Miss Piggy" of the Muppets—appealing but surely pig-headed. And "Pigpen," from the comic strip "Peanuts," is forever dirty.

Although in the modern urbanized world, far from the backyard pig-sties of our ancestors, hogs are raised and slaughtered in places sequestered from human habitation, they continue to stick to our ribs mentally. We need them to express otherwise ineffable concepts. "Piggy banks" are still in vogue. These money-pigs, which undoubtedly originated from the greedy and miserly image attributed to hogs, are given to children to encourage thrift and are often presented to newborns as gifts. "For centuries, the pig was the prime source of stockpiling for winter, so it was natural to choose the animal as a symbol of saving" (Nielsen 1978:44). Although usually money can be withdrawn from other types of banks without destroying them, piggy banks are ultimately smashed open when they have been filled with pocket money. "It is a rite that, although bloodless, commemorates the flesh and bone sacrifice of the pig, stuffed with good things and then, poor thing, ripped to pieces" (Bonera 1990:81).

In everyday life we use expressions like porker, hogwash, male chauvinist pig, gas hog, road hog, living high on the hog, happy as a pig in muck, going hog wild, piggish, and crying like a stuck pig. There are fascist pigs and Nazi pigs; prostitutes and policemen are called pigs. People are cautioned that "You can't make a silk purse out of a sow's ear." Not long ago, a mother was arrested for child abuse when she placed her seven-year-old son on a bench in front of his home with his hands tied behind his back and his face painted blue. He wore a cardboard pig nose and a sign on his chest read: "I am a dumb pig. Ugly is what you will become every time you lie and steal." The boy had stolen $25 worth of merchandise (Woman 1988:1). Recently President Bush's "tough talk" regarding his rival, Ross Perot, included his angry admonishment to voters against choosing an unknown candidate: "There's too much at stake to buy a pig in a poke" (Keen 1992:6A). This common metaphor refers to the risk once inherent in buying a pig that was carried to market in a sack, called a poke, without exam-

The Sacred Bee, the Filthy Pig, and the Bat Out of Hell

ining the contents. The buyer had to beware of unknowingly purchasing a runt, or even a cat, that an unscrupulous dealer might substitute for the pig (Funk 1985:105–106).

The Bat Out of Hell

Whereas pigs are creatures of the earth, bats are perceived as denizens of the realm below—the underworld, or hell. Like pigs, bats were objects of taboo for the Israelites: "These you shall regard as detestable among the birds. They shall not be eaten; they are an abomination: . . . the stork, the heron of any kind, the hoopoe, and the bat" (Leviticus 11:13, 19). The bat, here categorized as a bird, has always seemed an anomaly, defying classification with its mammalian body combined with the avian ability for flight, thus becoming a source of anxiety, suspicion, and dislike. Other characteristics have enhanced this reputation, causing the ambiguous creature to remain mysterious, the object of fear even today. According to biblical prophecy, on the day of God's judgment "people will throw away to the moles and to the bats their idols of silver . . . and of gold" and flee into "caverns of the rocks" and "clefts in the crags" (Isaiah 2:20–21). Idolaters would thereafter find the images they once worshiped fit only to be found "in those neglected places, where moles and bats have their abode" (Fisher 1972:30).

Bats' nocturnal habits ensure that they are seldom seen clearly or at close range. Their dark color, strange-looking heads, faces that may have menacing expressions, sharp teeth (in some cases), and enormous wings out of proportion to their bodies make them frightful. Their habitations in dark caverns, empty barns, or abandoned houses and their hanging upside down during the day lead us to associate bats with gloom and weirdness. Their uncanny ability to fly through darkness using unique powers of "radar" seems unnatural and sinister. The fact that one member of the species, the vampire, sucks blood, even though all others subsist on insects or fruit, gives the bat an evil connotation.

Expressions such as batty, bats in the belfry, going on a bat, blind as a bat, and like a bat out of hell are commonly used and exemplify human fascination with bats. Batman, himself an anomaly with his dual life of mortal man and supernatural hero that reflects his namesake's double nature as mouse and bird, is currently a prominent figure in comics and for the second time ranks as a top box-office attraction at the movies. In *Batman Returns*, the hero's main antagonist is "Penguin," a man-beast who takes his identity from a bird that does not fly but rather walks upright on land and swims like a fish—also an anomalous creature. Bram Stoker's *Dracula* remains a popular classic and is frequently recreated in film and drama. An early and influential indictment of the bat is found in Dante's *Inferno*, the great dramatization of Christian doctrine. Lucifer, the fallen angel, known as Dis, Satan, or the Devil, is seen and described by the traveler through hell. The ugly "emperor of the despondent kingdom" has three faces, and "beneath each face of his, two wings spread out / as broad as suited so immense a bird: / I've never seen a ship with sails so wide. / They had no feathers, but were fashioned like / a bat's; and he was agitating them, / so that three winds made their way out from him" (1982:313).

Whenever a bat is seen, even in the present day, many women evidence strong fear that the creature will entangle itself in their hair. As recently as 1991, a newspaper reported a person's childhood experience in which a woman's hair had to be cut in order to dislodge a bat (Tatem 1992:6). The reality of this phenomenon, however, is undoubtedly a great deal rarer than the dread of it, for bats are skillful at avoiding any obstacle to their flight. The persistent and ubiquitous association of bats with female tresses should not be pragmatically dismissed, as one zoologist has recently done, as merely attributable to the behavior of bats which, swooping after insects, come close to people's heads, evoking an exaggerated response (McCracken 1992:16). In fact, the tradition traces back to the New Testament teaching of Paul, who proclaimed that women must cover their heads in church (I Corinthians 11:5–6, 10–13). In the pertinent passages, "because of the angels" refers to "cosmic demonic powers against whom protection is needed" (Metzger and Murphy 1991:NT 241). Specifically, these powers are the spirits that were alleged to be attracted to or controlled by unbound female hair. The conviction that bats will fly into a woman's hair stems from the ancient notion that women's hair attracts demons. Consequently, Paul ordered that women conceal their hair while in church (Walker 1988:314, 362). The belief that bats are irresistibly attracted to women's hair—and,

The Sacred Bee, the Filthy Pig, and the Bat Out of Hell

once entangled there, can be released only by a pair of scissors wielded by a man (Cavendish 1970:228)—is in keeping with this feminine association. The widespread belief that having a bat entangled in the hair causes baldness (Hill and Smith 1984:177) adds to the dread. The idea of bats becoming entangled in a woman's hair was recently given vivid expression in *Batman Returns*. During one dramatic scene of that film, "The Ice Princess," a scantily clad and sexually seductive woman with a fluffy coiffure, falls to her death from the top of a building, after being horrified by bats that flew into her hair.

In Western culture, bats have long been identified as demons, depicted in Christian art with bat wings. A once-prevalent belief, still extant in some areas of Europe, held that human souls take the form of bats when they leave the body during sleep, which is the reason bats are not seen by day when people are awake. The association with souls led naturally to a belief that the pagan dead might also become bats flying about in search of a means of rebirth—or the blood that, according to Genesis 9:4, signified the "life" of all creatures (Walker 1988:362).

Religious views of the bat as demon are traceable to perceptions about the animal's habits. Edward Topsell (1572–1625), a British parson who wrote both as a churchman and a naturalist, included the bat in his treatise on avian species. He observed that the bat desires darkness, does not abide in the light of the sun, and flies in secret and solitary corners. Thus bats resemble devils, who love darkness more than light, and fly from Christ, the true light of the world. Bats' wings are made of skin with no feathers—just as devils are "incorporeall," flying without the feathers of humane desires and motivated by lust. Topsell points out that the bat has four feet, and yet flies, and is neither skillful at flying nor at moving on foot. Likewise devils are neither good fliers like angels nor good walkers like men. Moreover, no other birds have teeth like bats—and this makes them resemble devils, who are more greedy for revenge than other creatures. Bats lay no eggs, but bring forth their young without the delay of brooding and hatching—like devils who so speedily perform their evil work, working malice as soon as they are born. Sinners too are comparable to these animals, for lecherous persons, adulterers, wait for twilight for their activities. Moreover, bats will eat the oil from the church lamps—just as the envious rob good people of the grace and praise that belong to them. Bats stick to walls and live in secret holes of houses and chimneys—just as proud persons advance themselves above their neighbors by usurping the goods, lands, and favors of other people (Harrison and Hoeniger 1972:61).

Until relatively recently in the Western world, the "night air" was greatly feared as the abode of evil beings, and homes were closely shuttered after sundown to prevent any intrusion. Because bats thrive and fly about in the night air, they took on a malignant image by association. Bats are said to be symbolic of demons that cohabit with women at night, for both avoid the light of day. Because they sleep upside down, bats are enemies of the natural order, like evil spirits (Matthews 1986:18). Christians have called the bat "'the Bird of the Devil,' an incarnation of the Prince of Darkness." As a hybrid of the bird and the rat, the bat is associated with duplicity and hypocrisy (Cooper 1978:18). The bat's seemingly supernatural ability to hunt prey in total darkness gives it a frightening reputation as a creature with occult powers, a mysterious capacity making it the object of hatred. So tenacious is the notion that the devil assumes the shape of a bat that some people, even today, burn bats alive over a candle or crucify them by nailing them to their doors (Cavendish 2, 1970:226; Mercatante 1974:181).

For some of these same reasons, Christianity and Western culture associate the bat with black magic and witchcraft. In certain areas, "if the bat is observed to rise, and then descend again earthwards, you may know that the witches' hour is come—the hour in which they have power over every human being not shielded from their influence" (Opie and Tatem 1989:14). Most unfortunate for the strange but gentle bat, human beings see a similarity between its dark wings draped about its body when at rest and the cloak of a witch. Common belief held that witches transformed themselves into bats. Along with black cats and toads, bats were also thought to be witches' familiars. In contrast to heavenly bees, "God's birds," bats became known as demonic "witches' birds." Bats were once nailed to barns to repel witches, just as I have observed modern-day ranchers fastening magpie corpses to sheds "as a warning to the others to stay away." Bat's blood was an ingredient in recipes for witches' flying ointment, since witches wanted to acquire the animal's unique ability to fly at night without bumping into anything. Wool of bat was included in the brew created by the witches in

The Sacred Bee, the Filthy Pig, and the Bat Out of Hell

Macbeth, and reportedly wings or entrails were added to witches' cauldrons as well. The belief that if a bat flies near you someone is trying to bewitch you adds to batphobia (Barth 1972:53; Radford and Radford 1974:33; *Encyclopedia* n.d.: 202; Potter 1991; Lasne and Gaultier 1984:89).

Thomas Bewick, the British naturalist who denigrated the pig, presented a negative view of the bat as well, stressing its "middle nature" between animal and bird and calling it an "imperfect animal." He noted that it showed "extreme awkwardness" in walking and lack of ease in the air, where its flight is "laboured and ill directed," meriting the name "Flittermouse" (1807:510). Aesop's fable about the bat features its beast/bird dichotomy and expresses the ambiguity of its status (Mercatante 1974:181). Many diverse cultures, such as the Fipa of Africa described by Roy Willis, view the bat in the same light (1974:47–48). Because of its uncertain position between mammal and bird, the bat has been perceived as a hermaphrodite or as representing androgyny. Anxiety regarding its proper category makes it impure and gives it malevolence. Bats' membranous wings with claws associate them with dragons (Matthews 1986:18; Cooper 1978:18; Wootton 1986:75).

For diurnal human beings, death and disaster seem to occur most frequently at night—hence the nocturnal bat is an evil omen associated with death. Traditionally, encountering a bat is a forewarning of death or misfortune. Belief holds that if a bat hits against a person's window or enters his house, there will be a death in the family (Hill and Smith 1992:159; Opie and Tatem 1989:14). The popular vampire legends, in which bats emerge from graves at night to suck human blood, have added greatly to the bat's deadly image, even though the vampire bat's range is restricted to certain Neotropical areas. Only recourse to a crucifix (or garlic) can protect the potential victim against being bitten. The bat's preference for holes, caves, vaults, dungeons, towers, and deserted places gives it an association with chaos, disorder, and dirt. Its image as a "flying rodent" causes the bat to share the reputation of rats and mice as foul polluters. The idea that bats are filthy and commonly carry diseases has added greatly to people's fear and loathing of the animals, especially since rabies can be transmitted by vampire bats. Scientists point out, however, that in the last thirty years only ten people in the United States have died of rabies contracted from bats—far fewer than are killed annually by bee stings and dog attacks. And although the only other disease bats transmit to humans is histoplasmosis, fear of the flying mammals as disease-ridden vermin is deeply entrenched and risks of contact are greatly exaggerated (Johnson 1985:42; Tuttle 1988:17; Halton 1991:9–10).

Knowledge that bats play a vital ecological role in consuming insects, pollinating plants, and dispersing seeds does not redeem the species in the eyes of the general public. Other benefits—providing meat in Africa and Asia, producing guano for fertilizer and gunpowder, helping scientists to develop an ultrasonic orientation system for blind people through analysis of bat echolocation—are less well known. Varieties of bats do not have common names like birds, because they are unfamiliar to the average person. Yet present-day people who may never see a bat (except as depicted in Halloween decorations) frequently employ it to express concepts of perceived "batness." A gossipy or mean elderly woman is an old bat—a metaphor that recently received reinforcement from a prominent psychiatrist specializing in fears and phobias who revealed that bats symbolize "the mother as devouring ogre, the bad mother who is going to destroy something" (Kennedy 1992:2). An idle loafer is said to be batting around, a prostitute may be called a bat, and a brothel the bat-house. Blind as a bat (erroneous because bats actually see well) describes the way the nocturnal animal stares in the daytime. Batting one's eyelashes relates to the bat's fluttering movements. Bats in the belfry expresses the essence of craziness, as the bell tower of a church suggests a person's muddled head through which bats fly erratically as they might do in a church steeple.

Biophilia and Animal Symbolism

Edward O. Wilson defines biophilia as "the innate tendency to focus on life and lifelike processes," noting that human beings "learn to distinguish life from the inanimate and move toward it like moths to a porch light." He argues that "to explore and affiliate with life is a deep and complicated process in mental development." And according to Wilson, "human beings live— literally live, if life is equated with the mind—by symbols, particularly words, because the brain is constructed to process information almost exclusively in their terms." Humankind is "the poetic species," and "the symbols of art, music, and language freight power well beyond their outward and literal meanings" (1984:1, 74). Thus it is because "life of any kind is infinitely more interesting than almost any conceivable variety of inanimate matter" (84) that people so inevitably turn to the animal kingdom for symbolic expression.

The power of biophilia is manifest not only in direct interactions between people and animals but also through the process of symbolizing through animals. Indeed, for a large share of the population in the modern industrialized world, relationships with animals as they are symbolically perceived have to a great extent replaced interactions with their living counterparts. Moreover, commonly held beliefs about a particular animal, rather than personal experience, generally determine the character of interactions with its species. Interpretation of an animal's behavior in metaphorical terms can result in the creature being classified as "good" or "evil"—with consequent effects on the preservation or destruction of the species. The symbolizing process can enhance positive affiliation, resulting in preservation, or it can cause alienation of that animal from the human sphere with consequent destruction.

From the numerous animals in the world that are laden with symbolic significance, I have selected for consideration three that are commonly used in everyday thought and conversation as repositories of shared concepts and values. They represent species that are viewed in particular ways in Western society, and the roots of perceptions about them are deeply entrenched in Christian tradition. These cultural forces give power to the analysis of the animals' symbolic roles, providing a lens through which preconceived ideology determines the collective view of bees, pigs, and bats. The symbolic meanings of animals depend on the cultural context. Bees, for example, had a fearsome image in the Old Testament, with emphasis on their aggressiveness and stinging capacity, until Christianity invested them with gentleness and purity. In many New Guinea tribes, pigs are the favorite animal, greatly esteemed, with none of the negative connotations assigned to them in Western society. Bats in China are associated with good luck and happiness. Even in cultures that have an almost uniform view of certain animals, there are exceptions that prove the rule. Certain people harbor phobias and hatred of bees. Champions of pigs, as exemplified by

G. K. Chesterton, see them as "very beautiful animals," the lines of a fat pig being "the loveliest and most luxuriant in nature," with "the same great curves" as we see "in rushing water or in rolling cloud" (Ash 1986:12). The great naturalist W. H. Hudson considered pigs "the most intelligent of beasts"—their unique "disposition and attitude toward all other creatures, especially man," makes them view us democratically "as fellow-citizens and brothers" (1923:295–296). Poet Emily Dickinson, whose unorthodox view of darkness and light inverted the traditional metaphor so that darkness was valued rather than feared as symbolic of chaos, saw bats as beneficent, associated not with evil but with the special freedom provided by night (see Barker 1987; Allen 1983:40–42).

Many diverse circumstances—biological, behavioral, and political— influence animal symbolism. On certain Pacific islands, for example, where bats are very large and live conspicuously in the treetops like birds, people do not fear and persecute them as in Europe and America where they are small, evasive, mysterious, and nocturnal. In some Pacific island legends, bats are even depicted as heroes (Tuttle 1988:17; Givens 1990:61). Although, as Leach points out, in Western society virtually all insects are hated as "evil enemies of mankind" and are liable to ruthless extermination, the bee is the notable exception—the one credited with "superhuman powers of intelligence and organization" (1975:40). The Nazis elevated the pig, a predominant food animal, to a sacred or sacrificial status, separating their ideology from that of the Jews they hated. Nazi farm propaganda proclaimed that the pig occupied "first place in the cult of the Nordic peoples." Unlike the Semites, who do not understand the pig and reject it, Nazi doctrine recognized that the pig "has drawn its originality from the great trees of the German forest" (Arluke and Sax 1992:12).

It is significant that the human mind chooses certain characteristics of the animal and ignores others. Symbolizing with animals is not simply a rational process that views a species holistically: it involves deeper levels of consciousness as well as external stimuli. The use of the bee as a model for Christianity, for example, although deeply entrenched as an apt metaphor, becomes incongruous if accepted in totality. For the religion places particular emphasis on the individual—something that honeybee society does not do. The bees' killing of hive members who are not needed is certainly antithetical to the Christian ethic of preserving life at all cost, even for

The Sacred Bee, the Filthy Pig, and the Bat Out of Hell

people no longer useful to society. Bats, on a pragmatic level, could be viewed as pollinators and insect destroyers with "sonar" power superior to that of humankind rather than ghoulish vermin flapping up out of hell. Hogs might be seen as intelligent beings whose form and behavior have been shaped to fill the human desire for an economical source of protein rather than as repositories for greed and lust.

Natural history observations may be a starting point, but they are strongly molded by cultural constructs and by our need to affiliate with the rest of creation through metaphor. Signifying by means of animals takes place at deep levels of human consciousness, emanating from the same type of psychic experience as myth, poetry, and religion whose language is also symbols. The human is not the wholly rational, materialist being as is sometimes assumed, and symbolizing through animals emanates from the realm of the collective unconscious, as do myths and folklore, expressing truth that is deeper than observable fact. This is the same domain of knowing where the "bee's wisdom" lies and where animals communicate with humanity in terms of universal understanding. "What the bee knows," according to Travers, is "a wisdom of the ages, a time-honoured wisdom from a timeless source. Unlike the ephemeral knowledge of modern man, the bee's wisdom is one with the fount of all things and transcends the mirage of man's achievement." It is "the wisdom that sustains our passing life" (1989:90, 306).

Animal symbolism is biophilia in that it represents another step in the age-old search for "man's place in nature." Through such symbolizing, there is a kind of merging—animals take on human qualities and humans take on animal qualities. Antithetically, the process of symbolizing with animals also makes use of and preserves the separateness that exists between people and animals. Cognitively, we balance the alternatives in the question that determines all human / animal interactions: are the other life-forms like us or different? Though they seem closely related to us, animals are viewed as the quintessential Other, a means of identifying what characteristics make us human and weighing those qualities according to an imposed system of values. In a chaotic world, human beings seek their own identity through reference to the alternate domain of the beasts.

As Lévi-Strauss has noted, the central problem of anthropology is "the passage from nature to culture" (1963:99). Through animal symbolism, key issues in the nature / culture dichotomy are explored and given concrete expression. As Wilson phrases this process in *Biophilia*, animals are "agents of nature translated into the symbols of culture" (1984:97). Using animal symbolism enables us to confront the beast, relate to it, and, while remaining human, "take on beasthood" and "see our former selves as strange, as from the outside" (Willis 1974:66). Thus we are able to measure and evaluate ourselves, probing the enigma of what it means to be human.

Symbolized animals are ideal instruments for examining the key human issue of the individual versus society. Our strong need to cope with the dilemma of the social order makes us worship the bee, whose priority, clearly, is responsibility toward the group. We vilify the pig, whose behavior—if it were human—would be classed as antisocial, its impugned greed and sensuality raising the specter of too much individualism that could destroy social cohesion. And because of habits unfamiliar to people—being nocturnal, avoiding human society, hanging upside down—bats are the extreme example of defying the human social order: they become demons from hell. The human demand for classification and the need to have all creatures conform to a set pattern is pervasive. The holy bee is worshiped because of its classificatory logic and its reflection of the qualities we perceive in and desire for our own species, whereas the pig with no cud who eats both animals and plants and the bat who combines mouse and bird are despised as anomalous. In the case of the pig, it is so crass and offensive to aesthetic social norms that it is a metaphor for the absurd. Not only in churches, but also in ubiquitous contemporary statues, illustrations, and throughout popular culture, pigs ironically play musical instruments. Thus the saying "when a pig plays a flute" signifies an improbable action (Clark 1968:15). "In a pig's eye," another expression for something not likely to happen, traces back to prohibition days when the peephole in a speakeasy was known as a "pig's eye" and getting admitted to the saloon was extremely difficult (Tuleja 1987:168–169).

The universality of animal symbolism throughout the world and over eons of time indicates the profound significance of this inherent form of biophilia. Vestiges of the ancient beliefs of our ancestors retain their place in our minds, inextricably interwoven into the human condition because we are evolutionarily and physically, as well as aesthetically, spiritually, psychologically, and emotionally, tied to our animal kin. The old ways of per-

The Sacred Bee, the Filthy Pig, and the Bat Out of Hell

ceiving the universe, obliterated in their literal manifestations by the destruction of gods and spirits and the impact of science and technology, persist in human thought. The concepts of animism and totemism still fill the world with spiritually linked animals—creatures that are never neutral but are good or evil, pure or polluted, friendly or hostile. Literal reincarnation and metamorphosis of human into animal and animal into person persist on a symbolic level. In the imagination, animals are embodiments of gods or demons. Actual scapegoats like the Hebrews' sacrificial goat are replaced by modern-day conceptual and verbal sacrificial beasts who are burdened with our images of ourselves. Our pejoratively anthropocentric outlook gives us an imperative to impose order on our universe, fitting animals into categories that we feel they should occupy, rather than acknowledging what they are and opening our minds to their true natures.

One of the most urgent issues in this present era of ecological crisis is the hostile relationship that so often exists between humankind and animals with the consequent destruction of so many species. By analyzing the ways in which symbolic attributes become attached to animals we may begin to understand why in some circumstances people cherish and protect animals while in other cases they are devalued and destroyed. This kind of inquiry also has implications for elucidating the issue of animals' intrinsic value versus their worth as measured by human exploitation. Thus it has important bearing on the concept of reciprocal relationships as they should exist between people and other species. Understanding the symbolization of animals as part of the biophilia hypothesis, which involves the synthesis of alternative modes of thought with science, can play a vital role in contributing to the establishment of more harmonious relationships with all life on earth.

Speculations

Language continually adapts to changing conditions in an evolutionary process that has been compared to biological evolution. It is difficult to predict the ways in which our diminishing interactions with the natural world and different perceptions of animals will affect future expressions of cognitive biophilia. In pondering the development of animal symbolism in the years to come, many unanswerable questions arise. If we continue our current policy of destructiveness toward nature, does this mean that language will contain fewer and fewer symbolic references to animals—with consequent impoverishment of thought and expression? Or is our species so intimately connected to other forms of life that we will continue, or even increase, our inherent cognitive relationship to natural forms, even while we proceed to distance ourselves physically from them? As the domesticated sphere that we have created inexorably expands at the expense of the wild, will people eventually draw linguistic imagery exclusively from those animals that are controlled and near at hand? Or will the human mind nostalgically persist in imaging the faraway wild realm and using that now-cherished but forever lost kingdom as a measure of the diminished and artificial universe?

Ironically, as countless life-forms disappear from the earth, many people experience a fuller consciousness of nature and acquire deeper aesthetic appreciation for the nonhuman realm. These newly awakened feelings will likely result in an increased utilization of animal symbolism in literature and art as well as in everyday language. An important factor, too, is the recent surge in biological and behavioral knowledge about animals and the widespread dissemination of this information to the public. Through the media, especially television, the accessibility of animal lore to the average citizen has reached unprecedented heights, even as the animals themselves are swiftly disappearing. Growing realization of the ways in which other species seem to share human characteristics may enhance empathetic responses that make animals more prominent in the thought processes of people who are otherwise remote from them. Such developments could well lead to an increasing use of animal images and symbols that will greatly enrich the language of the next millennium, keeping alive the cognitive biophilia that arose with the first awareness of human affinity to the rest of nature.

REFERENCES

Achenbach, J. 1992. "My President as a Dog." *Providence Journal-Bulletin*, 21 April, p. A11.

Allen, M. 1983. *Animals in American Literature*. Urbana: University of Illinois Press.

Arluke, A., and B. Sax. 1992. "Understanding Nazi Animal Protection and the Holocaust." *Anthrozoos* 5:6–14.

Ash, R. 1986. *The Pig Book*. New York: Arbor House.

Atwood, J. 1982. "The Appeal of Pigs." *US Air* 4(9):452–458.

Baring-Gould, S. n.d.[ca. 1912]. *A Book of Folk-Lore*. London: Collins' Press.

Barker, W. 1987. *Lunacy of Light: Emily Dickinson and the Experience of Metaphor*. Carbondale: Southern Illinois University Press.

Barth, E. 1972. *Witches, Pumpkins, and Grinning Ghosts*. New York: Clarion Books.

Barth, F. G. 1991. *Insects and Flowers: The Biology of a Partnership*. Princeton: Princeton University Press.

Bewick, T. 1807. *A General History of Quadrupeds*. Newcastle upon Tyne: Edward Walker.

Bonera, F. 1990. *Pigs: Art, Legend, History*. Boston: Little, Brown.

Bouissac, P. 1990. "The Profanation of the Sacred in Circus Clown Performances." In *By Means of Performance*, edited by R. Schechner and W. Appel. New York: Cambridge University Press.

Bowman, S., and L. Vardey. 1981. *Pigs: A Troughful of Treasures*. New York: Macmillan.

Britt, K. 1978. "The Joy of Pigs." *National Geographic* 154:398–415.

Cansdale, G. S. 1970. *All the Animals of the Bible Lands*. Grand Rapids: Zondervan.

Cavendish, R., ed. 1970. *Man, Myth and Magic*. 24 vols. New York: Marshall Cavendish.

Charbonneau-Lassay, L. 1991. *The Bestiary of Christ*. New York: Parabola Books.

Chetwynd, T. 1987. *A Dictionary of Symbols*. London: Paladin Grafton Books.

Clark, J. D. 1968. *Beastly Folklore*. Metuchen, N.J.: Scarecrow Press.

Clausen, L. W. 1962. *Insect Fact and Folklore*. New York: Collier Books.

Cooper, J. C. 1978. *An Illustrated Encyclopaedia of Traditional Symbols*. London: Thames & Hudson.

———. 1985. *Symbolism: The Universal Language*. Wellingborough, Northamptonshire: Aquarian Press.

Coudert, J. 1992. "The Pig Who Loved People." *Reader's Digest* 140:150–154.

Dante, A. 1982. *The Divine Comedy: Inferno*. New York: Bantam Books.

Domacinovic, V., and V. Tadic. 1991. "Relationships with Bees and Sheep in Yugoslavia." Paper presented at the World Veterinary Congress, Rio de Janeiro, Brazil.

Douglas, M. 1976. *Purity and Danger*. London: Routledge & Kegan Paul.

Encyclopedia of Magic and Superstition. n.d. n.p.: Black Cat.

Ferguson, G. 1961. *Signs and Symbols in Christian Art*. New York: Oxford University Press.

Fisher, J. 1972. *Scripture Animals*. Princeton: Pyne Press.

Funk, C. E. 1985. *A Hog on Ice and Other Curious Expressions*. New York: Harper & Row.

"Fur, Fins and Fasts." 1992. *Economist* 322:93.

Givens, K. T. 1990. "Going Batty." *Modern Maturity* 33:60–64.

Graham, J. M., ed. 1992. "Newsnotes." *American Bee Journal* 132:259.

Griffin, D. R. 1981. *The Question of Animal Awareness*. New York: Rockefeller University Press.

Halton, C. M. 1991. *Those Amazing Bats*. Minneapolis: Dillon Press.

Harris, M. 1974. *Cows, Pigs, Wars, and Witches: The Riddles of Culture*. New York: Random House.

Harrison, T. P., and F. D. Hoeniger, eds. 1972. *The Fowles of Heaven or History of Birds*. Austin: University of Texas Press.

Hedgepeth, W. 1978. *The Hog Book*. Garden City: Doubleday.

Henderson, W. 1879. *Notes on the Folk-Lore of the Northern Counties of England and the Borders*. London: W. Satchell Peyton and Co.

Hill, J. E., and J. D. Smith. 1992. *Bats: A Natural History*. Austin: University of Texas Press.

Hudson, W. H. 1923. *The Book of a Naturalist*. New York: Hodder & Stoughton.

Jay, R. 1987. *Learned Pigs and Fireproof Women*. New York: Villard Books.

Johnson, S. A. 1985. *The World of Bats*. Minneapolis: Lerner Publications.

Keen, J. 1992. "'Tough Talk': A Preview of Bush's Strategy." *USA Today*, 22 June, p. 6A.

Kennedy, D. 1992. *Living Things We Love to Hate*. Vancouver: Whitecap Books.

Lasne, S., and A. P. Gaultier. 1984. *A Dictionary of Superstitions*. Englewood Cliffs, N.J.: Prentice-Hall.

Lavine, S. A., and V. Scuro. 1981. *Wonders of Pigs*. New York: Dodd, Mead.

Lawrence, E. A. 1982. *Rodeo: An Anthropologist Looks at the Wild and the Tame*. Chicago: University of Chicago Press.

Leach, E. 1975. "Anthropological Aspects of Language: Animal Categories and Verbal Abuse." In *New Directions in the Study of Language*, edited by E. H. Lenneberg. Cambridge: MIT Press.

Leach, M., ed. 1949. *Funk and Wagnall's Standard Dictionary of Folklore, Mythology and Legend*. 2 vols. New York: Funk & Wagnalls.

Lévi-Strauss, C. 1963. *Totemism*. Boston: Beacon Press.

Longgood, W. 1985. *The Queen Must Die*. New York: W. W. Norton.

McCracken, G. F. 1992. "Bats and Human Hair." *Bats* 10:16.

MacLennan, D. A. 1966. *Revell's Minister's Annual*. Westwood, N.J.: Fleming H. Revell.

Maeterlinck, M. 1954. *The Life of the Bee*. New York: New American Library.

Matthews, O., trans. 1986. *The Herder Symbol Dictionary*. Wilmette, Ill.: Chiron Publications.

Mercatante, A. S. 1974. *Zoo of the Gods*. New York: Harper & Row.

Metzger, B. M., and R. E. Murphy, eds. 1991. *The New Oxford Annotated Bible*. New York: Oxford University Press.

Miley, A. L. 1992. *Christian Education Resources Catalog*. San Diego: Christian Ed Publishers.

Nielsen, R. G. 1978. *The Pig in Danish Culture and Custom*. Copenhagen: Malchow Bogtryk.

Norman, J. 1990. *Honey*. New York: Bantam Books.

Opie, I., and M. Tatem. 1989. *A Dictionary of Superstitions*. New York: Oxford University Press.

Parkhill, J. M. 1980. *Honey*. Memphis: Wimmer Brothers.

"Pet Pigs Are Becoming More Popular." 1986. *Bay Window*, 29–30 October, p. 4.

Potter, C. 1991. *Knock on Wood*. Stamford, Conn.: Longmeadow Press.

Radford, E., and M. A. Radford. 1974. *Encyclopaedia of Superstitions*. London: Book Club Associates.

Ransome, H. M. 1986. *The Sacred Bee*. Burrowbridge: BBNO.

Reynolds, B. 1992. "King Verdict: Watershed Case Can Change Black Lives." *USA Today*, 1 May, p. 15A.

Russell, G. K. 1991. "Arousing Biophilia: A Conversation with E. O. Wilson." *Orion* 10:9–15.

Scott, A. 1980. *A Murmur of Bees*. Oxford: Oxford Illustrated Press.

Shaler, N. S. 1904. *Domesticated Animals*. New York: Scribner's.

Sillar, F. C., and R. M. Meyler. 1961. *The Symbolic Pig*. London: Oliver & Boyd.

Sparks, J. 1982. *The Discovery of Animal Behaviour*. Boston: Little, Brown.

Stratton-Porter, G. 1991. *The Keeper of the Bees*. Bloomington: Indiana University Press.

Tatem, M. 1992. "Bats Out of Hell?" *Newsletter of the Folklore Society* 14:6.

Towne, C. W., and E. N. Wentworth. 1950. *Pigs from Cave to Corn Belt*. Norman: University of Oklahoma Press.

Travers, P. L. 1989. *What the Bee Knows*. Wellingborough, Northamptonshire: Aquarian Press.

Tuleja, T. 1987. *The Cat's Pajamas*. New York: Fawcett Columbine.

Tuttle, M. D. 1988. *America's Neighborhood Bats*. Austin: University of Texas Press.

van Loon, D. 1980. *Small Scale Pig Raising*. Charlotte, Vt.: Garden Way.

Virgil. 1982. *The Georgics*. New York: Penguin Books.

Walker, B. G. 1988. *The Woman's Dictionary of Symbols and Sacred Objects*. New York: Harper & Row.

Warshaw, R. 1986. "Pets to the Rescue!" *Woman's Day*, 23 December, pp. 82–83, 113, 115.

White, L. 1973. "The Historical Roots of Our Ecological Crisis." In *Western Man and Environmental Ethics*, edited by I. G. Barbour. Reading, Mass.: Addison-Wesley.

Willis, R. 1974. *Man and Beast*. New York: Basic Books.

Wilson, E.O. 1984. *Biophilia*. Cambridge: Harvard University Press.

"Woman in Court After Dressing Son Like Pig." 1988. *USA Today*, 13 September, p. 1.

Wood, J. G. 1880. *Wood's Bible Animals*. Philadelphia: Bradley, Garretson.

Wootton, A. 1986. *Animal Folklore, Myth and Legend*. New York: Blandford Press.

Zeuner, F. E. 1963. *A History of Domesticated Animals*. New York: Harper & Row.

Part Five

EVOLUTION

CHAPTER 11

God, Gaia, and Biophilia

Dorion Sagan and Lynn Margulis

EVER SINCE THE human species evolved some 4 million years ago it has been expanding, first by nomadic hunting and gathering, then, in civilized times, by agriculture and industry. As our numbers have grown we have changed the environment. At this late juncture we have come to see that there is no way we can expand indefinitely without imposing indefinite unpleasant changes on the environment. We have walked on enough concrete, smelled enough air pollution, eaten enough processed food to realize that the sort of comfort afforded us by technology in the long run differs from, and is less sustainable than, the green fruit tree paradise of our simian ancestors. Our evolution has brought us beyond a point of no return.

The new high regard for earth is not unlike the feelings of customers at a fabulous small restaurant that is getting big. As word spreads of the restaurant's outstanding quality, more and more people come until the restaurant expands, new management comes, and the restaurant is no longer superb and small but *lousy*, perhaps even part of a national chain. Likewise, owing primarily to the fortuitously evolved manipulative skills of certain grass-walking African mammals, humans have always fed well on the victuals of earth. We have transformed local ecologies into technology, killed animals for clothing and meat, grown plants for food, shaped rocks and trees into shelter and tools. Like the small restaurant that was so good it became lousy, earth was so paradisiacal and susceptible to technological plunder that we plundered it—and are now faced with the results. Unlike the restaurant, we have no place else to eat.

In this essay we attempt to show how our technological plundering of the planet has forced us to revalue our biological connections to other species and living beings. This revaluation is forcing us to see the collusion in our way of life of traditional Western religion, which has provided an impetus for our technological plundering. Moreover, this same Judeo-Christianity still undergirds the assumptions of much "secular" science.

The renewed focus on the positive aspects of our connections to other living things has lately been called biophilia, from the Greek words for love and life. But as we can see from the use of the word *lousy* above—an adjective derived from a parasitic clinging insect—our connections to other life-forms are not always positive. In fact, the emotional palette of our responses to life-forms is rich, labile, and complex. Specific life-forms "push our buttons"—they elicit strong, relatively constant responses varying from disgust (maggots, bacterial infection), care (kittens, puppies), horror (spiders, snakes), awe (tigers), and well-being (magnolia trees, actinobacteria with their woodland scent) to longing or envy (birds in flight). As E. O. Wilson has suggested in his coining of the term biophilia, our intrinsic love for life can be used to help preserve crucial reserves of planetary biodiversity. He has further suggested that our positive affections, such as our appreciation for lush greenery, may be inbred—genetically based on the importance such early life-forms held for us. Other sensations, such as our instinctive avoidance of butyl mercaptan, the noxious ingredient in skunk spray, seem to benefit other organisms by keeping us away. We are, like many insects and other mammals, manipulated by our love of sweets and fresh colors, which over the millennia have induced us, for example, to eat cherries and hence act as couriers of the immobile cherry tree's seeds.

The point is that there is no simple biophilia, no unconditional, unchanging love for members of other species. Some men love racing cars and, indeed, may be attracted to the curvaceous bikini-clad women advertisers portray with such cars. We are attracted to bright colors, as well, an attraction whose application to painted automobiles comes long after the evolutionary crucial biophilia of primates to trees with brightly colored fruits. So not only is our love for life impure, not only do we have mixed feelings toward other life-forms, but our affection is also changeable, plastic.

With such complexities, such an admixture of feelings both positive and negative, and subtler states in between, a mixture which can moreover be changed and applied to more recent technological objects, it is difficult to speak monolithically of biophilia, a simple love of life. Perhaps it would be better to speak of prototaxis—the generalized tendency of cells and organisms to react to each other in distinct ways. Ivan E. Wallin defines prototaxis in *Symbionticism and the Origin of Species* as the "innate [that is, genetic] tendency of one organism or cell to react in a definite manner to another organism or cell." Let us think then of both positive and negative biophilia (sometimes called biophobia) as aspects of global prototaxis. The principle of prototaxis ought to be perceived as intrinsic to living beings, all of which have distinct lineages and combinations of genes. Like Wallin's profound conclusions on the role of symbiosis in the origin of species (Mehos 1992) and in embryogenesis, this notion of prototaxis is not well known.

Biologists define pioneering species as those which spread rapidly throughout an environment but quickly saturate it and reach their limit. Pioneer species are the first to come—like the customers in the restaurant parable. Although the term usually applies to plants, a case can be made that human beings, combined with our technology, are the global equivalent of a pioneer species. We may now have reached our saturation point, the limits of our growth; if so, we may be detecting signals from our living environment that it is no longer able to support continuous growth.

In the pioneering stage people told themselves stories that made it seem as if it were our destiny to endlessly plunder the natural environment, converting animals, plants, and rocks into extensions of ourselves. In retrospect these stories, which center, in the West, on the monotheistic conceit that humanity is Numero Uno for whose benefit God has made all other life-forms, were the rallying cry of a nomadic tribe. But the tribe, having become sedentary on all the continents of the globe, is no longer nomadic. Nonetheless, as often happens in cultural evolution, information contin-

God, Gaia, and Biophilia

ues to flow long after it is useful. Moreover, this data lag can be seen not only in the prescientific histories of religion, but in the scientific sagas that replaced or supplemented them. And, of course, the most compelling of these scientific supplements is the story of evolution.

But evolution no more evolved from nothing than God did. It, too, appeared within a social setting and cultural milieu. A telling marker of what might be called cryptotheism—a lingering of theological thought in scientific discourse—can be found in much present-day evolutionary biological, ecological, and environmental discourse. This marker is the prevalence with which even the most Darwinian of naturalists reserve some favored trait to distinguish humanity from the rest of life on earth, the rest of what was once called "Creation." Thus we are told by turns that humans are uniquely superior due to our upright posture (allowing us to think of ourselves as literally "above" other species), our opposable thumb (man the tool user), our linguistic abilities (man the symbol user, the storyteller), our superanimalistic soul (Descartes' ploy), our self-awareness, our moral superiority (even in the absence of God), or one of the most recent and desperate euphemisms: our "big brains."

Even Stephen Jay Gould, an ardent foe of the idea of progress in evolution (1980), would have us believe (and he is by no means alone) that all other organisms on the planet are shackled to the ancient system of natural selection whereas humans, and humans alone, can evolve through "cultural selection." Of course, Darwin's very term, "natural selection," was coined in comparison to the "artificial selection" of animal breeders. Darwin wanted to show the evolution of all species from a common ancestor. But he also had to make evolutionary theory palatable to a monotheistic populace. The acceptance of evolutionary theory required that it take over many functions of Judeo-Christianity—and in doing so compromised from the start the potential for a biophilia which would have seemed natural considering the kinship Darwin demonstrated between human and other life-forms.

If we believe that other animals have feelings, that we have no intrinsic superiority over them but are part of a global nexus of life, we are confronted with a moral crisis. This is the crisis of the animal rights groups and those who believe that humans are compromising the welfare of planetary life. As our growth and exploitation of resources force us to reconsider our relationship with other life-forms, we may find new value in systems of beliefs either dismissed by Christianity or absorbed by monotheism. The animism, theriomorphism (totem worship), pantheism, and polytheism that preceded the advent of monotheism as Judaism, Christianity, and Islam may contain powerful sources for present and future action and reflection. Culturally, biophilia and biodiversity are scientifically sanctioned catchwords calling for us to attend seriously to nature and our responses to nature—forms of attention already more fully developed in traditions less nomadic and technologically expansive than those of the West. If the love of life and the preservation of biodiversity are to become planet-scale education projects, Western countries should certainly lead the way—and by example, not by preaching. Ethically speaking, the West, which has led the way in environmental destruction, has the greatest obligation to restore biodiversity.

Yet nature is already saved and, moreover, largely out of our hands. If once we thought all organisms were for our benefit, and later we thought we could with bombs kill off all life on the planet, it is once again a mark of our hubris to think that we may now save the biological world. It is true that the current rate of extinction on the surface of the earth is comparable to major losses—the so-called mass extinctions—of life in prehistory. Indeed, the current rate of extinction is estimated to be the greatest since the end of the Cretaceous. A planetary catastrophe (implicated in the demise, among many other life-forms, of all the dinosaurs) may well have been caused by a bolide (meteorite, comet, or planetoid) landing offshore the Yucatan peninsula in Mexico.

The recent mass extinctions are claimed to differ, however, in that they arise not from an outside force but from within, as the result of human expansion. Some would have us think that the wreaking of such havoc on the environment is unparalleled in earth's long history. This is a kind of negative theology making us, if not God's chosen ones, then his prodigal sons; in any case, as good guys or bad, we remain the stars of the evolutionary show. The deflating fact, however, is that we have been preceded in our massive ecocide by other life-forms.

Because of the limited materials on the earth's surface, organisms have been competing for resources, polluting environments, and feeding on un-protected corpses and living bodies for over 3 billion years. The whole changeover of the atmosphere—from an anaerobic one suited for organisms poisoned by oxygen to an oxygen-rich one suitable to our ancestors—occurred as the result of a pollution crisis. Before we bow down in fear to our shadows as the grim reapers of evolution, let us remember that the Chinese ideogram for crisis combines the sign of "danger" with that of "opportunity" and recall, too, that other organisms have dangerously altered the planetary environment before us. Two billion years ago cyanobacteria, newly evolved microorganisms that used the hydrogen of water for photosynthesis, plunged the biosphere into crisis mode. Their "waste"—the free oxygen that sent thousands of varieties of organisms to early graves—altered the previous planetary habitat forever. From the point of view of anaerobes, the global environment was ruined. But for the oxygen-tolerant and oxygen-respiring forms among which are to be counted our remote bacterial ancestors, this ecocide, this destroying of the planetary home, made life possible.

Chaos mathematics, disequilibrium thermodynamics, and complexity studies have shown how certain structures, which seem fragile, amorphous, or dangerously out of balance, are as often as not at a bifurcation—a turning point or critical juncture on the way to still more complex structures. Planet Earth with its global human-fostered technology may presently be undergoing such a difficult transition period. The case history of cyanobacteria is worth thinking about when people, scientists among them, sound the alarms for us to gather round and "save the planet." By innovatively using light to split water, and rampantly growing wherever they could, cyanobacteria altered the atmosphere and poisoned large numbers of its inhabitants, not least of all themselves. Our hunting of animals for food, our razing of trees in lush species-rich Amazonia, and our urbanization of landmasses have also degraded the environment in a major way. People have every right to care about such degradation and loss of species, to fight against it and organize Brazilian mutual funds or whatever it takes to preserve biodiversity. There are, as many have pointed out, aesthetic, pharmaceutical, genetic, historical, and other reasons for saving the environment. The most important of these, and least often mentioned, may be the relationship of certain lush regions of the earth and the present biogeo-chemical regime—not just global climate, but global chemistry—that supports human beings.

But let us not kid ourselves into thinking we are saving life on earth as a whole. For all we know the demise of human beings may accelerate the appearance of some new complexity as far beyond primate intelligence as primate intelligence is beyond rodent responsiveness. After all, without the decline of the reptiles, mammals might never have been able to come into their own. So let us cut through the salvationist hyperbole and see that talk of saving the world really means saving that part of the planetary environment which has *traditionally and comfortably* supported human beings. It is fine to urge the salvation of the environment in which our species first flourished, but in fact even this cannot be done. Any return to green pastures, flowering fruit trees, bubbling brooks, and rolling glades will be a turn not of the circle but of the spiral. Or, as the Buddhists say, all beings are already saved.

Although the loss of charismatic large animals such as elephants, giraffes, and tigers from the surface of the planet would represent a tragedy comparable, on a smaller scale, to the murder of members of one's own family, it is not true that our rapidly multiplying, change-engendering life-form is the only one ever to cause mass extinctions of fellow organisms. Only a sort of well-wishing, or perhaps a deep guilt combined with an equally deep repression, can make us forget life's inescapably murderous legacy. One hears a Christian, even a Puritan, echo in the talk of our need to save the planet. In fact, we cannot stop evolution. We can, and probably should, try to stop certain global human activities among which may be counted overuse of plastics, rain forest destruction, and soil erosion. But to think that by doing so or not we are either going to kill off life on earth or save it is a form of unscientific self-aggrandizement. Such egotism smacks of the dated Christian notion of people being one step above the beasts and two steps (after the angels) below God. In terms of biophilia and biodiversity, we believe it is better to think of ourselves as all just a part of Gaia and not even, in any way, the most important part.

What is Gaia? Although memorizable phrases may be inadequate and specious we can try to convey the power of Gaia as principle and being.

God, Gaia, and Biophilia

First of all, on the cultural level, as a conscious taking of the name of the ancient Greek earth goddess and mother of the Titans, Gaia disturbs, perhaps even cancels out, the lingering theology of an external male god who has made humanity in his image and then narcissistically countenanced us to use the rest of creation to be fruitful and multiply ourselves. Roughly, Gaia is the nexus and nest, the global life and environment, the planetary surface seen as body rather than place. Recognizing prototactic living organisms such that they, in their patchy environments, themselves become selective agents is essential to the Gaian view of life on earth. The 3 to 30 million species of protoctists (protists: ciliates, foraminifera, algae, amoebae, and their larger descendants), fungi, animals and plants, and the entire bacterial continuum of gene-exchanging microbes together with their physical surroundings prevent the rampant exponential growth of populations: simply put, Gaia is Darwin's natural selector. All of these organisms have a tendency for population explosion. That this enormous population potential fails to be reached is Darwin's lesson. There are checks upon growth at all times throughout the life cycles of all organisms. Gaia, the sum of the interacting organisms of the biosphere, checks growth and therefore acts as the natural selector.

The Gaia hypothesis claims that, on earth, the atmosphere-hydrosphere, surface sediments, and all living beings together (the biota) behave as a single integrated system with properties more akin to systems of physiology than those of physics. The traditional Darwinian view is a linear scheme in which organisms are affected by the environment and the environment in turn is the result of chemical and physical forces. This linear scheme may owe much to the Victorian era of science in which Darwin worked, an era in which, to make evolution acceptable to a religious populace, Darwin had to give it a credible, detailed mechanism. Since the most respected science of the time was the physical discoveries of Isaac Newton, Darwin tried to portray evolution as the result of blind principles and mechanical interactions, just as Newton had portrayed gravity. Gaia has a different view of the environment. It is seen less as matter interacting blindly than as a superordinated collection of living things. The environment, far from being a static backdrop influenced only by physical and chemical forces, is highly active and biologically modulated.

The environment is an integral part of the Gaian system of the living earth as seen from space. The Gaia hypothesis asserts that the temperature and aspects of the chemical composition of the earth's surface are directly regulated by the metabolic, growth, and reproductive activities of a vast biota. Gaia theory, first formulated by British inventor and atmospheric chemist James E. Lovelock in the late 1960s, has been developed in the scientific literature for more than twenty-five years (Margulis and Lovelock 1989). Recent forays into Gaia science have been boosted by continued space exploration: views of the entire globe from orbit in comparison with other planets greatly influence all of us: clearly life on the planet is some kind of interacting unity. If symbiosis is defined as the living together in protracted physical continuity of different kinds of organisms then, as Hinkle (1992) asserts, Gaia is simply symbiosis seen from space.

In its stronger forms, the Gaia hypothesis claims that the mean global temperature, the composition of reactive gases in the atmosphere, and the salinity and alkalinity of the oceans are not only influenced but regulated, at a planetary level, by the flora, fauna, and microorganisms. This regulation, as we have seen, is not completely homeostatic. It is not like the thermostat of a house set at a single temperature for all time. It is homeorrhetic—regulated around what systems engineers call a moving set point, a set point which can change, as when global oxygen rose from a trace gas to a major constituent of the earth's atmosphere some 2 billion years ago. If we look at the development of a human body, from fertilized egg through blastula and embryo to child and adult, it becomes clear that the regulation of living systems is far more complex and fascinating than anything so far engineered. The chemical reactions of a physiological system, unlike those of an inert physical (geological or geochemical) system, are under active biological control. In the absence of the global physiology postulated by Gaia, variables such as global mean temperature, atmospheric composition, and ocean salinity would be deducible directly from Earth's position in the solar system. These aspects of the planetary surface, responding to changes in the energy output of the sun, would conform to the known physical and chemical laws.

Yet an examination of Earth's surface shows that such aspects vary widely from what would be expected based on the principles alone of physics, chemistry, and other nonbiological sciences. These principles predict that Earth should have reached a chemical steady state with carbon dioxide and nitrogen as compatible gases, as on Mars or Venus, for example. Chemically, however, Earth is extraordinarily anomalous: oxygen, methane, and hydrogen coexist in the atmosphere; carbon dioxide is in decorative carbonate rocks instead of in the air; iron is found in huge bands from kilometer-wide to micron-scale patterns; ancient gold is intertwined with long stretches of organic carbon in locales few and far between: Witwatersrand, South Africa, and Michepecoten, Ontario. Such planetwide disparities are what led Lovelock to propose the Gaia hypothesis that the earth is a physiological system.

The Gaia hypothesis has been criticized because of its controversial claim that the earth behaves like a living being. Some believe that Gaian views lend credence to the idea that Earth—the global biota in its gaseous and aqueous environment—is a single gigantic organism. Since this notion resonates with ancient beliefs and, relative to Western secularism, leads to a radical reenchanting of the world, it has come in for suspicion, especially from the Neo-Darwinian biologists whose nonchemical view of life Gaia threatens to make irrelevant by comparison. Nonetheless, an organism-like response of the planetary environment and its biota is clearly detectable—a behavior distinguishing Earth from Mars, Venus, Mercury, and any outer planet or its moons. The evidence in support of the Gaian idea that the earth's surface behaves as a macrobody includes the realization that the atmosphere is an extension of the biota. If the earth's surface were not covered with oxygen-emitting bacteria, algae, and plants, as well as methane- and hydrogen-producing bacteria and countless other organisms, its atmosphere would long ago have degenerated to the same carbon dioxide–rich steady state that today can be found on Mars and Venus (Margulis and Olendzenski 1991).

Another strong argument for Gaia comes from astrophysical models of the evolution of stars. Early in its history the sun was some 30 to 40 percent cooler than it is at present. Yet fossil evidence shows that life has existed since just after the earth's formation. (The Earth–Moon system is 4.6 billion years old and the first fossil communities, domed rocklike structures called stromatolites, left their record at least 3.9 billion years ago.) The more than 3 billion years of life's tenure on earth can be verified by the fossil evidence of such bacterial communities. Since organisms survive only within the limited temperature range in which water is a liquid (o to 100°C), the global mean temperature of earth apparently has not strayed outside these bounds since life's inception. The sun's increase in luminosity implies the surface temperature of earth should have increased correspondingly—but it hasn't, which suggests that the biota, cybernetic interactions of life with its environment, in a word, Gaia, has behaved in ways that have cooled the planet. The literal impact of meteors on global temperatures is becoming increasingly well known, as the debris, spewing into the atmosphere and blocking sunlight, makes it colder. But alterations in temperature are not evidence against global temperature control; perturbations are precisely what physiology deals with, reacts to, senses, and counters or attempts to counter. Although complex and certainly not understood in detail, the biologically assisted removal of greenhouse gases such as carbon dioxide and methane has probably played an important role in global thermoregulation. Temperature regulation may be a geochemical accident, as traditionally assumed, but it seems at least equally likely that exponentially growing populations of gas-producing organisms could exercise a global control on planetary temperature. The possibility seems less remote as people are beginning to accept the sometimes surprising behavior of complex systems. After all, we should recall, the human body is a complexity of cells.

Gaian ideas have been fruitful in inciting communication about the earth as system among scientists who usually do not study each other's work. Gaian ideas have generated new hypotheses relating to the control of global variables that are currently being explored. One such idea suggests that oceanic salt and acidity levels are also actively sensed and stabilized by the biota. Salts, especially sodium chloride, delivered continuously to the marine realm by rivers, should have accumulated by now in the oceans to levels (greater that $0.6M$ sodium chloride) far exceeding those permissible to any ocean life except the most halophilic (salt-loving) bacteria. Yet the earth's oceans have remained hospitable to fish, plankton, and many other life-forms for hundreds of millions of years at least. The relative constancy of ocean salinity (at 3.4 percent) suggests that the waters continuously undergo some form of desalination. The Gaia hypothesis suggests that salt

God, Gaia, and Biophilia

control on the global scale is analogous to that within the human body. Biological homeostasis might be accomplished by myriad interacting mechanisms—all products of the evolutionary process. Ocean salt regulation may even be achieved, at least in part, by the formation of evaporite flats. We know these structures result from activities of microbial communities and we know they can tie up great quantities of salt. Lovelock (1988) has even argued that life has influenced the movement of continental crust to the tropical regions, where rapid evaporation occurs. If this is the case then even plate tectonic movement is encompassed within the sprawling realm of life.

Gaia has evolved by prototaxis coupled with continuously checked exponential growth. Earth's atmosphere maintains an anomalous amount of oxygen (about 20 percent) in the presence of gases that react with it; the surface atmosphere has a mean mid-latitude temperature of 18°C; the pH of the lower atmosphere and oceans is slightly greater than 8. All of these values have been relatively constant for millions of years, and all are within ranges permissive to life. Such persistent and drastic differences between Earth and its neighbores reinforce the Gaian view of planet Earth and its recognition that biota and environment—biosphere—form one planet-wide homeorrhetic system. Prototaxis of the individual components leads the system to respond with alacrity to tendencies of the physical and chemical surroundings toward excursions beyond the limits to life. One predictable response includes the rapid growth of populations of metabolically and morphologically distinctive organisms whose interactions stabilize the whole.

Biodiversity is essential, therefore, to the physiology of the planet and perhaps we "biophiliacs" sense this. Sensitivity (and therefore prototaxis), biodiversity, and exponential growth rates of populations are intrinsic to Gaian physiology, but therein lies the rub. Gaia persisted long before people described or even worshipped her. Gaia, with or without humans, is likely to generate more diversity and continue to persist long after the extinction or speciation of humans, perhaps even after the atmosphere is depleted of the carbon dioxide needed to cool itself in the face of an increasingly luminous sun. Gaia, radiating forms of diversity as yet only dimly conceivable to us, may even survive the predicted explosion of the sun into a red giant, a final magnificent sunset which will boil away earth's oceans.

Let us try to come to grips with this evolutionary becoming that swamps the human species no less than the march of generations tramples an individual animal's life span. Evolution is a planetary phenomenon of thermodynamic disequilibrium: powered by the sun and, so far as is known, confined, until very recently, to the surface of the earth. (One of us would argue that *Apollo, Soyuz, Viking, Mariner, Voyager,* and other such missions represent the beginnings of a planetary budding, organic in nature, that will culminate in the extravagant reproduction of offspring biospheres; Sagan 1992.) The strongest argument for biophilia (and for the hypothetical outcome of biophilia's disciplined practice, biodiversity) is not ethical. Our reaction to other life-forms may be highly negative—as it is with cockroaches, spiders, maggots, snakes, rats, indeed virtually any organisms that reproduce rapidly or threaten to harm our person. Biophobia and biophilia are part of a finely differentiated prototaxis that extends throughout not only the animal but also the plant, fungal, protoctist, and bacterial kingdoms. Although plants, for example, do not have emotional reactions, their chemistry, their smells and visual attributes, draw to them and keep away certain very specific others. Fungi, too, elicit strong emotional responses through chemistry alone, as in the human aversion to toadstools.

The presence of biophilia suggests we not only love birds and flowers but also have an inbred contempt, distaste, and perhaps hatred of certain other life-forms. Even if we were to obey Kant's categorical imperative and treat all beings, starting with humans, as ends rather than means, cultivation of biophilia in the broad sense would lead us not to preserve biodiversity but only favored plants and animals. "All organisms are equal," we seem sometimes to want to say in the discourse on biodiversity, "yet some animals are more equal than others." Not surprisingly these "more equal" beings are often large mammals either like us or like those found in the savanna in which humanlike primates first evolved. One of the reasons for the decline of the aesthetically pleasing and emotionally resonant beasts such as African elephants and Bengal tigers is that human beings in our agricultural prowess have found shortcuts in the trophic line. From the vantage point of the charismatic vertebrates, and our love for them, this is very sad. But from the viewpoint of an evolving biosphere it may be analogous to the "trimming" that goes on in an expanding corporation.

If we were truly serious about saving *all* other organisms, we would follow Jainist principles and filter our water to save the paramecia. We would surgically implant chloroplasts in our skin in order to photosynthesize ourselves and not uproot lettuce or carrot plants. We certainly would not cavalierly flush away our solid wastes that serve as a breeding ground for *E. coli* and other gut bacteria. This *reductio ad absurdum* shows the hypocritical element implicit in the rhetoric of ecological salvation. In fact, part of the reason a predator like the Bengal tiger is so physically arresting is that it feeds at the top of the trophic chain; it is a carnivore, a killing machine, a king unfairly taxing plant and animal pawns. It has been said that all great poems contain an element of cruelty. Perhaps the same may be said of animals in the biosphere.

Nor is the strongest argument for biophilia practical. Preserving the Amazonian rain forest may serendipitously preserve a tree or insect species from which we can derive a valuable new drug or food or fiber. Such economic incentives may make the difference for a pragmatist, an industrialist trying to reduce quality to quantity on the spreadsheet of profit. For us, however, the strongest argument for a directed biophilia leading to a general if not all-encompassing biodiversity has to do with survival—not the abstract ethical survival of all sentient entities, but our own survival, the preservation of a certain quality of human life.

All life on earth is a unified spatiotemporal system with no clear-cut boundaries. Encouraging our biophilia, preserving blocks of biodiversity before they are converted to concrete skyscrapers and asphalt parking lots, is a way of enhancing the possibility that human beings will persist into the future. This future may be indefinite, as some few species do not become extinct but "scale back" and become symbiogenically attenuated and reintegrated into new forms of life and patterns of living organization. If we consider, for example, the ancestral oxygen-respirers that evolved into the mitochondria of all plants, animals, and fungi, we would have to say that this mitochondrial "species," codependent as it is, has resisted extinction, surviving and spreading (and still going strong) in multifarious forms for some 2,000 million years. Humanity seems to have been presented with an opportunity, rare in evolution, to do likewise. By allying ourselves more closely with once distant life-forms, by affiliating ourselves biophyletically, not only with the plants and animals whose ongoing demise weighs so heavily at present on our memory, but also with the waste-recycling, air-producing, and water-purifying microbes we as yet take largely for granted, we may be able to aid in the flowering of earth life into the astronomically voluminous reaches of space.

Like the ecocidal rampage exacted by the violently fast spread of ancient strains of photosynthetic bacteria, our expansion across the surface of the planet has created environmental havoc, and wholesale biological destruction, in our wake. Like those cyanobacteria we have polluted, we have murdered, we have slaughtered with laughter and pride. Like them, we are not good or evil. Like them, the planetary changes effected by our explosive population growth have prepared the way for strange new living things.

In a process of negative feedback not unlike that illustrated by the expanding small restaurant, the worldwide propagation of human beings has led to a planet ever more inhospitable to human life. As earth becomes increasingly polluted and overcrowded, as the global commons of atmosphere and ocean are spoiled as surely as the common grazing areas of small towns were once destroyed for all, self-sufficient environments at home and abroad, in space and beneath the ocean, become more attractive. Future human settlements may be like Arizona's Biosphere II: materially closed but informationally and energetically open systems of bio-affiliated life-forms that replenish water and air and indefinitely return their wastes into food, clothing, and shelter. Such biodiversity-containing artificial biospheres, materially separated from the global ecosystem, could also persist in orbit or on the surface of other planets without the need for resupply from earth. Because the tendency of all life is to reproduce, and in so doing spoil the environment, such enclaves would provide insurance against global environmental deterioration. Artificial biospheres and closed ecological systems are analogous to aerobes that flourished in the wake of the environmental destruction wreaked by the cyanobacterial spread 2 billion years ago. And, crucially dependent for their existence on both biodiversity and biophilia, they represent the only currently imaginable means of completely recycling wastes into food away from earth. This suggests yet again that the greatest level of living organization yet to evolve is Gaia.

Although Gaia's biodiversity is currently spread across the planet, the thought experiment of biospheres shows how Gaian biodiversity may be "individuated" or concentrated into independent units: Gaian offspring. One criticism leveled against Gaia is that earth cannot possibly be an organism, since it has no little ones. Yet the creation of recycling chambers with humans, such as Biosphere II, currently housing eight humans, food species, and technology near Oracle, Arizona, represents the first wave of an ultimately natural process of Gaia producing little ones. In far more sophisticated, as yet almost inconceivable forms, perhaps such systems will preserve biodiversity after humanity and the death of the sun. From the human perspective biospheres are communities, but from the Gaian perspective they are propagules.

The "technology" needed to cut the material umbilicus to earth, to truly migrate and live independently in space, is nothing other than other life-forms. Only select samples of earth's biodiversity, natural systems with soil bacteria, recycling fungi, food species, and many other organisms can support people in space. Probably nothing else so clearly illustrates that Gaia is not just a metaphor. To survive in space we require thousands of other living beings, entire ecosystems. They are not lower but *essential* to a life-form of which we are a mere part. Ultimately, we may be a dispensable part. Moreover, despite the great technological accomplishments of the human species, we are not yet close to recreating photosynthesis in the laboratory, let alone miniaturizing it as cellular life does. Gaia's photosynthesis, nitrogen fixation, and other chemical production and waste management abilities are still far ahead of modern technology.

Can we, as humans, destroy the environment we love and yet remain hopeful and festive? Population growth has decimated the earth. An ecologically correct alternative is to rally the peoples of the earth together into an enforced state of stasis, one in which population growth and exploitation of the living environment for human ends are tightly controlled. Certainly what most environmentally minded persons advocate—conservation—itself seems to accord with the precepts of the Judeo-Christian tradition, from the rhetoric of salvation to responsible stewardship over nature. But even assuming that the nationalistic economies of the world could be convinced of the dangers of growth (a doubtful proposition), even assuming that the world's governments could be persuaded to confine themselves to their borders and leave other nations alone, can one truly imagine such retrenchment enduring indefinitely? Would not the stage be set for defectors?

Life on earth is a complex, fractally individuated, chemical system whose basis is a mostly green layer of photosynthetic matter as bacteria, algae, and plants. This layer makes its own nutrition from air, water, and sun. This layer continues to grow and tempt any life-forms that would "cheat" and make use of it (or each other) rather than build themselves from scratch. What with solar radiation impinging on the surface of the earth, and its storage in the sediments as energetically exploitable matter, it seems inevitable that "unfair players," either cheating bands of humans or new species of organisms, will evolve, willing to transgress the enlightened growth-curbed policies of any hypothetical ecologically correct humans. Conservation on an evolving planet is ultimately a lost cause. Truly considered, this is a very difficult, even a dangerous, thought—indeed, most would rather not think it, as it seems to admit of no solution save a fruitless resignation to the endless murderous quality of life in an energetic universe. Maybe other beings have thought similar thoughts, and that is part of the natural antipathy, revulsion, and embarrassment—all forms, by the way, of biophobia—we sometimes feel face to face with our "ancestors," be they an unhip parent in polyester leisure suit, the fornicating apes Bishop Wilberforce could not admit were his relatives, or the microbial gunk in the sewer. And yet over against this instinctual distaste there is awe that we have come from such and are going—where? Often the strength and the weakness of something can be one and the same. The Judeo-Christian ethical perspective is a mental safety net protecting us against the onset of a Dionysian nature madness induced by a lack of guidelines. But it can also be an iron gate barring access to visions of the future as well as a clear grasp of biology's amoral status quo.

Once we disabuse ourselves of the ecologically correct inheritance of Jean-Jacques Rousseau's liberal nostalgia for a pristine (good, unpolluted, tranquil) past which in fact never existed, we will be in a better position to appreciate our present situation as mere humans trying to survive within a biosphere that our own agricultural and technological manipulations have irreversibly altered. Our very self-centeredness has led us to reproduce without concern for the environment around us. But now our past has caught up with us. We are stuck in the delicate position of having to undo our ecological karma. At the same time, as Gary Nabhan and Sara St. Antoine point out in Chapter 7, the epidemic global spread of technological humanity has whittled away the oral traditions of native cultures with specialized knowledge of local ecologies. Elsewhere in this volume, Paul Shepard focuses our attention on the cultural narcissism of the human species which prefers its wild animals caged and has rendered domestic pets into genetic "goofies" incapable of independent survival. One should note, however, that many other organisms in the history of life have been rendered chronically dependent as a result of interspecific alliances. Although the ancestors of mitochondria were free-living, independent organisms, their descendants are totally incapable, even in nutrient media, of survival outside the host cell. Thus a movement from, say, free-living wolves to urban dogs dependent on regular servings of pet food may be lamented, but it is hardly unique.

Indeed, if global biospheric relations are undergoing a major reorganization due not so much to the *interference* of humanity (this would again be the epitome of the shallow-ecological view, since it keeps people apart from nature) but rather the development within the biosphere of the human phenomenon, it is perfectly natural for us as sentient beings to feel distress in the presence of such sweeping changes. What is in question, however, is the assumption that we know that the planet is sick and can fix it by bringing it to some sort of environmental stasis. Without being dismissed as technophiliac, we would like to suggest that the decline in species diversity may be balanced by an increase in technological diversity—a trade-off that may ultimately enhance the longevity of the biosphere.

In 1973 the Soviet biologist M. M. Kamshilov (1976) performed a controlled experiment in which he added harmful phenolic acid to a series of laboratory communities, each more complex than the last. The first ecosystem consisted only of bacteria—the only kingdom of life whose members are varied and biochemically versatile enough to completely recycle foods into wastes without the aid of members of other kingdoms. By themselves the bacteria were able to break down the phenolic acid, but not as quickly as the more complex systems. The second vessel, which contained not only bacteria but aquatic plants, was able to neutralize the toxic acid more rapidly than the bacterial ecosystem. Following in this trend, the third system, to which was added mollusks, was even more effective. And the fourth, which incorporated fish, mollusks, plants, and bacteria, removed the phenolic acid at a quicker rate still. Notice that the model systems that recycled fastest were not simply the most complex assemblages but those that incorporated more recently evolved organisms into a base of ancient life-forms.

The appearance of dramatically new life-forms may cause an initial period of destabilization and discomfort as they rapidly spread. But for a newly evolved life-form to survive in the long run it must integrate itself into the global ecosystem of which it forms an increasingly large part. The global ecosystem is far bigger and more metastable than any single life-form, including the most disruptive. This statement applies emphatically to technological humanity—a species now confronting with greater responsibility than ever before (out of sheer necessity) the consequences of its pioneer stage of rapid proliferation and settlement. If this is the case, then the present concerns for the environment need no more signify planetary pathology than they indicate robust global health. Indeed, they may be more like the pains of some strange animal which, in sensing the culmination of its difficult pregnancy, takes conscious care to eat well and procure extra rest.

REFERENCES

Gould, S. J. 1980. *The Panda's Thumb*. New York: Norton.

Hinkle, G. 1992. "Undulipodia and Origins of Eukaryotes." Ph.D. thesis, Boston University, Department of Biology.

Kamshilov, M. M. 1976. *Evolution of the Biosphere*. Moscow: Mir Publishers. Original experiments described (in Russian) in M. M. Kamshilov, "The Buffer Action of a Biological System," *Zhurnal Obshchei Biologii* 34(2) (1973).

Lovelock, J. 1988. *The Ages of Gaia*. New York: Norton.

Margulis, L., and J. Lovelock. 1989. "Gaia and Geognosy." In M. B. Rambler,

364 L. Margulis, and R. Fester (eds.), *Global Ecology: Towards a Science of the Biosphere*. Boston: Academic Press/Harcourt Brace Jovanovich.

Margulis, L., and L. Olendzenski, eds. 1991. *Environmental Evolution*. Cambridge: MIT Press.

Mehos, D. C. 1992. "Ivan E. Wallin and His Theory of Symbionticism." In appendix of L. N. Khakhina, *Concepts of Symbiogenesis: A Historical and Critical Study of the Research of Russian Botanists*. New Haven: Yale University Press.

Sagan, D. 1992. "Metametazoa: Biology and Multiplicity." In J. Crary and S. Kwinter (eds.), *Incorporations*. New York: Zone.

Wallin, I. E. 1927. *Symbionticism and the Origin of Species*. Baltimore: Williams & Wilkins.

CHAPTER 12

Of Life and Artifacts

Madhav Gadgil

I AM, I GUESS, a confirmed biophilic. My pleasantest memories are all of encounters with wild animals in their natural settings: like the evening some fifteen years ago when I was standing in a watchtower overlooking a waterhole in the Bandipur tiger reserve in South India. Just as the sun was about to set a herd of some forty elephants came tramping up to the pond. In the lead was an enormous cow and close on her heels an even more imposing tusker. The cow was apparently in heat and the tusker mounted her as she waded into the water. And at that moment, no fewer than twenty-two other male elephants, from youngsters just beginning to sprout tusks to big males only slightly smaller than the tusker, sorted themselves into eleven couples of even size and jostled back and forth, arrayed around the margin of the waterhole, as the pair in the center mated. This drama lasted a full fifteen minutes. It left an indelible impression on my mind.

Artifacts as a Selective Force

Reading Edward Wilson's fascinating speculations on biophilia (1984) prompted me to reflect on my own love for the natural world. For this is not a trait widely shared by my kinfolk. If there is a learning rule that inclines humans to love natural diversity, very few among the urban middle classes of India actually come to do so. My father was a rare exception and was fond of watching birds and trekking in the mountains. He had acquired these interests, not from his family, but from friends as a young student at Cambridge University in England. I learned to love nature enough from him to want to take up the study of the natural world as my lifelong avocation. Only then did I come into contact with the tribals and peasants and herders of India, who loved the natural world as much as I did.

My own son picked up these interests at an early age and became an avid naturalist. But when he was fourteen I bought a personal computer. My son's attention was very soon riveted on this complex "organism." For while he enjoyed watching frogs and lizards and birds, he could make the computer do all sorts of fascinating things. His exploratory instincts found an even more attractive outlet in learning about and then fiddling with the computer than they had in observing the natural world. Soon he was hooked on the world of artifacts, in part at the cost of his fascination for the natural world.

Hominids have been populating their world with such complex entities for 2 million years. That means 100,000 generations, surely a period long enough for the artifacts to have acted as a selective force in their own right. Artifacts have been playing an increasingly significant role in governing human existence over this long time interval. For our ancestors, digging sticks must have meant ability to get at tubers underground, cooking hearths freedom from parasites infecting meat, and a floating log access to an island free of predators even before our own species, *Homo sapiens*, appeared on the scene 200,000 years ago. It is plausible, then, that humans are programmed to learn to be interested in a well-controlled fire just as they are known to be attracted to natural landscapes with water. Of course, a great many artifacts are very recent and humans are unlikely to have evolved appropriate tendencies of attraction or aversion toward them. Thus humans appear to have an innate aversion to closed spaces or spiders, but not to handguns or frayed electric wires (Chapter 3 in this volume). But, by and large, it is plausible that humans may have evolved innate attention or aversion responses to a variety of artifact configurations, just as they may have evolved such responses to living organisms.

Mimicking Life-Forms

Our animal ancestors are equipped with an exploratory drive, a natural curiosity toward unusual configurations in their environment. Such a drive may be at the common root of the human attraction or aversion to other living organisms, as well as artifacts. The simplest artifacts are pieces of stone or bone shaped to form a cutting edge or a piercing point. But when humans began to fabricate more complex artifacts, they tended to relate them to the living creatures in their surroundings. The earliest human-made figurines are icons of women and animals, as are the earliest paintings. Dagger handles have faces of tigers and lions on them, as have pillars of buildings. Birds have inspired people to build flying machines. Human fascination for complex entities, whether artifacts or living organisms, may thus have a common basis.

Evolution: Cultural and Biological

Notably, the cultural evolution of human artifacts shows many parallels with the organic evolution of living forms. More and more complex living organisms of larger and larger size have with time made their appearance on the evolutionary stage (Bonner 1988; Wilson 1992). The earliest living organisms were microscopic, single-celled, without well-defined cell organelles. To these have been added successively macroscopic organisms—among the largest ever being the modern kelps, whales, and redwood trees. These organisms are themselves composed of millions of cells, each with a number of well-defined cell organelles. So have the earliest artifacts been bits and pieces of stones, small flakes, devoid of any internal structure. To

Of Life and Artifacts

these have slowly been added pyramids and skyscrapers made up of thousands of stones and bricks, iron and windowpanes, electric wires and telephone lines. And just as modest-sized humans are in some ways the most complex of organisms in terms of the incredible number of linkages their neurons make, so we now have modest-sized computers each with an exceedingly complex internal organization.

More and more complex organisms appearing on earth have tended to colonize a larger and larger range of habitats. Life probably began in warm, shallow pools of water on the surface of the ooze. It then burrowed in the mud and migrated to deeper seas. Later still, organisms came out on dry land and flew up in the air. So were artifacts first confined to tropical savanna habitats of early hominids. Later they permitted people to penetrate rain forests and colder latitudes, to reach oceanic islands, and, finally, to penetrate outer space.

With time the total diversity of living organisms has been continually increasing despite continual extinctions of individual species and several episodes of megaextinctions. So has the diversity of artifacts been continually on the increase (Petroski 1992). Consider just one category: transport vehicles. The earliest such vehicles were logs floating on water, later scooped into dugout canoes. Then were added rafts of reeds, bullock carts, and horse chariots. Today they span deep-sea submersibles and nuclear submarines, oil tankers and rowboats, bicycles and dogsleds, bulldozers and bullet trains, hang gliders and supersonic jets, satellites and beach buggies. Japanese car manufacturers produce a new model of car every month.

As the size, complexity, variety, and ubiquity of living organisms have increased, so too have their mass and the fluxes of energy and matter driven by them. Woody land plants arose only some 400 million years ago, although life itself originated more than 3.5 billion years ago (Cowen 1990). But their mass now exceeds that of all other kinds of living organisms. Warm-blooded vertebrates, birds and mammals, came into existence even later, only 250 million years ago or so. But they metabolize and therefore drive material and energy fluxes at rates far higher than any organism of comparable weight. Plants cultivated by humans are in a way artifacts. They came on the scene only about 10,000 years ago, well after our species appeared some 200,000 years ago. But today the mass of cultivated plants is of the same order as the mass of natural vegetation.

The mass of human-constructed buildings in metropolitan centers far exceeds the mass of the most luxurious forest. On a worldwide scale the mass of the buildings may be only one order of magnitude lower than that of woody plants. Humans increasingly drive energy and material fluxes of the world, thanks to their control of a large mass of artifacts. One estimate puts at 40 percent the human appropriation of the entire terrestrial photosynthetic production in the world, apart from tapping a substantial fraction of past photosynthetic reproduction sequestered as fossil fuels (Vitousek et al. 1986). The rate of consumption of energy by artifacts like rocket engines far exceeds the rate of metabolism of any animal.

Causes of Artifact Evolution

This evolution of artifacts has been driven by a variety of competitive interactions both within and between human groups. Primitive agriculturists with their axes, digging sticks, and earthen storage jars added to both the variety and complexity of artifacts employed by hunter-gatherer societies. These implements enabled them to maintain higher population densities and create surpluses of food that sustained them for several months without devoting any time to gathering or cultivation activities. They could therefore effectively wage aggression against hunter-gatherers and came gradually to replace them, and their simpler artifacts, over much of the world. The process of complex artifacts helping societies wage aggression against others continues to this day: in the recent Gulf War the American victory was primarily based on access to more complex electronic devices. Such military advantages generate forces that prompt societies all over the world to acquire as well as continue creation of ever more complex devices of aggression.

Corporate human groups based on a specific function such as processing or transport of materials, energy, or information have come to play an increasingly important role in human societies. Such groups are continually involved in intergroup competition over scarce resources. Access to

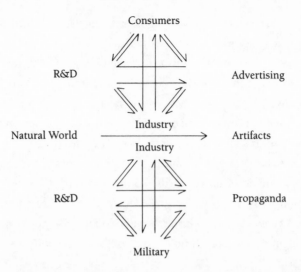

FIGURE 12.1. A positive feedback cycle—involving growing demands for consumer goods and military hardware, industrial production, persuasion through advertising and propaganda, and research and development—is responsible for production of artifacts at the cost of the natural world at an ever accelerating pace.

complex artifacts plays a critical role in determining the outcome of such competition. Thus the British textile mills caused the extinction of the Indian handloom weaving industry because of their advanced technology, in turn prompting the establishment of textile mills in India. Similarly Japanese fishing fleets with their complex coordination and on-board canning and other facilities have outcompeted less sophisticated fishing concerns.

Within human groups possession of ever more complex artifacts is more and more a symbol of higher social status—whether compact disc players, cellular phones, or electric toothbrushes. So corporate groups go on producing ever more complex artifacts and spending ever greater resources on advertisement to persuade people that the possession of more and more complex artifacts is indeed related to a higher social status. These processes have today created very powerful feedback cycles that promote the production of ever larger quantities of more and more complex artifacts, inevitably at the cost of the natural world (Figure 12.1).

As a result, the mass of human-made buildings and roads, as well as human-controlled fields, plantations, and aquaculture ponds, increases daily. So too does the concentration of human-induced gases such as CO_2, CO, CFCs, and nitrogen oxides in the atmosphere and hydrosphere, along with concentrations of human-synthesized chemicals such as pesticides and chemical fertilizers in the hydrosphere and the biosphere.

Community of Beings

As the artifacts have blossomed, human societies have increasingly come to transfer to them the feeling of affinity earlier directed toward the natural world. This feeling of affinity is an extension of the affinity among members of human groups. Kinship and reciprocity comprise the glue that holds human societies together. Humans help others, even at a cost to themselves, because other members of the society are blood relatives or can be expected to reciprocate in the future.

Societies of hunter-gatherers and subsistence agriculturists extended this notion of kinship and reciprocity toward nonhumans as well. For them rivers and streams could be mothers, antelopes and bears brothers. They lived as members of a community of beings, with other humans, landscape elements, trees, and animals all part of a fellowship (Martin 1978; Gadgil and Berkes 1991). Such nonhuman fellow members provided many benefits to humans: water, shelter, food. Humans reciprocated by affording them protection from excessive interference and offering them gifts. Thus a stretch of stream might never be fished; thus a tree might be offered a fowl before being cut. Such protection or offer of gifts could be related to an attribution of the quality of sacred to streams, mountain peaks, groves, or individual animals. A whole range of genuinely effective conservation measures was implemented through such practices. Thus on the hill chain of Western Ghats in South India, as much as 6 percent of the land was covered by a dispersed network of sacred groves covering all habitat types from swamps and gallery forests to stunted scrub on wind-swept hilltops. Even today, when the network is greatly reduced, sacred groves protect the northernmost stands of *Dipterocarpus* and rare habitats like *Myristica* swamps. In the same region all trees of the genus *Ficus*, now recognized to be keystone resources, are given nearly total protection against felling, as are all primate species protected against being hunted. Especially vulnerable stages, such as birds nesting at a heronry or fruit bats roosting in daytime, are protected though they may be otherwise hunted (Gadgil and Berkes 1991).

Worshiping Artifacts

Human societies have extended this relationship of kinship and reciprocity and attribution of sacred qualities to artifacts as well. Hindus have special holy days on which artifacts that play a key role in their subsistence strategies are worshiped by being offered flowers, food, and gifts. On these days, peasants venerate their plows, traders their account books, warriors their swords and shields, weavers their looms, truck drivers their trucks, and indeed metallurgists their scanning electron microscopes. And everybody worships their canoes or bullock carts or bicycles or cars. Many Indians still pray to wick lamps or fluorescent tubes before turning them on every evening. Although different in form, Americans revere their automobiles in the same spirit. On May Day, communists used to worship their tanks and missiles in Red Square in Moscow. India's first prime minister, Jawaharlal Nehru, a staunch secularist, termed hydroelectric projects and steel mills the temples of modern India.

Supplanting Life-Forms

As artifacts have come to play an increasingly significant role for human societies, they have usurped the place that natural elements and living organisms once played as objects of human reverence. Such a transition has been a significant element of the Middle Eastern religions of Judaism, Christianity, and Islam. As Christianity spread over Europe more than a thousand years ago sacred groves of oak were cut down to be replaced by churches whose tall spires mimicked a forest. In recent decades Christian missionaries have shot sacred lemurs in Madagascar, to demonstrate the frailty of indigenous religious beliefs, and cut down sacred trees and groves in tribal tracts of northeastern India.

The major religions of the East, Buddhism and Hinduism, have taken a less adversarial position with respect to veneration of natural elements, plants, and animals. But even where these religions are in ascendance nature worship has been slowly on the retreat. For worship of nature is a matter of a direct relationship between the worshiper and a concrete, highly localized sacred object, perhaps a peepal (*Ficus religiosa*) tree. The worshipers are local inhabitants who make offerings to the tree spirit on their own and consume the chicken or the goat that may be sacrificed. In these transactions there is no surplus available for usurpation by a professional priest. That can materialize only if the tree spirit is identified with some deity of the Hindu pantheon, calling for a more elaborate worship for which the paid services of a professional priest become necessary. Professional Brahmin priests have therefore been active throughout India identifying the diverse nature spirits as forms of some Hindu deity. Once this happens the spirit which may be in the form of a tree or an anthill or a rock is established as a deity in an iconic form. Soon the icon is housed inside a temple building, often destroying in the process the original sacred tree or grove. Indeed, the Brahmin priests often perform special rituals to pacify the original nature spirit angered by the cutting down of its tree abode. They also deal with timber contractors involved in marketing the wood, receiving a commission from them.

As communication and market forces reach out to more and more remote tracts of India, one therefore sees in action a process of desacralization of nature and a shift of worship from natural elements to man-made icons in formally constructed temples. Notably these temples become more and more complex in form, as do the transport vehicles in which the deities are taken out in a procession on special holy days. Once there were just bullocks and elephants, then came elaborate wooden chariots, now they take the form of limousines. An increasingly unified world is thus rapidly losing its ancient heritage of respect for life in all its variety.

Of Life and Artifacts

As a culmination of this process, the world is now entering a new phase, perhaps the final stage of replacing the diversity of living forms with a multiplicity of artifacts. For advances in molecular biology now permit us to transform life-forms themselves, erasing the boundary between living organisms and artifacts. Such engineered life-forms will soon become the artifacts of greatest significance in determining the outcome of competition among business enterprises. These enterprises will then attempt to create conditions under which (1) they possess superior abilities to create newer, more complex engineered life-forms while at the same time trying to ensure that (2) the ability of potential competitors to create more complex engineered life-forms is depressed as much as possible. The ability of a business enterprise to create superior engineered life-forms would be enhanced by its access to the store of natural biodiversity that is their raw material. At the same time, the competitive ability of other enterprises would be effectively depressed if they have access to as little of such natural biodiversity as possible.

All business enterprises are therefore likely to strive to establish exclusive control over natural biodiversity. Three major instruments would be used for this purpose. One involves accumulation of as much natural biodiversity as possible in *ex situ* storage. The second involves legal devices such as patenting of knowledge and engineered life-forms. The third involves ensuring that nobody else has physical access to natural biodiversity. The last would logically imply deliberate extermination of any natural biological population harboring a given element of biodiversity as soon as that element has been brought under an enterprise's own *ex situ* control— or, even more frighteningly, appears likely to pass under some other enterprise's *ex situ* control.

Flowering of Diversity

Humanity must find ways of averting such a catastrophe. This would require controlling the blind forces of competition that permit a corporation or a state to acquire an edge over others at the cost of the natural world.

Evolution

Such a moderation of competition is likely only in a far more peaceful and equitable global society.

Humanity could then continue to produce more and more complex artifacts while at the same time imposing less and less impact on the natural world by ever more efficient use of material and energy resources. Indeed, living organisms could inspire the development of such artifacts. Biosensors that effectively detect low concentrations of specific molecules use chemical reactions involving enzymes or antibodies. Neural computation has its roots in attempts to build machines mimicking biological structures that exhibit elementary cognition. And Freeman Dyson (1989) makes the provocative suggestion that genetically engineered organisms could open up unparalleled opportunities for a further flowering of diversity. After all, living organisms have been colonizing a greater and greater range of environments over their evolutionary history. Starting, perhaps, in warm shallow seas, they have moved to the deepest ocean trenches, to the driest of land areas, to the highest of mountain peaks. But so far as we know, they are confined to a narrow film at the surface of the earth. With new technologies at hand, humans could deliberately design life that could colonize space. It may then evolve on its own, gradually diffusing outward throughout the universe. In the process it could undergo a further, immeasurably greater, diversification than has been possible on the surface of the earth.

I believe, therefore, that humans possess a fascination for all sorts of complex entities, living and nonliving, natural and artificial. Such a fascination may have been favored during the course of evolution since complex entities—other human beings, other living creatures, artifacts, elements of landscape—must have played a significant role in human lives. Biophilia may be one manifestation of this generalized love for complexity. Hunter-gatherer and horticultural societies treat all complex objects as other "beings"; humans are members of such a "community of beings." Reciprocity is an important basis of human sociality, and such societies extend the notion of reciprocity toward other nonhuman members of the community: trees, monkeys, springs, or mountain peaks. This reciprocal relationship involves offerings such as food and guarantees of protection against human interference. The notion of "sacred" may have arisen out of

Of Life and Artifacts

such a relationship, initially with natural entities, later with man-made artifacts.

Just as a greater and greater diversity of more and more complex living organisms has developed during the course of biological evolution, so have humans come to fabricate a greater and greater diversity of more and more complex artifacts over the course of cultural evolution. Humans have extended notions of reciprocity to these artifacts; they have also attributed qualities of sacredness to them. Technological progress has conferred an ever growing role on the artifacts in domination of other human groups as well as animate and inanimate nature. Indeed, the diversity of living creatures has been beating a retreat as a larger and larger mass of an ever increasing variety of artifacts has come to populate the earth. As this process has gone on, people have increasingly transferred their love and veneration for natural entities to the artifacts. Thus sacred groves have come to be replaced by churches and temples; thus chariots and cars have become objects of love, if not veneration, in place of elephants and bulls. Love for life has been giving way to love for artifacts.

With the advent of molecular biology, humans are beginning to create an entirely new kind of artifact: the genetically engineered living organism. This development is ushering in a new era in which forces of competition among states and corporate groups may in the long run prompt a liquidation of all natural diversity. Humans must then urgently find ways of nurturing natural diversity even as artifacts continue to evolve. If we succeed in this effort, we may not only maintain natural diversity but perchance engineer new organisms that could undergo a tremendous diversification as they colonize the rest of the universe.

ACKNOWLEDGMENTS

I am grateful to V. D. Vartak and M. D. Subash Chandran, who have taught me much over the years of sacred trees, groves, and people/nature relationships. The Ministry of Environment and Forests, Government of India, has supported my research through a series of long-term grants.

REFERENCES

Bonner, J. T. 1988. *The Evolution of Complexity by Means of Natural Selection*. Princeton: Princeton University Press.

Cowen, R. 1990. *History of Life*. Boston: Blackwell.

Dyson, F. 1989. *Infinite in All Directions*. New York: Harper & Row.

Gadgil, M., and F. Berkes. 1991. "Traditional Resource Management Systems." *Resource Management and Optimization* 8:127–141.

Martin, C. 1978. *Keepers of the Game: Indian-Animal Relationships and the Fur Trade*. Berkeley: University of California Press.

Petroski, H. 1992. "The Evolution of Artifacts." *American Scientist* 80:416–420.

Vitousek, P. M., P. R. Ehrlich, A. H. Ehrlich, and P. A. Matson. 1986. "Human Appropriation of the Products of Photosynthesis." *Biosciences* 36:368–372.

Wilson, E. O. 1984. *Biophilia*. Cambridge: Harvard University Press.

———. 1992. *The Diversity of Life*. Cambridge: Harvard University Press.

Part Six

ETHICS

AND

POLITICAL

ACTION

CHAPTER 13

Biophilia, Selfish Genes, Shared Values

Holmes Rolston III

Two central features of Edward O. Wilson's work are selfish genes and biophilia (1975a, 1984a). Perhaps more than any other living biologist, he has sought an ethics that is, both subjectively and objectively, based on biology. We moral agents, human *subjects* who act, must have morality based in our genes. And those who are the focus of concern, the *objects* or beneficiaries of our moral behavior, are not simply other humans but plants and animals. This ethics is based in a love for all forms of life: biophilia. So the chief exponent of selfish genes reaches toward a more comprehensive ethics, one even including ants.

Hence the puzzle: can we get biophilia from selfish genes? If so, well and good. If not, must we choose one or the other? Here I will propose a theory that describes what is going on and prescribes what ought to be. We can start with selfish genes but will have to expand progressively outward until we end with the whole Earth. By a series of ever more extensive hookups we will weave the selfish genes into global natural history. Philosophically this is a study in integration and identity in natural history.

Among sociobiologists, Wilson is notable for his ardent environmentalism. In *Biophilia*, with the subtitle *The Human Bond with Other Species*, he urges "an advance in moral reasoning . . . to create a deeper and more enduring conservation ethic." And he insists: "The only way to make a conservation ethic work is to ground it in ultimately selfish reasoning—but the premises must be of a new and more potent kind." Wilson worries about only "a surface ethics" and continues: "It is time to invent moral reasoning of a new and more powerful kind . . . a deep conservation ethic [based on] biophilia. . . . The more the mind is fathomed in its own right, as an organ of survival, the greater will be the reverence for life for purely rational reasons" (1984a:126, 138–140).

In sum: "To the degree that we come to understand other organisms, we will place a greater value on them, and on ourselves" (1984a:2). Wilson struggles both to keep and to break out of a selfish conservation ethic. He hopes to place great value on other organisms and we find that promising. We will return to the vocabulary of value. We want to get values in the right places—whether by placing them there, by finding them in place, or by sharing them. We hope to put selves in their places, as well, and thereby to put into place an environmental ethics.

Selfish Genes

In Wilson's classic *Sociobiology*, we are introduced to "The Morality of the Gene" on the first page (1975a:3). The genes, he says, hold culture on a leash: "Human behavior—like the deepest capacities for emotional response which drive and guide it—is the circuitous technique by which human genetic material has been and will be kept intact. Morality has no other demonstrable ultimate function" (1978:167). He continues: "Human emotional responses and the more general ethical practices based on them have been programmed to a substantial degree by natural selection over thousands of generations. . . . The deep structure of altruistic behavior . . . is rigid and universal" (1978:6, 162–163).

"Morality, or more strictly, our belief in morality," says Wilson, "is merely an adaptation put in place to further our reproductive ends. . . . In an important sense, ethics . . . is an illusion fobbed off on us by our genes to get us to cooperate" (Ruse and Wilson 1985). "Human beings function better if they are deceived by their genes into thinking that there is a disinterested objective morality binding upon them, which all should obey. We help others because it is 'right' to help them and because we know that they are inwardly compelled to reciprocate in equal measure. What Darwinian evolutionary theory shows is that this sense of 'right' and the corresponding sense of 'wrong,' feelings we take to be above individual desire and in some fashion outside biology, are in fact brought about by ultimately biological processes" (Ruse and Wilson 1986:179). Bluntly put, ethics results in genetic fertility; that is its deepest explanation.

Can biophilia be such "an illusion fobbed off on us by our genes to get us to cooperate" for our reproductive advantage? If all human behavior is a "technique by which human genetic material has been and will be kept intact," if "morality has no other demonstrable ultimate function," then we must "ground it in ultimately selfish reasoning" and all morality will be "the morality of the gene." This requires asking whether genes can be moral or immoral. Are there selfish genes that keep biophilia on a leash?

We do not ask about the morality of the liver or endoplasmic reticulum, for organs and organelles cannot be moral agents. But genes do code for life (for livers, cells, and organismic behavior as a whole) and perhaps there can indeed be a morality of genes. Genes govern the process; they are not simply products, and maybe there is some selfishness in the executive program. It is logically essential to the ordinary concept of selfishness that some entity (a "self") act in its own interests in an arena where peer entities (other "selves") have interests that can be acted for or against. We must be able to identify one self among other selves where the result of behavior benefits one and costs others. For a selfish gene, the contrasting class would be other genes located within or without a particular individual organism. Gene A benefits; gene B loses. Otherwise the possibility of selfish behavior lapses.

It is essential to any censurable selfishness that the agent acting selfishly has an option. Ought implies can; ought not implies can do otherwise. Since genes have no such behavioral options, we may not be dealing with censurable selfishness, but rather with a compulsive selfishness governed by the genes as they determine (but do not choose) behavior. Already we see that we must be circumspect about selfishness in genes.

Questions arise whether one gene can act against the interests of other genes that coinhabit the same organism. Or against the interests of genes inside other organisms. We must locate genes in their communities, their ecosystems. Biological phenomena take place at multiple interconnected levels—from the microscopic genetic through the organismic to the ecosystemic, bioregional, and planetary. Bigger networks are superposed on smaller, and these on lesser networks still; we descend from global scales to those in nanometer ranges. When we locate a gene in such a fishnet of fishnets of fishnets, it is difficult to think what it would mean for a single gene to operate "selfishly" in any biological sense (much less in any moral sense). Identity becomes a complex, multilevel phenomenon.

Part of the problem is that the benefits and costs accrue at a level different from that at which the gene immediately acts. There is a genetic level of *coding* and an organismic level of *coping*. Structure and metabolism both are genetically controlled; the genotypic level is doubly cross-wired to the phenotypic level. One gene may affect numerous phenotypic traits (pleiotropy); a single morphological or behavioral trait may depend on the contribution of many genes (polygeny). Many genes are epistatic (affect one another's effects).

In the functioning organism, proteins of thousands of different kinds, made on different genes and delivered to their vital sites, must all thereafter coact with the rest of the somatic materials and metabolic processes with which the organism manages to cope. Regulatory genes switch on and off the structure-producing and enzyme-producing genes. So a "selfish" regulatory gene can only be expressed in the phenotype if it switches on and off appropriately some structural or other gene—presumably too a "selfish" gene but one that can, in turn, be selfish only subject to the operation of a regulatory gene.

Though Wilson believes in selfish genes, he also knows that "real selection, however, is not directed at genes but at individual organisms, containing on the order of ten thousands of genes or more" (1975a:70). No gene is fit by itself; it has fitness only in the company it keeps. If a gene has a "self" to be selfish about, this too is only in the company it keeps. Hence these allegedly selfish genes are already set in the context of sharing, even before we pass outside the boundaries of the individual organism. One cannot be very selfish if one's fate is blended and interlocked with that of a hundred thousand others. Especially with behavior, which involves complex neural, cognitive, and muscular activities, the whole organism is involved, interacting with its environment.

A genetic reductionist approach sees the organism as nothing but an aggregation of genes and their outputs, each gene being individually "selfish," a kind of bottom-up approach. But the truer picture is a top-down approach: the organism is a whole, a synthesis, and codes its ways of coping in the genes, which are analytic units of that synthesis, each gene a cybernetic bit of the program that is the specific form of life. A gene exists in the microworld of coding, though its output functions sooner or later in the ecological macroworld of coping. We are having trouble seeing how any one gene is in any position to act selfishly—as though this could mean in its "own" interests separately from the interests of other genes or separately from the interests of the organism in which it is embedded.

Selfish Selves

We next turn to the organism facing its outer environment. Although many genes and their products coordinate into one integrated organism within the skin, facing outward life is lived as a singular individual. The organism is on its own. At this point natural selection does operate to select the better-adapted fits, those coded for the best coping. Now "selfish" behavior becomes more plausible. Behavior is a characteristic of the organism, not of this or that gene.

Again we must ask whether there is an identifiable entity (a "self") that can act in its own interests in an arena where peer entities (other "selves") have interests that can be acted for or against. In the case of a selfish organism, the contrasting class will be other organisms, either of the same or other species. The network of its coordinated parts comes to integrated unity in the organism as a whole. That is the "self." Such organisms frequently behave so as to benefit themselves at cost to others. One chicken grabs a grain of corn in the barnyard, and others do not get it. So it certainly seems that an organismic self can act against the interests of other organismic selves. Neither plants nor animals have intentions about this matter in the reflective sense required of moral agents. Except possibly for certain higher animals, it is not possible for them to do otherwise. So it is still not clear that selfishness is an appropriate label to apply to genetically based behavior and performance, where there are no options. But at least we can see how one organism can gain while other organisms lose. And so, perhaps, we have a biological analog to, a precursor of, selfishness.

We need to reflect on "selves," vital to the question of identity in natural history. Life requires reproduction, and that requires genes. Life also requires an inside and an outside, an organism that has separated itself from its environment. The definition of life—as we know it on Earth in any case—really includes this definition of self from nonself. There must be some kind of a cell, some defining envelope. After that, a cell, an organism, can take in nutrients from the environment and sequester them for its own uses. Our biology has to be arranged so as to keep us apart, though we must immediately add that our biology has to relate us to others with whom we are interdependent. The conservation of self-identity by a semipermeable organism is the larger truth within which we must interpret this alleged selfishness. The individual must live in an environment with which it must be in constant exchange. Self-identity means self-defense, self-stability, self-integrity.

"Self" is often a psychological concept, an ego, so we must be clear here that "self" is a biological concept. Plants and paramecia have no subjective life, though they defend objective selves. In all the advanced species of natural history, those with immune systems, the self is a singularity. In some forms of life, selves are clones of each other. Sometimes selves are histocompatible. But this is not true past the earlier levels of evolutionary history. After that, nature began to make idiographic selves. Indeed, the degree of idiosyncrasy in nature is quite remarkable. There is historical particularity in Earth's natural history, right down to the biomolecular level.

In organisms without moral capacity, we make a category mistake if we let "selfish" have a moral meaning. But in nature, there are selves—biological organismic identities to be preserved. Such a self-impulse cannot in itself be a disvalue. Quite to the contrary, this self-impulse is just the life impulse, the principal carrier of biological value. An organismic self is not a bad thing, nor is the defense of it. The system evolves organisms that attend to their immediate somatic needs (food, shelter, metabolism) and reproduce themselves in the very next generation. In the birth-death-birth-death system a series of replacements is required. The organism must do this; it has no options; it is "proper" for the organism to do this (Latin *proprium*, one's own proper characteristic). Somatic defense and genetic transmission are the only conservation activities possible to most organisms; they are necessary for all, and they must be efficient about it.

If there is some disvalue, this must lie in an overextension or aberration of the self-impulse. When a subordinate monkey relinquishes a feeding site to a dominant, the dominant may be said to have "selfishly" taken over. Or males may "selfishly" dominate females or defend territories. But if we strike out the negative moral overtones and replace them with positive self-preservation, what is going on? The monkey with the superior genes gets fed and bred; the monkey with the inferior genes does not, or at least not first. What is so disvaluable about that? Should it rather be the other way round—that the inferior genes get nourished and propagated and the superior ones do not?

Some sociobiologists can be quite emphatic about this organismic selfishness, which, they think, we humans are born into. Richard Dawkins reaches this conclusion: "We are survival machines—robot vehicles blindly programmed to preserve the selfish molecules known as genes. Let us try to *teach* generosity and altruism, because we are born selfish" (1976:ix, 3). George Williams complains: "The process and products of evolution are morally unacceptable . . . and justify an . . . extreme condemnation of nature. . . . Brought before the tribunal of ethics, the cosmos stands condemned. The conscience of man must revolt against the gross immorality of nature. . . . Natural selection . . . can honestly be described as a process for maximizing short-sighted selfishness." Behavior such as that of the dominant monkey, says Williams, is "not only selfish in some theoretical sense but patently pernicious. Only the morally and intellectually dishonest could label it otherwise." Williams urges: "An unremitting effort is required to expand the circle of sympathy for others. This effort is in opposition to much of human nature" (1988:383–385, 392, 437). Not only must humans get an ethic from outside biology, they must defeat their biology with it.

Many thinkers have concluded that humans are born selfish, and some influential ones find little possibility of altruism in the deeper sense, short of some kind of redemption of the nature we inherit biologically. These views have sometimes claimed to be scientific, as in Freud's psychoanalysis or Skinner's behaviorism; but even before the rise of science they were just as intensely advocated by Luther, Calvin, Aquinas, Augustine, Saint Paul, Jesus, and Gautama Buddha. The novel discovery here locates the nature of our bondage to selfishness in genetic determinants.

Some sociobiologists think that this discovery frees us from such bondage. But Wilson, as cited earlier, does not seem to think we can escape that leash. Rather we will have to find moral altruism if and only if some kind of altruism can be found within the constraints of the selfish genes. We get a clue how this may be so when we notice that the higher organisms, which "behave" and "act," often cooperate with one another. They mate in pairs and rear their offspring, they hunt in packs, they nest in colonies, they give alarm calls, they lead each other to food and share it. How are we to explain this behavior?

Inclusive Selfish Genes

To answer we must go down to the genetic level and consider kinship from the "selfish" gene's-eye view. When geneticists become sociobiologists, though they continue to suppose there are selfish genes, they insist that when we ask about "my genes" we have to enlarge the scope of "my" and go up to the family level by the same logic that goes down to the genetic level. From the gene's-eye view, since a gene is an information bit, a gene is present in all cells where there are copies of it. A particular gene is copresent in myriads of cells within any one individual but likewise may be copresent in relatives—copies within kin in a different skin. Facing out, we find that we are sometimes facing in, finding ourselves in others. Expanding the concept of the self to include this "inclusive fitness" (Hamilton 1964), the survival and reproduction of a relative are partly equivalent in evolutionary effect to one's own survival and reproduction. Animals, including humans, are evolved not only to reproduce the genetic materials in their own bodies by creating and assisting their descendants, but also to assist copies of their genes that reside in collateral relatives. Assistance to a relative will be favored if the benefit to the relative, proportioned to the degree of relationship, exceeds the cost to the donor.

Consider a baboon on sentry duty. He is not getting anything to eat while others are eating. But seen in terms of inclusive fitness, the distribution of benefits is the reverse of what it first seems. The sentry duty reduces what Wilson calls personal fitness (better: individual fitness). But it does not reduce inclusive fitness. The dominant male's genetic self, arising from his genetic type, is copresent in those he guards. He has one-half of a self in offspring, one-eighth of a self in first cousins, and so on. If we add these up, and adjust for risks and probabilities, the "selfish" benefits distributed elsewhere exceed the losses to the whole self within himself. So he is really defending his enlarged reproductive self when he risks his individual organismic self.

Here we are reaching an odd sort of selfishness, too, just as conceptually odd as we found selfishness to be at the genetic level. An individual's fitness is shared with kin—more and less with mother, father, sisters, brothers, children, cousins, uncles, aunts, nieces, nephews—all those blood relations in whom there are partial copies of "my genes," of whose genes "my genes" are partial copies. "Inclusive" mellows "fitness" from the skin outward. It does not matter whether the descendants (gene copies) are mine immediately, as a result of my individual fitness, or in my family, my inclusive fitness. If I fail to reproduce, it is just as well to have copies transmitted over there in my cousins.

Now we have clouded the seeming clarity of having located an idiographic "self" that can be selfish. It is not just the organismic, somatic self (the one protected so zealously by the immune system) that counts; it is the reproductive (genetic) self. In relatives, a self acts to preserve shared genes even if the self is not the one to perpetuate them. You can insist, if you wish, that these are still selfish genes, partial copies of oneself over there in daughter or nephew, uncle or cousin. But this is a strange kind of selfishness smeared out into a network of family relationships. Really, there is no reason to prefer a reductionist explanation of such behavior, "nothing but self-ishness," especially when labeling it pejoratively with a term borrowed from more complex human moral failure.

We are dealing with vital competence in an animal that has no duty. It is more plausible to interpret such behavior as self-defense, self-actualizing proper to every animal life, a defense not only of somatic self but of familial and specific forms of life. The "self" is not so much isolated and singularly preserved as it is fragmented and redistributed, mingled with other "selves," likewise shuffled. We are really dealing with in-common genes, in which any one family member participates, such as the dominant male baboon or a juvenile he protects. We are dealing with a heritage. We find a much expanded "selfishness" that becomes indistinguishable from family—one that shares most genes with conspecifics, a self stretched into community.

Humans have evolved with this animal heritage, and therefore we can interpret much human cooperation with the same theory applied to primates—for example, "altruistic" acts when a family member risks danger to protect his kindred. But this is no "killjoy" explanation of human ethics reduced to animal selfishness. Rather if human ethics originated here, this is really a quite promising origin of ethics in values already shared in pre-moral animal behavior. Inclusive fitness, where "my" becomes "our," is a welcome precursor to ethics, although we must be clear about what additionally emerges with its elevation into altruistic moral concern.

Sexuality and Self

This "our" of shared genes is a widening circle. Consider sexuality. Few phenomena are more pervasive throughout natural history than sexuality; few have proved more challenging to interpret from the point of view of selfish genes. The idiographic self cannot survive alone but has to mate. Sexuality requires male animals to couple defense of blood family to the nonkindred genetic lines of female mates. But this includes not only mates of a particular male but also those nonkindred female "others" who mate his kin, because "his" fractional genetic selves also couple with outside lines. The female mammal does have to tolerate another—the fetus in her womb, only half her own. So a particular self's inclusive fitness (genes in his or her kin) become entwined with much "alien" fitness in the bloodlines of mates. The individual somatic self is smeared out into the family and entwined with the community. The self is checked by sexuality.

Outbreeding individuals mate with others who have different genes. The human individual, falling in love, urged to reproduce, cannot love self alone but loves self in family, a family initiated by union with a genetically unrelated other. Typically husband and wife do not have recently shared genes. Nor can brother marry sister, nor cousin close cousin, without inbreeding depression. When the genes go through just that phase of the life cycle where the fully selfish gene might wish to construct a faithful copy of itself, or at least to protect partial copies of itself in relatives, there is chopping up and reshuffling, as though to bar genetic fidelity as the only rule in the game. The system insists on variation. It is hard to be selfish if one is a genome and must be split in half at every reproduction.

Sexually reproducing organisms cannot make identicals; offspring must be *others* (Latin: *alteri*) and in this sense sexual reproduction is by necessity "altruistic." Organisms can only make similars, similars with differences, and such variations over evolutionary time are as critical as the similarities. It is not possible, of course, for an organism to make other-very-differents; it can only breed after its kind. Only in asexual reproduction can an organism make identicals, clones, but asexuals are disadvantaged over evolutionary time. There is not enough variation and no way to crossbreed discoveries. Pure replicators, making only identicals, do well enough in the short term or in little-changing environments; but in the long haul and in complex environments, they go extinct (Maynard Smith 1978).

Thus an organism arrives in the world as a beneficiary of past variations, and it inhabits a natural system in which it can cope only if it can make variant copies of itself. Insofar as they are copies, the organismic history is inherited; insofar as they are variants, history is generated anew. The organism is itself a product of history, but its "self" cannot continue long somatically: it dies. And it cannot replicate itself except as it also generates otherness, copies with variance. Sexuality is a key to this variance. It breaks up at the same time that it creates unique biological identity. The self cannot continue except by dividing to unite with an alien self; its selfishness is limited by a required sharing. Selves over time inhabit a breeding community. Each new generation of idiographic selves is born of complementarity.

Yet these other selves are not all that other. Humans are "all of one blood" in the species sense. The man and the woman, like any mating pair, must have enough in common to interbreed; they share far more in their biochemistries than they differ in their idiosyncrasies. Within the human population there are many alleles at many loci, and one human can only carry a few of these. In this sense two individual humans may differ by hundreds of genes. At the same time, genetic studies show a remarkable uniformity from one human population to the other. Only 15 percent of the variation within human blood types exists between groups, whereas 85 percent of the variation is shared across groups (Lewontin 1972). For most genes, differences between populations are of frequency only; the genes themselves tend to be the same in population after population from the equator to the arctic circle. Where there are differences in alleles, it is difficult to link such differences with any survival benefit. Despite intensive study, there are less than half a dozen such genes known. Genes for dark skin provide protection in sunny climates. The sickle-cell gene gives resistance to malaria in malarious regions. But there are hundreds of other human blood group polymorphisms that do not make any known difference to reproduction rates.

Some geneticists believe that most of these differences are due to genetic drift and are neutral to selection. It is difficult to think that such genes could defend themselves "selfishly," since selection does not act on them. Other differences (as with dark skin) may formerly have made more difference than they do now; they may be relict genes. Whatever the explanation for the differences, humans around the globe have enough in common genetically to interbreed. If we are thinking about the genes that make ribosomes, Golgi apparatus, erythrocytes, acetylcholine molecules, or stem cells for B and T lymphocytes, whatever distinctive mix there is of these alleles in the husband's body, they are more similar to than different from the genes that do those things in the wife's body. Most of my genes are nonrival with most genes in most other humans.

From this perspective the fifty-fifty male / female split was a misperception; there is no more than a fraction of a percent difference between us; we have 99.444 (!) percent of our genes in common with everybody else, sort them uniquely though we do. Those genes of hers that seemed alien a moment ago are mostly my genes after all—or, the other way round, my genes are hers. After all, my wife too has hemoglobin in her veins and an opposable thumb on her hands, as do all "alien" humans around the globe. There are only four blood types as far as transfusion is concerned. Where most genes are involved it is difficult to think of alien genes in any other human. All 5 billion humans have copies of genes mostly like the copies I share with them. The differences between us, if we must compete about these, all turn on a trifling fractional percent and a different turn of the genetic kaleidoscope. It is really only the relatively idiosyncratic genes about which we are quarreling.

The other side of the picture is that each idiographic self is really a cluster of bits and pieces borrowed from, inherited from, all over everywhere, copies of which are still present with us side by side in relatives. We are composites. For it is not so much that our genes are heterogeneous, as is our combinatorial package. Most of my genes are not unique to myself at all, nor even to my family; to the contrary, they are common to conspecifics. These in-common genes, insofar as they affect behavior as well as determine structure, will be pushing me to cooperate with any and all fellow species members, and they with me. Or perhaps they will be neutral to behavior that differentiates between members of my species, since they are copresent in all.

We do not want a pejorative picture of a world laden with selfishness from the genes on up if the selfishness is really theory-laden and in the eye of the beholder. We might be viewing wild nature through a human prism—fooling ourselves that this is objective hard science when it is really just a subjective way of framing the problem. In this case the theory is not revealing anything about values in nature; it is just confusing us. Selfishness is indeed real—we experience it in culture—but we do not want to speak as though animals and genes were ethical agents in conditions of only superficial similarity. That kind of science has become almost animistic, mistakenly ascribing personal characteristics to natural things that are

Biophilia, Selfish Genes, Shared Values

incapable of such characteristics. The immorality is not there in nature; it is in our theoretical habiliment. Theories are like suits of clothes: they do have to fit the data more or less, but a great deal depends on how you want to dress things up.

Satisfactory Fitness

No genes, no organism; but also no ecology, no organism. Genes and self are quite surrounded by their environs. All three levels are vital: genes, organisms, natural history. If we are going to see the whole picture, we must next place the self in an ecosystem where it has a satisfactory fitness. The skin is a surface of exchange with the environment, and what is outside is as vital to life as what is inside. The world offers resources and accepts our wastes, recycling them. Interdependence and dependence are as true as selfish genes and organism. The environment is something that is outside and, we might say, over against us, but also it is our life support, not something that we are against or that is against us. Self-actualizing is essentially the protection of individual biological identity in a world where life is maintained by the orderly control of what passses through membranes. An educated geneticist must be an ecologist.

From the perspective of selfish genes, "foreign" means any molecule not coded for by the organism's DNA. Everything in the environment is foreign. But from the perspective of ecology, the organism inhabits a niche; the environment is its domicile, its "home" (the root of ecology, Greek *oikos*). An organism without a habitat is soon extinct. Life, skin in, has to protect a self. Life, skin out, has to fit the organismic identity into an ecosystemic integrity. Selves survive a little while; but all the while, really, the ecosystem in which this self lives is the fundamental unit of development and survival. An organism is a member of a species; its self-identity is smeared out into family and kind; that was the previous point. The present point is that an organic self, a member of a species, is what it is where it is. There are no organisms, period; there are only organisms-in-ecosystems.

Although conflict is part of the picture, the organism is selected for a situated environmental fitness beyond an inclusive fitness. For several decades biology has emphasized the survival of those with better-adapted fit into their ecological communities. A bear-organism fits its forest community—as surely as its organs fit together to organize a bear, as surely as the genes program and defend that organization. There are differences: the heart and lungs are close-coupled in a way that bear and forest are not. An ecosystem is often weakly coupled. Still: no forest, no bear. Unity is admirable in the organism, but the requisite matrix of its generation is the open, plural ecology.

Most of the relations between organisms are networks of interdependence and tolerance. This includes eating each other and being eaten. It also includes, if there are to be idiographic selves with identity, standoff relations. Organisms must defend territories and offspring. There can sometimes arise adversary relations. But the bigger truth is ecological: every organism is connected to and dependent on many others. Joining this holistic biological picture with a philosophical perspective, we must find a place for both idiographic self-defense and community dependency in tandem.

To some, ecosystems are little more than stochastic processes. A seashore, a tundra, is a loose collection of externally related parts. Much of the environment is not organic at all (rain, groundwater, rocks, nonbiotic soil particles, air). Some is dead and decaying debris (fallen trees, scat, humus). These things have no organized needs; the collection of them is a jumble, hardly a community. Each self defends its own life and there is only fortuitous interplay between organisms. An ecosystem is a matter of the distribution and abundance of organisms, how they get dispersed here and not there, birthrates and deathrates, population densities, moisture regimes, parasitism and predation, checks and balances. There is really not enough centered process to call community. There is only catch-as-catch-can scrimmage for nutrients and energy.

Even if we think of animals and plants as selfish, we still may respect them because each defends an organized biological identity. An ecosystem is the necessary habitat for this self-defense, but an ecosystem itself has no genome, no brain, no self-identification. It does not defend itself against injury or death. It is not irritable. An oak-hickory forest has no self to defend. So it can begin to seem as if concern for ecosystems is secondary after all, instrumental to the defense of organismic selves.

Biophilia, Selfish Genes, Shared Values

But to say that and nothing more is to misunderstand ecosystems. The organism is selected for a situated environmental fitness. There is a crucial element of struggle, but it is equally important to see this element contained in community. Ecological science emphasizes how there is a biological sense in which deer and cougar cooperate, defend their selves though they may; and the integrity, beauty, and stability of each is bound up with their coactions. Predator and prey, parasite and host, grazer and grazed, require a coevolution where both flourish, since the health of the predator, parasite, grazer is locked into the continuing existence, even the welfare, of the prey, host, or grazed.

The community connections, though requiring adaptive fit, are looser than the organismic coactions. But this does not mean they are less significant. Internal complexity, a self, arises to deal with a complex, tricky environment, the world as foil of self. The skin-out processes are not just the support, they are the subtle source of the skin-in processes. Everything will be connected to many other things, sometimes by obligate associations, more often by partial and pliable dependencies; and, among other components, there will be no significant interactions. There will be shunts and crisscrossing pathways, cybernetic subsystems, and feedback loops, functions in a communal sense. The system is a kind of field with characteristics as vital for life as any property contained within particular organisms. The individual and species (the genetic line) and its environment are not in fortuitous contrast or accidental aggregation; the ecosystem is the depth source of individual and species alike.

In the current debate among biologists about the levels at which selection takes place—individual organisms, populations, species, genes—the recent tendency to move selective pressures down to the genetic level forgets that a gene is always emplaced in an organism that is emplaced in an ecosystem. The molecular configurations of DNA are what they are because they record the story of a particular form of life in the macroscopic, historical ecosystem. What is generated arises from molecular mutations, but what survives is selected for adaptive fit in an ecosystem. We cannot make sense of molecular life without understanding ecosystemic life. The one level is as vital as the other.

Sometimes it is even held that organisms—or their biochemical molecules: proteins and genes—are real whereas ecosystems are just collections of interacting individuals, epiphenomenal aggregations. This too is a confusion. Any level is real if there is significant downward causation. Thus the atom is real because that pattern shapes the behavior of electrons; the cell is real because that pattern shapes the behavior of amino acids; the organism is real because that pattern coordinates the behavior of hearts and lungs; the community is real because the niche shapes the morphology and behavior of the foxes within it. Genes are the coding for coping in ecosystems; this makes them what they are where they are, and it makes ecosystems as real, as ultimate, as any genetic self.

"A thing is right," concluded Aldo Leopold, "when it tends to preserve the integrity, stability, and beauty of the biotic community. It is wrong when it tends otherwise" (1968:224–225). Leopold urged a "land ethic" that embraces concern for individual plants, animals, and persons but fundamentally loves and respects biotic communities. An environmental ethics needs biophilia for what Leopold called the land. Selves there are; but selfishness is difficult to maintain when the self gets spread through kin and kind, mated with other selves, and, now, extended into the landscape one inhabits, an interconnected web of life. There is some exaggeration in the deep ecologist's "the world is my body." But for those intoxicated by selfish genes, it is a sobering thought.

Reciprocal Altruism

At the next level of the evolution of cooperation, we complicate the picture with reciprocal altruism (Trivers 1971). Serving their self-interest, animals may help each other out, now oblivious to close kinship. There are certain things it is difficult or inconvenient for a baboon to do for itself (backscratching) which others can conveniently do for it; and it can reciprocate for them (scratch their backs). So it is to the mutual advantage of social primates to backscratch for each other. At this level, genetic relationships make no difference; a foreign backscratcher will do as well as a brother. So one baboon may be inclined to scratch the back of another, subject only to the likelihood that the second will reciprocate later when the first gets an itch.

In a cooperative society, animals can lower their risks. A vervet monkey will give an alarm call. Any other monkey, related or not, can interpret the call and benefit from it, while the caller puts himself at some risk by identifying his location to the predator. But on a later occasion, if the caller himself is unaware of a nearby predator and is alerted by the call of some more distant monkey, perhaps one quite outside his own family line, his life is saved. When unaware of a nearby predator, a monkey is at high risk of losing everything; when calling to alert others to a predator that he has spied at a distance, a monkey is at comparatively low risk. Because of this asymmetrical risk factor—a little cost versus a lot of benefit—both parties can, overall, lower their risks by helping each other out. Each gains individual somatic fitness and is more likely to live to reproduce than if neither gives alarm calls. Reciprocal altruism raises somatic and genetic fitness without any need to introduce genetic or inclusive fitness: kin selection.

When reciprocal altruism is working well, there are no losers on long-term average, although there are short-term losers on occasion. The baboon scratching another's back is not getting anything to eat while he is backscratching; the alarm-calling monkey is momentarily at some risk, gaining nothing by this particular call. But generally each gains back more than was lost, although benefits and losses may, on statistically rare occasions, be maldistributed. "Selfishness" makes some sense when one wins and another loses, but it is difficult to think what selfishness means in a win / win situation when one "self" has a self-interest that coincides with that of another. Mutual backscratchers may each be acting in their self-interest, but there is nothing selfish about helping each other out to the mutual advantage of both.

Where there is memory and a capacity to discriminate between individuals, remembering who reciprocates and who does not, reciprocal altruism can evolve where kinship is marginal and in doubt, so that the benefited other may (or may not) be both kin and reciprocator. A strategy dubbed "tit-for-tat"—cooperating initially, never thereafter refusing to reciprocate if the other does, refusing to cooperate when and so long as the other refuses to cooperate, and restoring cooperation at once if the other ventures it—can get established in a population, remain established, and resist invasion by various other strategies, particularly by noncooperation (Axelrod and Hamilton 1981). (Caution: These are mathematical game models, run on computers, which are neither biological nor ethical. Whether they map anything going on in the real world of genetics and ecosystems, much less that of personality and morality, needs more discussion.)

All this is said to be enlarged "selfishness." But in the same way that "inclusive" fitness is not a very selfish kind of fitness, or that a satisfactory fitness in an ecosystem webs the self into an ecosystem, reciprocal altruism is not as "selfish" as alleged. The "self" is getting coupled up to other selves willy-nilly. One way to approach this issue is resolutely to hang onto the central paradigm of "selfishness" and see all these others as being exploited by the original self. But it is just as plausible to see the self as being distributed out into the communal system, and we reinterpret what is happening by transposing to a communitarian paradigm.

The social system is entwining the self, as backscratcher or alarm caller, inseparably with the destinies of others—somewhat analogously to the way in which, earlier, the system embedded the fate of any one gene with the collective fates of myriads of others copresent in the genome of the integrated organism. This organism was in turn embedded in a family, its genes smeared out over kindred, and all these genes were interlocked sexually with mates. The organism got placed in a species line, in a breeding population, and further placed in a biological community on a landscape. Now, in social systems, the self is again being expanded, past those who are kindred, to all those of like kind with whom one interacts. Again, this is no problematic, ugly, ungodly, evil, or embarrassing precursor of ethics.

Adding in these genetically unrelated but socially related reciprocators with whom the individual interacts makes the picture all the more communitarian, since the genes of all these reciprocators are benefited with this coupling to "my" selfish genes. Just as it was earlier difficult to think of a "selfish" gene, owing to its inescapable organismic interlocking, it now becomes problematic to think of a selfish self. Willy-nilly there is reciprocation; there is community. If one insists on the word, the individual acts "selfishly" in his or her own interests, but "selfish" has been first stretched to cover benefits to father, mother, niece, nephew, cousin, children, aunts, uncles, and so on, and then stretched to cover benefits made to reciprocating

Biophilia, Selfish Genes, Shared Values

nonkindred others. The "my" that once seemed located from the skin in has been so much the further reallocated into a broadly scoped "our."

The evolutionary adventure is becoming less and less private and individualistic, more and more social and communal. The picture we are getting is one of benefits dispersed as much as benefits hoarded. Reciprocal altruism, though present even in animals, is not extensive there, however. Animal relationships are usually not sufficiently complex, enduring, or remembered to permit its elaboration (Wilson 1975a:120). Most organisms, living in rather local environments and narrow niches, are incapable of much reciprocity. Animals do not have much capacity to act or interact outside their own immediate sector of residence.

But humans can vastly expand the circle of reciprocal altruism, and this is the basis of all cultural cooperation. In all cultures, ancient and classical, people did not help just their blood relations; they helped other members of their tribe. Persons today cooperate at work, in politics, at school, in business, and so on with other persons with whom they have no known kinship except that they are all members of *Homo sapiens*. In modern nations, with trade by truck, mail, and telephone, they may never even see or know the names of these people. The small circles of reciprocal altruism in the animal world become national and international networks of cooperation. In this kind of behavior, judgments of kinship are irrelevant. This embedding of individual in society involves transmitting neural information superposed on genetic cybernetic systems. It involves language, artifacts, markets, computers, oil tankers, and jet planes.

The human person, already an integrated whole by concerted action of the genes from the skin in, already having found gene copies in kindred outside of the self, is all the more embedded in a community. Out-group cooperation can be just as beneficial as in-group cooperation. Brothers and cousins are nearby and can often help, but they are not likely to possess goods to which I do not myself also have some access. Foreigners have access to goods and skills I may need—and this is in fact what has happened in the modern world. The local self eats breakfast (coffee, orange juice, bananas) with resources drawn from 10,000 miles away and brought to oneself through the reciprocal cooperation of 10,000 persons (all those who had anything to do with getting breakfast here). Then one drives to work in a car made in Japan.

All this is a way of coping in the world, and such a propensity to cooperate must be coded in our genes. We could think of this as just glorified backscratching—with the additional complication, beyond what is instinctive from our animal heritage, that now we have to make judgments of the likelihood of reciprocation over greater distances and time spans. Those who can figure out these trade-offs and probabilities will live longer and reproduce more; their genotypes will be selected for.

Like animal cooperators, mutual human backscratchers may each be acting in their self-interest. But there is nothing selfish about helping each other out to the mutual advantage of both—not unless all self-interested actions are condemned as selfish. No ethical system, nor any religion, has ever condemned cooperation in which both partners gain. One of the two Great Commandments urges us to love others as we do ourselves. This injunction presumes self-love as an unquestioned principle of human behavior and urges us to combine this with loving others. If we can do this with overall loss to none, so much the better. We need not always love others instead of ourselves to fulfill this commandment.

The tit-for-tat strategy, though initiated at the nonmoral level, is not an immoral strategy if a moral agent were to continue it. It is an operational version of the Golden Rule, doing to others as you would have them do to you, while refusing to be taken advantage of. The strategy it displaces (dubbed "always defend") really means always defend your own self immediately, the only strategy possible to lower life-forms. There is nothing improper about this strategy at that level and indeed there is something impressive when in higher animals it evolves into: If possible, always cooperate in defending your values, but refuse to be a pushover for noncooperators, because this destabilizes the cooperative system. As before, values thereby become entwined in community—now the moral community superposed on what was before a biotic community.

This process suggests how reciprocal altruism may have evolved into ethical altruism in humans. It shows the enlarging of self-interests in cultural systems. This is an evolutionary development that makes interhuman

Biophilia, Selfish Genes, Shared Values

ethics possible. But plants and animals are not reciprocators, and we still need to know whether environmental ethics is possible. Let us return to the self that has been progressively enlarged into family, kind, ecosystem. Where and how do we place value now?

Shared Values

Can we describe the natural system more accurately and less pejoratively? Let us choose a positive axiological paradigm, rather than a negative ethical one, trying out a different interpretive gestalt. We can rewrite "selfishness" as "the conservation of intrinsic value." This too will be an interpretive scheme, a more plausible one we think, but if not it illustrates at least how the sociobiological account is itself interpretive. It will also help us find an environmental ethic, because we will get values in the right places, and human duties will follow accordingly.

Every organism must be self-projecting, pushing itself forward. But by the revised account, this process is not nasty; that is the beauty of life. Self-development, self-defense, is the essence of biology, the law of the wilderness, though there is also all that we have said about such a self being extended into family, kin and kind, niche and landscape. And there is still more to be said when culture emerges. An organism is the autonomous seat of its own life program, as rocks and rivers cannot be. The coping organism is coded at its information center. The genome is set to drive the movement from genotypic potential to phenotypic expression. Given a chance, these molecules seek organic self-expression.

They thus proclaim a life way. And with this an organism, unlike an inert rock, claims the environment as resource and sink from which to abstract energy and materials and into which to excrete them. Life thus arises out of earthen sources (as do rocks), but life turns back on its sources to make resources out of them (unlike rocks). Rocks do not give rise to other rocks; rivers do not reproduce themselves. But oaks make other oaks. An acorn becomes an oak; the oak rises from the ground and stands on its own.

So far we have only a description of the logic of life. We pass to value when we recognize that the genetic set is a normative set; it distinguishes between what *is* and what *ought to be*. The genome is a set of conservation molecules. The organism is an axiological, evaluative system. So the oak grows, reproduces, repairs its wounds, resists death. The physical state that the organism seeks, idealized in its programmatic form, is a valued state. *Value* is present in this achievement. *Vital* now seems a better word for it than *biological*. The living individual, taken as a point experience in the web of interconnected life, is per se an intrinsic value. A life is defended for what it is in itself, without necessary further contributory reference, although, given the structure of all ecosystems, such lives necessarily do have further reference. The organism has something it is conserving, something for which it is standing: its life. Organisms have their own standards, fit into their niche though they must. They promote their own realization while at the same time they track an environment. They have a technique, a know-how. Every organism has a *good-of-its-kind*; it defends its own kind as a *good kind*.

Bacteria, mice, and chimpanzees have projects of their own; each is a life-form to be defended for what it is intrinsically. To label this "selfish genes" is to misunderstand the biology and the metaphysics of what is going on. Every organism must project itself in the world. Instead of thinking of a ground squirrel, much less a single gene within a ground squirrel, as acting "selfishly," we will substitute the equally descriptive but nonpejorative acting "for its own sake" and even substitute the positive "to protect its intrinsic value." These are "axiological genes."

A gene is really an information fragment—and information does not have to be lost to be shared. It is really difficult to interpret selfishly the transmission of information. When that information overleaps death it would seem as appropriate to say that it has been "shared" (distributed) as that it has been "selfishly" reproduced (hoarded). Since a parental organism "donates" (distributes) information to offspring via genes, "altruism" is as easy an inference as is selfishness—if one insists on moral labels. Genes are no more capable of "sharing" than of being "selfish"—it must at once be said—where "sharing" and "selfish" have their deliberated, moral meanings. Since genes are not moral agents, they cannot be selfish and, equally, they cannot be altruistic. But they can transmit information. And if we are going to stretch a word employed in the moral world and make it serve in this amoral realm, then "share" is as descriptive as "selfish" and without the pejorative overtones. Genes do generate; they reproduce or communicate what they possess; they share (distribute in portions) their information, literally, although preconsciously and premorally. That places each gene where it belongs: on a commons in which it participates. What is selfish about dispersing vital information, sharing a value?

Natural history is not an evil scene driven by maliciously selfish genes. It is a wonderland of adaptive fit, a community of intrinsic values woven instrumentally into a systemic web. There is the conservation of intrinsic value, but this is not permitted to be an isolated thing: it is webbed into the family, the population, the species line, the ecosystemic community, the landscape, as an individual is given a place to live and a role to play in the valuable system. Intrinsic value is smeared out into instrumental and systemic value, no less than was the self smeared out into the whole. Values enjoyed have to be values shared.

We want a nonhumanistic, nonanthropocentric account, one unbiased by our morals. This is really a much better paradigm because there is no good reason to think that genes are selfish; there are no moral agents in wild nature even at the organismic level, much less the genetic one. But there is good reason to think that there are objective, nonanthropocentric values in nature and that these are defended and distributed by wild creatures in their pursuits of life in the midst of their entwined destinies. The axiological paradigm is the objective and natural one; the ethical ("selfish") paradigm is subjective and humanistic. We want to try to pass judgment on the value of nature for what it is in itself—with criteria appropriate to nature, that is, not with anthropocentric criteria. Let nature be what it is; do not fault it morally. Value it biologically; do not disvalue it ethically.

Suppose we cast the event, say, of a dominant monkey's feeding first in terms of values defended. What is of value here (the superior genome) gets transmitted, maintained through feeding and breeding, while what is of relative disvalue gets selected against. There is no moral agency at issue; what is at stake is value that is self-actualized. To ask these monkeys to behave as altruistic humans is to misunderstand the events and misvalue them accordingly. Read out the immorality, and the picture looks different. Take off the dark glasses and put on clear ones. It is a category mistake to describe (and censure) what goes on in wild nature with terms borrowed from culture and projected onto nature. There is nothing here particularly disvaluable that moral agents, when they come, will want to deplore and rectify in the animals—although nothing follows from this about how they should behave in culture. The alleged selfishness is really the conservation of value intrinsic to the organism in the only manner possible and appropriate to it.

All such contests at feeding and reproduction are endured for "selfish" advantage by males or females only in a problematic sense, since the somatic individual soon dies anyway. A better way of interpreting events is to say that the contest is to share genes. It is self-defense in one sense. But if males and females spend time, energy, and effort to reproduce, this is self-sacrificing in another sense. By those who resolve to see everything through selfish lenses, this will (rather confusingly) be called selfishness again, seen from the nonmoral genetic level. But we get a much clearer picture of what is going on if we interpret this as values being transmitted over generations.

Although the organism is engaged in a short-range reproduction of its kind, the systemic processes are neither short-range nor do they selfishly maximize only one kind. The evolutionary system is 3.5 billion years old; it has steadily produced new arrivals, replacements, and elaborations of kinds, going from zero to 5 or (or 10) million species, through 5 (or 10) billion turnover species in a kaleidoscopic panorama. Every organism, in the subroutines of this system, actualizes its own values and transmits them to the next generation (with variations). Apart from humans, to whom we next turn, that is all any organism has the capacity to do, a capacity of critical value. The result is quite a dramatic story—not just a long, long chain of "patently pernicious" short-sighted selfishness. The value account seems quite descriptively plausible, not at all "morally and intellectually dishonest."

Philosophers sometimes note that on close examination a seemingly bold hypothesis dies the death of a thousand qualifications. What happens to the seemingly bold hypothesis of selfish genes is that they live the life of ten thousand interconnections.

Biophilia on the Home Planet

Incremental quantitative changes can add up to a qualitative change. We start with night, add light bit by bit, and pass into day; the night is gone. We have started with selfish genes, added "other" values interconnection by interconnection, and passed over to valuing others. Humans can see these ten thousand interconnections and love this system of life in which they too are entwined. Humans, alone on the planet, can realize that they are kindred with all. Darwin taught us so, a century ago, from an evolutionary perspective; today, microbiologists confirm it. For structural genes, "the average human protein is more than 99 percent identical in amino acid sequence to its chimpanzee homolog" (King and Wilson 1975:112). Differences between the species lie largely in regulatory genes (Sibley and Ahlquist 1984).

Edward Wilson recognizes this as well: "We are literally kin to other organisms. . . . About 99 percent of our genes are identical to the corresponding set in chimpanzees, so that the remaining 1 percent accounts for all the differences between us. . . . Furthermore, the greater distances by which we stand apart from the gorilla, the orangutan, and the remaining species of living apes and monkeys (and beyond them other kinds of animals) are only a matter of degree, measured in small steps as a gradually enlarging magnitude of base-pair differences in DNA" (1984a:130). "At the biochemical level," he says elsewhere, "we are today closer relatives of the chimpanzees than the chimpanzees are of gorillas" (Ruse and Wilson 1986:176). Aren't these small steps gradually enlarging the self by degrees until the self is identified with more and more others?

Suppose we translate such genetic similarity into the vocabulary of selfish, kin-selecting genes. If a human (Jane Goodall) were to devote her life to saving chimpanzees, this would really be 99 percent selfish and only 1 percent altruistic, at least for structural genes. Likewise with the Siberian tigers to whom George Schaller is perhaps 95 percent related. And so on down the evolutionary lineage. We get a circle of alleged selfishness expanded several orders of magnitude past siblings and cousins, aunts and uncles. This is just as curious a big-scale selfishness as the narrow, constrained variety with which we started.

From the viewpoint of the gene that makes a cytochrome-*c* molecule, found in organisms ranging from yeasts to people, it is going to be difficult to locate much of a rival. Cytochrome-*c* molecules do evolve through various nucleotide substitutions, but they are comparatively stable molecules.

The primary structure is identical in humans and chimpanzees, which diverged about 10 million years ago; there is only one replacement between humans and monkeys, whose most recent common ancestor lived 40 to 50 million years ago (Dickerson 1971). The same is true from the viewpoint of genes that make adenosine triphosphate (ATP), biotin, riboflavin, hematin, thiamine, pyridoxine, vitamins K and B_{12}, or those involved in fatty acid oxidation, glycolysis, and the citric acid cycle, or those that make actin and myosin. The genetic code is essentially the same for all living organisms. The twenty amino acids are common to all.

Sometimes life lines, once independent, have fused into a single identity. Two of the most important processes energizing life on Earth use endosymbionts. One, involving mitochondria, powers animals; the other, with chloroplasts, powers plants; and, of course, plant power is the basis of animal power, including human power. In the full drama of natural history, identity is a multileveled, dynamic phenomenon. Biological identity is not so idiographic after all: it mingles with biological solidarity and is shared with the fauna and flora of the ecosystemic whole.

Such a vastly expanded kinship suggests that the better way to view this ever more extended inclusive self is to regard it as individuals residing within a community of shared values. Though an individual self, I, in my effort to survive, am not really pushing the line of a solitary individual at all: I live in a community on a front of shared family heritage, shared human heritage, shared primate, mammalian heritage, indeed shared biological heritage. Perhaps I still have some "inclusive fitness," carried genetically, fractionally, and which I particularly have to defend in my local niche. But from a gene's-eye view, if we take the 99 percent seriously then it seldom matters whether the genes are inside me, inside my cousin, or inside a chimpanzee. Indeed, it may not matter whether they are inside me or inside an oyster or an ant.

When we move from the microscopic level to the range of ordinary experience, selves have entwined destinies with the landscapes they inhabit. Maybe you do not feel all that related to chimpanzees and insist on being discriminating about your relationships. Maintain your distance from the other creatures as you wish, insist that the self is over against the world. But you cannot take the self out of the world. We continue to inhabit this home

Biophilia, Selfish Genes, Shared Values

planet, a relationship you cannot escape. A bumper sticker says, "Earth: love it or leave it." Since leaving is difficult, loving Earth is the only real option.

Perhaps we are genetically adapted to loving it. Natural selection could certainly select for loving that with which one has an entwined destiny: that could convey survival advantage. Biologists suppose that selection operates at the level of the individual; they prefer the lowest level possible. Most humans have inhabited local neighborhoods; they hardly knew they inhabited a planet. But they did know they dwelt on landscapes; they belonged to "countries," as they put it, and it seems quite plausible to think that humans could, over time, be selected to love their world.

This will be a flexible characteristic, however, since humans inhabit many different kinds of landscapes and often rebuild them to their liking. Animals have to take their landscapes ready to hand, as it were; they adapt their selves to them. Human selves can, in culture, rebuild their landscapes, more or less, adapting them to their preferences. This rebuilding too will convey survival advantage. Maybe there is some selection of those who love culture and conquer nature. Still, in the end, every culture remains set in an ecosystem. The human genetic destiny, if there is such a thing, can be expected to keep the self happy in its home place. People need to be natives, residents, as well as citizen-selves. Ethics is not so much ultimately selfish as self-involving. And when the full scope of these self-involvements is known, the planet is the self's ultimate survival unit. Inclusive fitness ends up being planetary fitness.

The opposite of selfishness is altruism, and we have been enlarging selfishness so that it becomes more altruistic, embracing an expanding circle of relationships. Have we not reached the point at which the circle comes to include genuine others—an altruism with universal intent? If so, the environmental ethicist is the ultimate altruist. We do reach a point where the quantitative expanding of self has reached a qualitative regard of a self for others with whom one is interconnected but whom one loves for what they are in themselves, not just for what they are for us. We cannot get off the Earth, out of the system, but we can get our identity enlarged to the whole. And then we see as what philosophers call ideal observers: those who see overall and not just from their narrow niche.

To try to see all ethics as nothing but extrapolated selfish genes might stunt humanity because it fails to realize the genuine human transcendence—an overview caring for others. Rather than using mind and morals as survival tools for defending the human form of life, mind forms an intelligible view of the whole and defends ideals of life in all their forms. Humans have oversight; they are worldviewers—today more than ever before. From this, morality follows as a corollary because of what humans can know and do—today more than before because of our increased knowledge and power. Humans can get "let in on" more value than any other kind of life. They can share the values of others and in this way become consummate altruists.

Animals have the capacity to see only from their niche; they have mere immanence. Humans can have a transcending view from no niche. It is not just our capacity to *say I*, to actualize an ego-self, but our capacity to *see others*, to oversee a world, that distinguishes humans. Skeptics and relativists may say that humans just see from another niche, and it is certainly true that when humans appraise soil or timber as resources, they see from within their niche. But humans also see other niches and the ecosystems that sustain niches; they study warblers or see Earth from space. No other species has such supersight, such spectacular oversight. What humans can do that nothing else can is recognize these intrinsic values for what they are, where they are, instrumentally woven into the ecosystemic Earth. Such value, which *is* present, *ought* to be preserved.

You can, if you insist, hang onto the old anthropocentric paradigm that we maintain such oversight lest we stunt humanity. But you are really failing to see the paradigm switch to a biocentric conviction arising from a love of life beyond self-love. The self has gotten deeper and deeper into its ecology; the shallow self is no more. This view is both radical, in that it goes to the roots, and conservative in that it conserves all life, not just human life extended. And it makes biophilia superbly possible. You can say, if you like, that what humans really want is the optimal (ideal) configuration of their world; indeed this is so. But this is to abandon the genetic leash; the criterion is no longer the maximum production of "my" offspring at all, nor even "our" human offspring.

But this, you may protest, is only theory. Who knows whether such an ethic can be lived? We need examples that verify this theory. Consider personal experience. When I donate money to promote an environmental cause to which I am committed—the fund for whales—I need not even know that I have genes. Or if I do, the genes be damned, so long as the whales are saved. I do want to convert other persons to my conservationist ideology, but their genetic relationship to me is immaterial. I enjoy knowing that the whales are safe in their marine ecosystems. Label that a "selfish" motivation if you must, but my enjoyment does nothing to increase my fertility. John Muir and David Brower, if anything, will have fewer offspring in the next generation on account of their time, energy, and effort spent in protecting Hetch Hetchy and Glen Canyon.

I do not expect whales, warblers, or grizzlies, much less forests and canyons, to reciprocate with mutual backscratching; the animals can do nothing to assist me (or any other humans) somatically or genetically. Insist that what I am really doing is identifying my "self" with the ecosystemic whole, or preserving my life support system, or whatever; this does not aggrandize the self or its genetic line so much as it stretches the "self" out into the community it inhabits, until the self has come to focus on not-self, on other selves that are good of their kind. Why not face up to the epistemic crisis? These are no longer selfish questions. They are questions whether each of the myriad "other" life-forms can be good-of-its-kind, good-in-its-kind-of-place, and about their all being in a good kind of place—and these add up to the question of well-placed goodness. Ethics is about optimizing these values.

Can you continue to insist that I do not really have a concern for the whales, warblers, or pristine forests, that I am only protecting my recreational opportunities? Or that I am only self-deceived and parading my beneficence so that other humans will laud me and assist my offspring, since I am an environmentalist hero? Surely it is better to say that the "self" has been elevated into genuine morality, where it can detect values outside itself, and come to embrace these values in freedom and love because it is right to do so. This is not naturalized ethics in the reductionist sense; it is naturalized ethics in the comprehensive sense.

If you are still unconvinced, let us close with an *ad hominem* argument addressed to Edward Wilson himself, who so superbly demonstrates what kind of love of life is possible in humans. We shall make him part of the evidence for our theory. Wilson claims: "Our societies are based on the mammalian plan: the individual strives for personal reproductive success foremost and that of his immediate kin secondarily; further grudging cooperation represents a compromise struck in order to enjoy the benefits of group membership" (1978:199). Can we make that claim self-referential?

Hardly. Because in the same breath he urges, as an interhuman ethics, the three primary principles. First: One ought to protect "the cardinal value of the survival of the human genes in the form of a common pool over generations." Second: One ought to "favor diversity in the gene pool as a cardinal value," for "of all the evils of the twentieth century, the loss of genetic diversity ranks as the most serious in the long run." Wilson fears a tragic loss of "the variety of human genes out of which endless new combinations can be drawn for the attainment of genius and further genetic evolution" (1978:196–199; 1980:61–62). Third: One ought to regard "universal human rights . . . as a third primary value" (1978:198). Sociobiology, Wilson concludes, is going to lead us to "a genetically accurate and hence completely fair code of ethics" (1975a:575). We hear hope in the man and commend him for it. But what we hear does not sound like grudging cooperation at all; it sounds, rather, like someone ardently defending the common sources that generate human life in all its diversity, producing a culture in which each person is worthy of respect as a matter of right. Meanwhile, it is hard to find this logic in the biology of the theory—which says that if it is genetic (maximizing one's own offspring, no matter what), then it cannot be completely fair (equity for all). One cannot be selfish about self and fair and at the same time give each his due by right, much less be altruistic toward any.

In environmental ethics, Wilson urges forming a human bond with other species, loving not only human diversity but biodiversity throughout the fauna and flora. He wants to stretch the self into a nobility of character that comes from a "generosity beyond expedience" (1984a:131) that he has himself embraced but cannot quite reach on the basis of his theory. He concludes that there ought to be a respect for life in which we value other forms of life as we do our own, a sort of Golden Rule in environmental ethics. The self-interest that an environmental ethic serves cannot be of the backscratching kind; the ants that Wilson wishes to protect are unlikely recip-rocators. Wilson confesses that, "in the end, the problem of wilderness preservation is a moral issue, for us and for our descendants," and he com-mends "species diversity as an ethical goal" (1984b). Here we are not deal-ing with a genetic determinism fobbing off illusions about why we are be-having so. We are dealing with an ethical "idea," an "ideal," a conviction detecting objective natural values present outside the self, outside culture, values that ought to be preserved.

Wilson asks: "What event likely to happen during the next few years will our descendants most regret?" His answer: "The one process now going on that will take millions of years to correct is the loss of genetic and species diversity by the destruction of natural habitats. This is the folly our descen-dants are least likely to forgive us" (1984a:121). If our descendants will judge it an all but unforgivable sin to destroy thousands of other species, this ca-tastrophe ought not to happen. Nor is it just our descendants' regret that we fear; it is life lost from this wonderland Earth.

No doubt these descendants will suffer losses in those species that do not survive. Their human quality of life may be at stake, but maximum re-productive success—the largest human population possible on Earth—is no criterion of this environmental ethic. Indeed, it is antithetical to it. Our human reproductive instincts must and ought to be replaced by biophilia and concern for environmental integrity. "To rear as many healthy children as possible was the long road to security," Wilson observes, "yet with the population of the world brimming over, it is now the way to environmental disaster" (1975b:50).

For those humans who can move outside their own pragmatic utilities and learn to appreciate the "mysterious and little-known organisms" with which we coinhabit this planet, "splendor awaits in minute proportions" (1984a:139). This, if you insist, is an enrichment of human welfare, but it has nothing to do with his or our fertility or selfish genes. None of this in-quiry about what humans ought to do in environmental ethics can be un-dertaken without being released from an ethics that is nothing but selec-tion for maximum production of human offspring. The one thing selfish genes do not do is promote diversity not their own. Rejoicing in the splendor of his planet, Wilson is finding it difficult to get biophilia out of selfish genes. That is because a single gene is really, so to speak, only a fragment of biophilia, a bit of life information. A gene is nothing much in and of itself; there is no self there to be selfish about. But these genes collectively, in their wholes, share and spin together the vital drama of life.

There is no need for a person with such an admirable love of life to retreat into a killjoy explanation of his love. Why not rise to a joyous explanation? The home planet is prolific with life, exuberantly projected up from the primeval ooze and mud, an emergent vitality expressed in 10 million species. The planet loves life and so do we. This is the evolutionary epic, and we are this love of life become conscious of itself. We do not want to depress life into nothing but selfishness, borrowing inappropriately a depressing category from human moral failure. We want to respect the life that has so marvelously expressed itself over evolutionary history, and reaching that respect will itself be an elevating moral achievement.

REFERENCES

Axelrod, Robert, and W. D. Hamilton. 1981. "The Evolution of Cooperation." *Science* 211:1390–1396.

Dawkins, Richard. 1976. *The Selfish Gene*. Oxford: Oxford University Press.

Dickerson, R. E. 1971. "The Structure of Cytochrome-*c* and the Rates of Molecular Evolution." *Journal of Molecular Evolution* 1:26–45.

Hamilton, William D. 1964. "The Genetical Evolution of Social Behavior, I and II." *Journal of Theoretical Biology* 7:1–52.

King, Mary-Claire, and A. C. Wilson. 1975. "Evolution at Two Levels in Humans and Chimpanzees." *Science* 188:107–116.

Leopold, Aldo. 1968. *A Sand County Almanac*. New York: Oxford University Press.

Lewontin, R. C. 1972. "The Apportionment of Human Diversity." *Evolutionary Biology* 6:381–396.

Maynard Smith, John. 1978. *The Evolution of Sex*. Cambridge: Cambridge University Press.

Ruse, Michael, and Edward O. Wilson. 1985. "The Evolution of Ethics." *New Scientist* 108 (17 Oct.):50–52.

———. 1986. "Moral Philosophy as Applied Science." *Philosophy* 61:173–192.

Sibley, Charles G., and Jon E. Ahlquist. 1984. "The Phylogeny of the Hominoid Primates, as Indicated by DNA-DNA Hybridization." *Journal of Molecular Evolution* 20:1–15.

Trivers, Robert L. 1971. "The Evolution of Reciprocal Altruism." *Quarterly Journal of Biology* 46:35–57.

Williams, George C. 1988. "Huxley's Evolution and Ethics in Sociobiological Perspective." *Zygon* 23:383–407 (and reply to critics, 437–438).

Wilson, Edward O. 1975a. *Sociobiology: The New Synthesis*. Cambridge: Harvard University Press.

———. 1975b. "Human Decency Is Animal." *New York Times Magazine*, 12 Oct., pp. 38–50.

———. 1978. *On Human Nature*. Cambridge: Harvard University Press.

———. 1980. "Comparative Social Theory." In Sterling M. McMurrin, ed., *The Tanner Lectures on Human Values, 1980*. Vol. 1. Salt Lake City: University of Utah Press.

———. 1984a. *Biophilia*. Cambridge: Harvard University Press.

———. 1984b. "Million-Year Histories: Species Diversity as an Ethical Goal." *Wilderness* 48(165):12–17.

CHAPTER 14

Love It or Lose It:

The Coming Biophilia

Revolution

David W. Orr

I have set before you life and death, blessing and cursing: therefore choose life, that both thou and thy seed may live.

—DEUTERONOMY 30:19

"NATURE AND I are two," filmmaker Woody Allen once said, and apparently the two have not gotten together yet.[1] Allen is known to take extraordinary precautions to limit bodily and mental contact with rural flora and fauna. He does not go in natural lakes, for example, because "there are live things in there." The nature Allen does find comfortable is that of New York City, a modest enough standard for wildness.

Allen's aversion to nature, what can be called biophobia, is increasingly common among people raised with television, Walkman radios attached to their heads, video games, living amidst shopping malls, freeways, and dense urban or suburban settings where nature is permitted tastefully, as decoration. More than ever we dwell in and among our own creations and are increasingly uncomfortable with the nature that lies beyond our direct control. Biophobia ranges from discomfort in "natural" places to active scorn for whatever is not man-made, managed, or air-conditioned. Biophobia, in short, is the culturally acquired urge to affiliate with technology, human artifacts, and solely with human interests regarding the natural world. I intend the word broadly to include as well those who regard nature "objectively" as nothing more than "resources" to be used any way the favored among the present generation see fit.

Is biophobia a problem like misanthropy or sociopathy? Or is it merely a personal preference, one plausible view of nature among many? Is it OK that Woody Allen feels no kinship with nature? Does it matter that a growing number of other people don't like it or like it only in the abstract as nothing more than resources to be managed or as television nature specials? Does it matter that we are increasingly separated from the conditions of nature? If these things do matter, how do they matter and why? And why have so many come to think that the natural world is inadequate? Inadequate for what?

At the other end of the continuum of possible orientations toward nature is biophilia, which E. O. Wilson defines as "the urge to affiliate with other forms of life."[2] Erich Fromm once defined it more broadly as "the passionate love of life and of all that is alive."[3] Both agree, however, that biophilia is not only innate but a sign of mental and physical health. To what extent are our biological prospects and our sanity now dependent on our capacity for biophilia? To that degree it is important that we understand how biophilia comes to be, how it prospers, what it requires of us, and how this is to be learned.

If biophilia were all that tugged at us, this book would be an unnecessary documentation of the obvious. But the affinity for life competes with other drives, including biophobia, disguised beneath the abstractions and presumptions of progress found in economics, management, and technology. My hypothesis about the biophilia hypothesis, then, is that whatever is in our genes, the affinity for life is now a *choice* we must make. Compared to earlier cultures, our distinction lies in the fact that technology now allows us to move much further toward total domination of nature than ever before. Serious and well-funded people talk about reweaving the fabric of life on earth through genetic engineering and nanotechnologies; others talk of leaving the earth altogether for space colonies; still others talk of re-shaping human consciousness to fit "virtual reality." If we are to preserve a world in which biophilia can be expressed and can flourish, we will have to decide to make such a world.

Biophobia: Its Origins and Consequences

In varying degrees humans have always modified their environment. I am persuaded that they generally intended to do so with decorum and courtesy toward nature. Not always and everywhere to be sure, but mostly. On balance, the evidence further suggests that biophilia or something close to it was woven throughout the myths, religions, and mindset of early humankind, which saw itself as participating with nature. In Owen Barfield's words, people once felt "integrated or mortised into" the world in ways that we do not and perhaps cannot.[4] Technology, primitive by our standards, set limits on what tribal cultures could do to the world; their myths, superstitions, and taboos constrained what they thought they ought to do. But I do not think that they *chose* biophilia, if for no other reason than there was no choice to be made. And those tribes and cultures which were biophobic or incompetent toward nature passed into oblivion through starvation and disease.[5]

Looking back across that divide, it is evident that tribal cultures possessed an ecological innocence of sorts because they did not have the power or knowledge given to us. We, in contrast, must choose between biophobia and biophilia because science and technology have given us the power to destroy so completely as well as the knowledge to understand the consequences of doing so. The divide was not a sharp break but a kind of slow tectonic shift in perception and attitudes that widened throughout the late Middle Ages to the present. What we call "modernization" represented dramatic changes in how we regarded the natural world and our role in it. These changes are now so thoroughly ingrained in us that we can scarcely conceive any other manner of thinking. But crossing this divide first required us to discard the belief that the world is alive and worthy of respect if not fear. To dead matter we owe no obligations. Second, it was necessary to distance ourselves from animals who were transformed by Cartesian alchemy into mere machines. Again, no obligations or pity are owed to machines. In both cases, use is limited only by usefulness. Third, it was necessary to quiet whatever remaining sympathy we had for nature in favor of hard data that could be weighed, measured, counted, and counted on to make a profit. Fourth, we needed a reason to join power, cash, and knowledge in order to transform the world into more useful forms. Francis Bacon provided the logic; the evolution of government-funded research did the rest. Fifth, we required a philosophy of improvement and found it in the ideology of perpetual economic growth, now the central mission of governments everywhere. Sixth, biophobia required the sophisticated cultivation of dissatisfaction which could be converted into mass consumption. The advertising industry and the annual style change were invented.

For these revolutions to work, it was necessary that nature be rendered into abstractions and production statistics of board feet, tons, barrels, and yield. It was also necessary to undermine community—especially the small community where attachment to place might grow and with it resistance to crossing the divide. Finally it was necessary to convert politics into the pursuit of material self-interest and hence render people impotent as citizens and unable to talk of larger and more important things.

To this point the story is well known, but it is hardly finished. Genetic engineers are busy remaking the fabric of life on earth. The development of nanotechnologies—machines at the molecular level—will create possibilities for good and evil that defy prediction. How long will it be until the genetic engineers or nanotechnologists release an AIDS-like virus? One can only guess. But even those promoting such technologies admit that they "carry us toward unprecedented dangers . . . more potent than nuclear weapons."[6] And immediately ahead is the transformation of human consciousness brought on by the conjunction of neuroscience and computers in machines that will simulate whatever reality we choose. What happens to the quality of human experience (or to our politics) when cheap and thoroughgoing fantasy governs our mental life? In each case untransformed nature pales by comparison. It is clumsy, inconvenient, flawed, difficult to rearrange. It is slow. And it cannot be converted to mass dependence and profits so easily.

Ethics and Political Action

Behind each of these endeavors lies a barely concealed contempt for unaltered life and nature, as well as contempt for the people who are expected to endure the mistakes, purchase the results, and live with the consequences, whatever those may be. It is a contempt disguised by words of bamboozlement like "bottom line," "progress," "needs," "costs and benefits," "economic growth," "jobs," "realism," "research," and "knowledge," words that go undefined and unexamined. Few people, I suspect, believe in their bones that the net results from all this will be positive, but most feel powerless to stop what seems to be so inevitable and unable to speak what is so hard to say in the language of self-interest.

The manifestation of biophobia explicit in the urge to control nature has led to a world in which it is becoming easier to be biophobic. Undefiled nature is being replaced by a defiled nature of landfills, junkyards, strip mines, clear cuts, blighted cities, six-lane freeways, suburban sprawl, polluted rivers, and Superfund sites, all of which deserve our phobias. Ozone depletion, meaning more eye cataracts and skin cancer, does give more reason to stay indoors. The spread of toxics and radioactivity does mean more disease. The disruption of natural cycles and the introduction of exotic species have destroyed some of the natural diversity that formerly graced our landscapes. Introduced blights and pests have destroyed American chestnuts and elms. New ones are attacking maples, dogwoods, hemlocks, and ashes. Global warming will degrade the flora and fauna of familiar places.[7] Biophobia sets in motion a vicious cycle that tends to cause people to act in a fashion that undermines the integrity, beauty, and harmony of nature—creating the very conditions that make the dislike of nature yet more probable.

Even so, is it OK that Woody Allen, or anyone else, doesn't like nature? Is biophobia merely one among a number of equally legitimate ways to relate to nature? I do not think so. For every biophobe others have to do that much more of the work of preserving, caring for, and loving the nature that supports biophobes and biophiliacs alike. Economists call this the "free-rider problem." It arises in every group, committee, or alliance, when it is possible for some to receive all the advantages of membership while doing none of the work necessary to create those advantages. Environmental freeriders benefit from others' willingness to fight for clean air which they

Love It or Lose It: The Coming Biophilia Revolution

breathe, clean water which they drink, the preservation of biological diversity which sustains them, and the conservation of the soil which feeds them. But they lift not a finger. Biophobia is not OK because it does not distribute fairly the work of keeping the earth or any local place.

Biophobia is not OK for the same reason that misanthropy or sociopathy are not OK. We recognize these aberrations as the result of deformed childhoods that create unloving and often violent adults. Biophobia in all its forms shrinks the range of experiences and joys in life in the same way that the inability to achieve close and loving relationships limits a human life. E. O. Wilson puts it this way:

> People can grow up with the outward appearance of normality in an environment largely stripped of plants and animals, in the same way that passable looking monkeys can be raised in laboratory cages and cattle fattened in feeding bins. Asked if they were happy, these people would probably say yes. Yet something vitally important would be missing, not merely the knowledge and pleasure that can be imagined and might have been, but a wide array of experiences that the human brain is peculiarly equipped to receive.[8]

Can the same be said of whole societies that distance themselves from animals, trees, landscapes, mountains, and rivers? Is mass biophobia a kind of collective madness? In time I think we will come to know that it is.

Biophobia is not OK because it is the foundation for a politics of domination and exploitation. For our politics to work as they now do a large number of people must not like any nature that cannot be repackaged and sold back to them. They must be ecologically illiterate and ecologically incompetent, and they must believe that this is not only inevitable but desirable. Further, they must be ignorant of the basis of their dependency. They must come to see their bondage as freedom and their discontents as commercially solvable problems. The drift of the biophobic society, as George Orwell and C. S. Lewis foresaw decades ago, is toward the replacement of nature and human nature by technology and the replacement of real democracy by a technological tyranny now looming on the horizon.

These are reasons of self-interest: it is to our advantage to distribute the world's work fairly, to build a society in which lives can be lived fully, and to create an economy in which people participate knowledgeably. There is a further argument against biophobia that rests not on our self-interest but on our duties. Biophobia is not OK, finally, because it violates an ancient charge to replenish the earth. In return for our proper use, the earth is given to humankind as a trust. Proper use requires gratitude, humility, charity, and skill. Improper use begins with ingratitude and disparagement and proceeds to greed, abuse, and violence. We cannot forsake our duties as stewards without breaking that trust. Neither can we forsake the duties of stewardship without breaking another trust to those who preceded us and those who will follow.

Biophobia is certainly more complex than I have described it. One can be both biophobic and a dues-paying member of the Sierra Club. It is possible to be averse to nature but still "like" the idea of nature as an abstraction. Moreover, it is possible to adopt the language and guise of biophilia and do a great deal of harm to the earth, knowingly or unknowingly. In other words, it is possible for us to be inconsistent, hypocritical, and ignorant of what we do.

But is it possible for us to be neutral or "objective" toward life and nature? I do not think so. On close examination, what often passes for neutrality is nothing of the sort but rather the thinly disguised self-interest of those with much to gain financially or professionally. For those presuming to wear the robes of objectivity, the guise, in Abraham Maslow's words, is often "a defense against being flooded by the emotions of humility, reverence, mystery, wonder, and awe."[9] Life ought to excite our passion, not our indifference. Life in jeopardy ought to cause us to take a stand, not retreat into a spurious neutrality. Further, it is a mistake to assume that commitment precludes the ability to think clearly and use evidence accurately. Indeed, commitment motivates intellectual clarity, integrity, and depth. We understand this in other realms quite well. When the chips are down, we don't go to doctors who admit to being neutral about the life and death of their patients. Nor when our hide is at stake do we go to lawyers who profess "objective" neutrality toward justice and injustice. It is a mistake to think that matters of environment and life on earth are somehow different. They are not. And we cannot in such matters remain aloof or indifferent without opening the world to demons.

Love It or Lose It: The Coming Biophilia Revolution

The Roots of Biophilia

We relate to the environment around us in different ways, with differing intensity, and these bonds have different sources. At the most common level we learn to love what has become familiar. There are prisoners who prefer their jail cell to freedom; city dwellers, like Woody Allen, who shun rural landscapes or wilderness; and rural folk who will not set foot in the city. Simply put, we tend to bond with what we know well. Geographer Yi-Fu Tuan describes this bonding as "topophilia," which includes "all of the human being's affective ties with the material environment."[10] Topophilia is less rooted in our deep psychology than it is in our particular circumstances and experiences. It is closer to a sense of habitat that is formed out of the familiar context of everyday living than it is a genuine rootedness in the biology and topography of a certain place. It is not innate but acquired. New Yorkers have perhaps a greater sense of topophilia than do residents of Montana. But Montanans are more likely to feel kinship with sky, mountains, and trout streams. Both, however, tend to be comfortable with what has become habitual and familiar.

E. O. Wilson suggests a deeper source of attachment that goes beyond the particularities of habitat. "We are," he argues, "a biological species [who] will find little ultimate meaning apart from the remainder of life."[11] We are bound to living things by what Wilson describes as an innate urge to affiliate which begins in early childhood and "cascades" into cultural and social patterns. Biophilia is inscribed in the brain itself, he says, expressing tens of thousands of years of evolutionary experience. It is evident in our preference for landscapes that replicate the savannas on which mind evolved: "Given a completely free choice, people gravitate statistically toward a savanna-like environment."[12] Removed to purely artificial environments and deprived of "beauty and mystery," the mind "will drift to simpler and cruder configurations" that undermine sanity itself.[13] Still, biophilia competes with what Wilson describes as the "audaciously destructive tendencies of our species" that seem also to have "archaic biological origins."[14] Allowing those tendencies free rein to destroy the world "in which the brain was assembled over millions of years" is, Wilson argues, "a risky step."[15]

Yet another possibility is that at some level of alertness and maturity we respond with awe to the natural world independent of any instinctual conditioning. "If you study life deeply," Albert Schweitzer once wrote, "its profundity will seize you suddenly with dizziness."[16] He described this response as "reverence for life" arising from the awareness of the unfathomable mystery of life itself. (The German word Schweitzer used, *Ehrfurcht*, implies more awe than the English word *reverence*.)[17] Reverence for life is akin, I think, to what Rachel Carson meant by "the sense of wonder." But for Schweitzer reverence for life originated in large measure from the intellectual contemplation of the world: "Let a man once begin to think about the mystery of his life and the links which connect him with the life that fills the world, and he cannot but bring to bear upon his own life and all other life that comes within his reach the principle of Reverence for Life."[18] Schweitzer regarded reverence for life as the only possible basis for a philosophy on which civilization might be restored from the decay he saw throughout the modern world. "We must," he wrote, "strive together to attain to a theory of the universe affirmative of the world and of life."[19]

We have reason to believe that this intellectual striving is aided by what is already innate in us and may be evident in other creatures. No less an authority than Charles Darwin believed that "all animals feel wonder."[20] Primatologist Harold Bauer once observed a chimpanzee lost in contemplation by a spectacular waterfall in the Gombe Forest Reserve in Tanzania. Contemplation finally gave way to "pant-hoot" calls while the chimp ran back and forth drumming on trees with its fists.[21] No one can say for certain what this behavior means, but it is not farfetched to see it as a chimpanzee version of awe and ecstasy. Jane Goodall and others have described similar behavior. It would be the worst kind of anthropocentrism to dismiss such accounts in the belief that the capacity for biophilia and awe is a human monopoly. In fact, it may be that we have to work at it harder than other creatures. Joseph Wood Krutch, for one, believed that for birds and other creatures "joy seems to be more important and more accessible than it is to us."[22] And not a few philosophers have believed with Abraham Heschel that "as civilization advances, the sense of wonder almost necessarily declines."[23]

Do we, with all our technology, still retain a built-in affinity for nature?

Love It or Lose It: The Coming Biophilia Revolution

I think so, but I know of no proof that would satisfy skeptics. If we do have such an innate sense, we might nevertheless conclude from the damage we have done to the world that biophilia does not operate everywhere and at all times. It may be, as Erich Fromm once argued, that biophilia can be dammed up or corrupted and can subsequently appear in other, and more destructive forms:

> Destructiveness is not parallel to, but the alternative to, biophilia. Love of life or love of the dead is the fundamental alternative that confronts every human being. Necrophilia grows as the development of biophilia is stunted. Man is biologically endowed with the capacity for biophilia, but psychologically he has the potential for necrophilia as an alternative solution.[24]

We also have reason to believe that people can lose the sense of biophilia. In his autobiography, Darwin admits that "fine scenery . . . does not cause me the exquisite delight which it formerly did."[25] It is also possible that entire societies can lose the capacity for love of any kind. When the Ik tribe in northern Uganda was forcibly moved from its traditional hunting grounds into a tiny reserve, their world, in Colin Turnbull's words, "became something cruel and hostile," and they "lost whatever love they might once have had for their mountain world."[26] The biophilia the Ik people may have once felt was transmuted into boredom and a "moody distrust" of the world around them and matched by social relations that Turnbull describes as utterly loveless, cruel, and despicable. The Ik are a stark warning to us that the ties to life and to each other are more fragile than some suppose and, once broken, are not easily repaired or perhaps cannot be repaired at all.

Much of the history of the twentieth century offers further evidence of the fragility of biophilia and of philia. Ours is a time of unparalleled human violence and unparalleled violence toward nature. This is the century of Auschwitz and the mass extinction of species, the age of nuclear weapons and exploding economic growth. Even if we could find no evidence of a lingering human affinity or affection for nature, however, humankind is now in the paradoxical position of having to learn altruism and selflessness—but for reasons of survival which are reasons of self-interest.

Ethics and Political Action

In the words of Stephen Jay Gould: "We cannot win this battle to save species and environments without forging an emotional bond between ourselves and nature as well—for we will not fight to save what we do not love."[27] And if we do not save species and environments, we cannot save ourselves who depend on those species and environments in more ways than we can possibly know. We have, in other words, "purely rational reasons" to cultivate biophilia.[28]

Beyond our physical survival, there is still more at risk. The same Faustian urges that drive the ecological crisis also erode those qualities of heart and mind that constitute the essence of our humanity. Bertrand Russell once put it this way:

> It is only insofar as we renounce the world as its lovers that we can conquer it as its technicians. But this division in the soul is fatal to what is best in man. . . . The power conferred by science as a technique is only obtainable by something analogous to the worship of Satan, that is to say, by the renunciation of love. . . . The scientific society in its pure form . . . is incompatible with the pursuit of truth, with love, with art, with spontaneous delight, with every ideal that men have hitherto cherished.[29]

The ecological crisis, in short, is about what it means to be human. And if natural diversity is the wellspring of human intelligence, then the systematic destruction of nature inherent in contemporary technology and economics is a war against the very sources of mind. We have good reason to believe that human intelligence could not have evolved in a lunar landscape devoid of biological diversity. And we have good reason to believe that the sense of awe toward the creation had a great deal to do with the origin of language and why early hominids *wanted* to talk, sing, and write poetry in the first place. Elemental things like flowing water, wind, trees, clouds, rain, mist, mountains, landscape, animals, changing seasons, the night sky, and the mysteries of the life cycle gave birth to thought and language. They continue to do so, but perhaps less exuberantly than they once did. For this reason I think it is impossible to unravel natural diversity without undermining human intelligence as well.

Can we save the world and anything like a human self from the violence we have unleashed without biophilia and reverence for the creation? All the arguments made by technological fundamentalists and by the zealots of instrumental rationality notwithstanding, I know of no good evidence that we can. We must choose, in Joseph Wood Krutch's words, whether "we want a civilization that will move toward some more intimate relation with the natural world, or . . . one that will continue to detach and isolate itself from both a dependence upon and a sympathy with that community of which we were originally a part."[30] The writer of Deuteronomy had it right. Whatever our feelings, however ingenious our philosophies, whatever innate gravity tugs at us, we must finally choose between life or death: between intimacy or isolation.

From Eros to Agape

We are now engaged in a great global debate about what it means to live "sustainably" on the earth. The word, however, is fraught with confusion—in large part because we are trying to define it before we have decided whether we want an intimate relation with nature or total mastery. We cannot know what sustainability means until we have decided what we intend to sustain and how we propose to do so. For some, sustainability means maintaining our present path of domination, only with greater efficiency. But were we to decide with Krutch and others that we do want an intimate relation with nature, to take nature as our standard, what does this mean? We must choose along the continuum that runs between biophilia and biophobia, intimacy or mastery, but how can we know when we have crossed over from one to the other? The choices are not always so simple nor will they be presented to us so candidly. The options, even the most destructive, will be framed as life-serving, or as necessary for a greater good someday, or as simply inevitable since "you can't stop progress." How, then, can we distinguish those things that serve life well from those that diminish it?

Biophilia is a kind of love, but what kind? The Greeks distinguished three kinds of love: *eros*, meaning love of beauty or romantic love aiming to possess; *agape* or sacrificial love that asks nothing in return; and *philia*, the love between friends. The first two of these reveal important aspects of biophilia, which probably begins as eros but matures, if at all, as a form of agape. For the Greeks eros went beyond sensuous love to include creature needs for food, warmth, and shelter as well as higher needs to understand, appreciate, and commune with nature.[31] But eros aims no higher than self-fulfillment. Defined as an innate urge, biophilia is eros: it reflects human desire and self-interest, including the interest in survival.

Biophilia as eros, however, traps us in a paradox. In the words of Susan Bratton: "Without agape, human love for nature will always be dominated by unrestrained eros and distorted by extreme self-interest and material valuation."[32] What we love only from self-interest we will sooner or later destroy. Agape tempers our use of nature so that "God's providence is respectfully received and insatiable desire doesn't attempt to extract more from creation than it can sustain."[33] Agape enlarges eros, bringing humans and the creation together so that it is not possible to love either humanity or nature without also loving and serving the other. Agape in this sense is close to Schweitzer's description of "reverence for life," which calls us to transcend even the most enlightened calculations of self-interest. Would not respect for nature do as well? I think not: it is just too bloodless, too cool, too self-satisfied and aloof to cause us to do much to save species and environments. I am inclined to agree with Stephen Jay Gould that we will have to reach deeper.

What, then, do we know about deeper sources of motivation—including the ways in which eros is transformed into agape—and what does this reveal about biophilia? First, we know that the capacity for love of any kind begins early in the life and imagination of the child. Perhaps the potential for biophilia begins at birth, as Robert Coles once surmised, as the newborn infant is introduced to its place in nature.[34] If so, the manner and circumstances of birth are more important than usually thought. Biophilia is certainly evident in the small child's efforts to establish intimacy with the earth—Jane Goodall, age two, sleeping with earthworms under her pillow,[35] for example, or John Muir "reveling in the wonderful wildness" around his boyhood Wisconsin home.[36] If by some fairly young age, however, nature has not been experienced as a friendly place of adventure and

Love It or Lose It: The Coming Biophilia Revolution

excitement, biophilia will not take hold as it might have. An opportunity will have passed and thereafter the mind will lack some critical dimension of perception and imagination.

Second, I think we know that biophilia requires easily and safely accessible places where it might take root and grow. For Aldo Leopold it began in the marshes and woods along the Mississippi River. For young E. O. ("Snake") Wilson it began in boyhood explorations of the "woods and swamps in a languorous mood . . . [forming] the habit of quietude and concentration."[37] The loss of places such as these is one of the uncounted costs of economic growth and urban sprawl. It is also a powerful argument for containing that sprawl and expanding urban parks and recreation areas.

Third, I think we can safely surmise that biophilia, like the capacity to love, needs the help and active participation of parents, grandparents, teachers, and other caring adults. Rachel Carson's relation with her young nephew caused her to conclude that the development of a child's sense of wonder required "the companionship of at least one adult who can share it, rediscovering with him the joy, excitement and mystery of the world we live in."[38] For children the sense of biophilia needs instruction, example, and validation by a caring adult. And for adults, rekindling the sense of wonder may require a child's excitement and openness to natural wonders as well.

Fourth, we have every reason to believe that love and biophilia alike flourish mostly in good communities. I do not mean necessarily affluent places. In fact, affluence often works against real community as surely as do violence and utter poverty. By community I mean, rather, places in which the bonds between people and those between people and the natural world create a pattern of connectedness, responsibility, and mutual need. Real communities foster dignity, competence, participation, and opportunities for good work. And good communities provide places in which children's imagination and earthy sensibilities root and grow.

Fifth, we have it on good authority that love is patient, kind, enduring, hopeful, long-suffering, and truthful, not envious, boastful, insistent, arrogant, rude, self-centered, irritable, and resentful (1 Corinthians 13). For biophilia to work I think it must have similar qualities. Theologian James Nash, for example, proposes six ecological dimensions of love: beneficence (kindness to wild creatures, for example); other-esteem, which rejects the idea of possessing or managing the biosphere; receptivity to nature (awe, for example); humility, by which he means caution in the use of technology; knowledge of ecology and how nature works; and communion as "reconciliation, harmony, koinonia, shalom" between humankind and nature.[39] I would add only that real love does not do desperate things and it does not commit the irrevocable.

Sixth, I think we know with certainty that beyond some scale and level of complexity the possibility for love of any sort declines. Beneficence, awe, reconciliation, and communion are not entirely probable attitudes for the poverty-stricken living in overcrowded barrios. With 10 or 12 billion people on the earth, we will have no choice but to try to manage nature, even though it will be done badly. The desperate and the hungry will not be particularly cautious with risky technologies. Nor will the wealthy, fed and supplied by vast, complex global networks, understand the damage they cause in distant places they never see and the harm they do to people they will never know. Knowledge has its own limits of scale. Beyond some level of scale and complexity the effects of technology, used in a world we cannot fully comprehend, are simply unknowable. When the genetic engineers and the nanotechnologists finally cause damage to the earth comparable to that done by the chemists who invented and so casually and carelessly deployed CFCs, they too will plead for forgiveness on the grounds that they did not know what they were doing.

Seventh, love, as Erich Fromm once wrote, is an art, the practice of which requires "discipline, concentration and patience throughout every phase of life."[40] The art of biophilia, similarly, requires us to use the world with disciplined, concentrated, and patient competence. To live and earn our livelihood means that we must "daily break the body and shed the blood of creation" in Wendell Berry's words. Our choice is whether we do so "knowingly, lovingly, skillfully, reverently . . . [or] ignorantly, greedily, clumsily, destructively."[41] Practice of any art also requires forbearance, which means the ability to say no to things that diminish the object of love or our capacity to work artfully. And for the same reasons that it limits the exploitation of persons, forbearance sets limits to our use of nature.

Finally, we know that for love to grow from eros to agape something like

Love It or Lose It: The Coming Biophilia Revolution

metanoia—the transformation of one's whole being—is necessary. Metanoia is more than a paradigm change. It is a change, above all, in our loyalties, affections, and basic character that subsequently changes our intellectual priorities and paradigms. For whole societies the emergence of biophilia as agape will require something like a metanoia that deepens our loyalty and affections to life and in time alters the character of our entire civilization.

The Biophilia Revolution

"Is it possible," E. O. Wilson asks, "that humanity will love life enough to save it?"[42] And if we do love life enough to save it, what is required of us? At one level the answer is obvious. We need to transform the way we use the earth's endowment of land, minerals, water, air, wildlife, and fuels: an efficiency revolution which buys us some time. Beyond efficiency, we need another revolution that transforms our ideas of what it means to live decently and how little is actually necessary for a decent life: a sufficiency revolution. The first revolution is mostly about technology and economics. The second revolution is about morality and human purpose. The biophilia revolution is about the combination of reverence for life and purely rational calculation by which we will *want* to be both efficient and live sufficiently. It is about finding our rightful place on earth and in the community of life; it is about citizenship, duties, obligations, and celebration.

There are two formidable barriers standing in our way. The first is the problem of denial. We have not yet faced up to the magnitude of the trap we have created for ourselves. We are still thinking of the crisis as a set of problems which are, by definition, solvable with technology and money. In fact we face a series of dilemmas which can be avoided only through wisdom and a higher and more comprehensive level of rationality than we have yet shown. Better technology would certainly help, but our crisis is not fundamentally one of technology: it is one of mind, will, and spirit. Denial must be met by something like a worldwide ecological perestroika predicated on the admission of failure: the failure of our economics which became disconnected from life; the failure of our politics which lost sight of the moral roots of our commonwealth; the failure of our science which lost sight of the essential wholeness of things; and the failures of all of us as moral beings who allowed these things to happen because we did not love deeply enough and intelligently enough. The biophilia revolution must come like an ecological enlightenment that sweeps out the modern superstition that we are knowledgeable enough and good enough to manage the earth and direct evolution.

The second barrier standing in the way of the biophilia revolution is one of imagination. It is easier, perhaps, to overcome denial than it is to envision a biophilia-centered world and believe ourselves capable of creating it. We could get an immediate and overwhelming worldwide consensus today on the proposition "Is the earth in serious trouble?" But we are not within even a light-year of agreement on what to do about it. Confronted by the future, the mind has a tendency to wallow. For this reason we can diagnose our plight with laser precision while proposing to shape the future with a sledgehammer. Fictional utopias, almost without exception, are utterly dull and unconvincing. And the efforts to create utopias of either right or left have been monumental failures, leaving people profoundly discouraged about their ability to shape the world in accord with their highest values. And now some talk about creating a world that is sustainable, just, and peaceful! What is to be done?

Part of our difficulty in confronting the future is that we think of utopia on too grand a scale. We are not very good at comprehending things at the scale of whole societies, much less that of the planet. Nor have we been very good at solving the problems utopias are supposed to solve without imposing simplistic formulas that ride roughshod over natural and cultural diversity. Except for certain anarchist varieties, utopianism is almost synonymous with homogenization. Another part of the problem is the modern mind's desire for drama, excitement, and sexual sizzle—which explains why we don't have many best-selling novels about Amish society, arguably the closest thing to a sustainable society we know. How do we fulfill the need for meaning and variety while discarding some of our most cherished fantasies of domination? How do we cause the "change in our intellectual emphasis, loyalties, affections, and convictions" without which all else is

Love It or Lose It: The Coming Biophilia Revolution

moot?[43] When we think of revolution our first impulse is to think of some grand political, economic, or technological change: some way to fix quickly what ails us. What ails us, however, is closer to home, and I suggest that we begin there.

The Recovery of Childhood

I began by describing biophilia as a choice. In fact it is a series of choices, the first of which has to do with the conduct of childhood and how the child's imagination is woven into a homeplace. Practically, the cultivation of biophilia calls for the establishment of more natural places—places of mystery and adventure where children can roam, explore, and imagine. This means more urban parks, more greenways, more farms, more river trails, and wiser land use everywhere. It means redesigning schools and campuses to replicate natural systems and functions. It means greater contact with nature during the school day, but also unsupervised hours to play in places where nature has been protected or allowed to recover.

For biophilia to take root, we must take our children seriously enough to preserve their natural childhood. But childhood is being impoverished and abbreviated, and the reasons sound like a curriculum in social pathology: too many broken homes and unloving marriages, too much domestic violence, too much alcohol, too many drugs, too many guns, too many things, too much television, too much idle time and permissiveness, too many off-duty parents, and too little contact with grandparents. Children are rushed into adulthood too soon, only to become childish adults unprepared for parenthood, and the cycle repeats itself. We will not enter this new kingdom of sustainability until we allow our children the kind of childhood in which biophilia can put down roots.

Recovering a Sense of Place

I do not know whether it is possible to love the planet or not, but I do know that it is possible to love the places we can see, touch, smell, and experience. And I believe, with Simone Weil, that rootedness in a place is "the most important and least recognized need of the human soul."[44] The attempt to encourage biophilia will not amount to much if we fail to create the kind of places where we might become deeply rooted. The second decision we must make, then, has to do with the will to rediscover and reinhabit our places and regions, finding in them sources of food, livelihood, energy, healing, recreation, and celebration.

Call it "bioregionalism" or "becoming native to our places." Either way it means deciding to relearn the arts that Jacquetta Hawkes once described as "a patient and increasingly skillful love-making that [persuades] the land to flourish."[45] It means rebuilding family farms, rural villages, towns, communities, and urban neighborhoods. It means restoring local culture and our ties to our local places where biophilia first takes root. It means reweaving the local ecology into the fabric of the economy and life patterns while diminishing our use of the automobile and our ties to the commercial culture. It means deciding to slow down—hence more bike trails, more gardens, more solar collectors. It means rediscovering and restoring the natural history of our places. And, as Gary Snyder once wrote, it means finding our place and digging in.[46]

Education and Biophilia

The capacity for biophilia can still be snuffed out by education that aims no higher than to enhance the potential for upward mobility—which has come to mean putting as much distance as possible between the apogee of one's career trajectory and one's roots. We should worry a good bit less about whether our progeny will be able to compete as a "world-class work force" and a great deal more about whether they will know how to live sustainably on the earth. My third proposal, then, requires the will to reshape education in a way that fosters innate biophilia and the analytical abilities and practical skills necessary for a world that takes life seriously.

Lewis Mumford once proposed the local community and region as the "backbone of a drastically revised method of study."[47] Study of the region would ground education in the particularities of a specific place and would also integrate various disciplines in accord with the "regional survey," including surveys of local soils, climate, vegetation, history, economy, and society. Mumford envisioned this as an "organic approach to knowledge" that began with the "common whole—a region, its activities, its people, its configuration, its total life."[48] The aim is "to educate citizens, to give them the tools of action," and to educate a people "who will know in detail where they live and how they live . . . united by a common feeling for their land-scape, their literature and language, their local ways."[49]

Something like the regional survey is required for the biophilia revolu-tion. Education that nourishes a reverence for life would occur more often out-of-doors and in relation to the local community. It would confer a basic competence in the kinds of knowledge that Mumford described a half cen-tury ago. It would help people become not only literate but ecologically lit-erate, understanding the biological requisites of human life on earth. It would confer basic competence in what I have called the "ecological design arts"—the set of perceptual and analytic abilities, ecological wisdom, and practical wherewithal essential to making things that fit in a world gov-erned by the laws of ecology and thermodynamics.[50] The components for a curriculum in the ecological design arts can be found in recent work in res-toration ecology, ecological engineering, conservation biology, solar de-sign, sustainable agriculture, sustainable forestry, ecological economics, energetics, and methods of least-cost, end-use analysis.

A New Covenant with Animals

The biophilia revolution would be incomplete without our creating a new relationship with animals—one, in Barry Lopez's words, that rises "above prejudice to a position of respectful regard toward everything that is dif-ferent from ourselves and not innately evil."[51] We need animals, not locked up in zoos, but living free on their terms. We need them for what they can tell us about ourselves and about the world. We need them for our imagi-nation and for our sanity. We need animals for what they can teach us about courtesy and what Gary Snyder calls "the etiquette of the wild."[52] The hu-man capacity for biophilia as agape will remain "egocentric and partial" until it can also embrace creatures who cannot reciprocate.[53] And, needing animals, we will need to restore wild landscapes that invite them again.

A new covenant with animals demands that we decide to limit the hu-man domain in order to establish their rights in law, custom, and daily habit. The first step is to discard the idea we got from René Descartes that animals are only machines incapable of feeling pain and to be used in any way we see fit. Protecting animals in the wild while permitting confine-ment feeding operations and most laboratory uses of animals makes no moral sense and diminishes our capacity for biophilia. In this respect I think Paul Shepard is right: to recognize animals and wildness is to decide to admit deeper layers of our consciousness into the sunlight of full consciousness again.[54]

The Economics of Biophilia

The biophilia revolution will also require national and global decisions that permit life-centeredness to flourish at a local scale. Biophilia can be suffocated, for example, by the demands of an economy oriented to accumulation, speed, sensation, and death. But economists have not written much about how an economy encourages or discourages love generally or biophilia in particular. As a result, not much thought has been given to the relationship between love and the way we earn our keep.

The transition to an economy that fosters biophilia requires a decision to limit the human enterprise relative to the biosphere. Some economists talk confidently of a fivefold or tenfold increase in economic activity over the next half-century. But Peter Vitousek and his colleagues have shown that humans now use or coopt 40 percent of the net primary productivity from terrestrial ecosystems.[55] What limits does biophilia set on the extent of the human enterprise? What margin of error does love require?

Similarly, in the emerging global economy in which capital, technology, and information move easily around the world, how do we protect the people and communities left behind? Now more than ever the rights of capital are protected by all the power money can buy. The rights of communities are protected less than ever. Consequently we face complex decisions about how to protect communities and their stability on which biophilia depends.

Biophilia and Patriotism

The decisions necessary to lead us toward a culture capable of biophilia are finally political decisions. But our politics, no less than our economy, has other priorities. In the name of "national security" or one ephemeral national "interest" or another we lay waste to our lands and the prospects of our children. Politics of the worst sort has corrupted our highest values, becoming instead one long evasion of duties and obligations in the search for private or sectarian advantage. "Crackpot realists" tell us that this is how it has always been and must therefore always be: a view which marries bad history to bad morals.

Patriotism, the name we give to the love of one's country, must be redefined to include those things which contribute to the real health, beauty, and ecological stability of our homeplaces and to exclude those which do not. Patriotism as biophilia requires that we decide to rejoin the idea of love of one's country to how well one uses the country. To destroy forests, soils, natural beauty, and wildlife in order to swell the gross national product, or to provide short-term and often spurious jobs, is not patriotism but greed.

Real patriotism demands that we weave the competent, patient, and disciplined love of our land into our political life and our political institutions. The laws of ecology and those of thermodynamics, which mostly have to do with limits, must become the foundation for a new politics. No one has expressed this idea more clearly than the former Czech president, Václav Havel: "We must draw our standards from our natural world. . . . We must honor with the humility of the wise the bounds of that natural world and the mystery which lies beyond them, admitting that there is something in the order of being which evidently exceeds all our competence."[56] Elsewhere he writes:

> Genuine politics . . . is simply a matter of serving those around us: serving the community, and serving those who will come after us. Its deepest roots are moral because it is a responsibility, expressed through action, to and for the whole, a responsibility . . . only because it has a metaphysical grounding: that is, it grows out of a conscious or subconscious certainty that our death ends nothing, because everything is forever being recorded and evaluated somewhere else, somewhere "above us," in what I have called "the memory of being."[57]

Beyond Utopia

Erich Fromm once asked whether whole societies might be judged sane or insane.[58] After the world wars, state-sponsored genocide, gulags, McCarthyism, and the "mutual assured destruction" of the twentieth century there can be no doubt that the answer is affirmative. Nor do I doubt that our descendants will regard our obsession with perpetual economic growth and frivolous consumption as evidence of theologically induced derangement. Our modern ideas about sanity, in large measure, can be attributed to Sigmund Freud, an urban man. And from the urban male point of view the relationship between nature and sanity may be difficult to see and even more difficult to feel. Freud's reconnaissance of the mind stopped too soon. Had he gone further, and had he been prepared to see it, he might have discovered what Theodore Roszak calls "the ecological unconscious," the repression of which "is the deepest root of collusive madness in industrial society."[59] He might also have stumbled upon biophilia. And had he done so our understanding of individual and collective sanity would have been on more solid ground.

The human mind is a product of the Pleistocene age, shaped by wildness that has all but disappeared. If we complete the destruction of nature, we will have succeeded in cutting ourselves off from the source of sanity itself. Hermetically sealed amidst our creations and bereft of those of The Creation, the world then will reflect only the demented image of the mind imprisoned within itself. Can the mind doting upon itself and its creations be sane? Thoreau would never have thought so, nor should we.

A sane civilization that loved more fully and more intelligently would have more parks and fewer shopping malls; more small farms and fewer agribusinesses; more prosperous small towns and smaller cities; more solar collectors and fewer strip mines; more bike trails and fewer freeways; more trains and fewer cars; more celebration and less hurry; more property owners and fewer millionaires; more readers and fewer television watchers; more shopkeepers and fewer multinational corporations; more teachers and fewer lawyers; more wilderness and fewer landfills; more wild animals and fewer pets. Utopia? No! In our present circumstances this is the only realistic course imaginable. We have tried utopia and can no longer afford it.

NOTES

1. Eric Lax, *Woody Allen: A Biography* (New York: Vintage Books, 1992), pp. 39–40.
2. Edward O. Wilson, *Biophilia* (Cambridge: Harvard University Press, 1984), p. 85.

3. Erich Fromm, *The Anatomy of Human Destructiveness* (New York: Holt, Rinehart & Winston, 1973), pp. 365–366.

4. Owen Barfield, *Saving the Appearances: A Study in Idolatry* (New York: Harcourt Brace Jovanovich, 1957), p. 78.

5. Jared Diamond, *The Third Chimpanzee* (New York: HarperCollins, 1992), pp. 317–338.

6. Eric Drexler, *Engines of Creation* (New York: Anchor Books, 1987), p. 174.

7. Robert Peters and J. P. Myers, "Preserving Biodiversity in a Changing Climate," *Issues in Science and Technology* 8(2) (Winter 1991–2):66–72.

8. Wilson, *Biophilia*, p. 118.

9. Abraham Maslow, *The Psychology of Science: A Reconnaissance* (Chicago: Gateway, 1966), p. 139.

10. Yi-Fu Tuan, *Topophilia* (New York: Columbia University Press, 1974), p. 93.

11. Wilson, *Biophilia*, p. 81.

12. Ibid., p. 112.

13. Ibid., p. 115.

14. Ibid., p. 118.

15. Ibid., p. 121.

16. Albert Schweitzer, *Reverence for Life* (New York: Pilgrim Press, 1969), p. 115.

17. See Joseph Wood Krutch, *The Great Chain of Life* (Boston: Houghton Mifflin, 1991), p. 160.

18. Albert Schweitzer, *Out of My Life and Thought: An Autobiography* (New York: Holt, Rinehart & Winston, 1972), p. 231.

19. Ibid., p. 64.

20. Charles Darwin, *The Descent of Man* (New York: Modern Library, 1977), p. 450.

21. Cited in Melvin Konner, *The Tangled Wing: Biological Constraints on the Human Spirit* (New York: Holt, Rinehart & Winston, 1982), p. 431.

22. Krutch, *The Great Chain of Life*, p. 227.

23. Abraham Heschel, *Man Is Not Alone: A Philosophy of Religion* (New York: Farrar, Straus & Giroux, 1990), p. 37; see also Konner, *The Tangled Wing*, p. 435.

24. Fromm, *Anatomy of Human Destructiveness*, p. 366.

25. Charles Darwin, *The Autobiography of Charles Darwin* (New York: Dover Books, 1958), p. 54.

26. Colin Turnbull, *The Mountain People* (New York: Simon & Schuster, 1972), pp. 256, 259.

27. Stephen Jay Gould, "Enchanted Evening," *Natural History* (September 1991):14.

28. Wilson, *Biophilia*, p. 140.

29. Bertrand Russell, *The Scientific Outlook* (New York: W. W. Norton, 1959), p. 264.

30. Krutch, *The Great Chain of Life*, p. 165.

31. Susan Bratton, "Loving Nature: Eros or Agape?" *Environmental Ethics* 14(1) (Spring 1992):11.

32. Ibid., p. 15.

33. Ibid., p. 13.

34. Robert Coles, "A Domain of Sorts," *Harpers Magazine* (November 1971). Reprinted in Stephen Kaplan and Rachel Kaplan (eds.), *Humanscape: Environments for People* (North Scituate: Duxbury Press, 1978), pp. 91–93.

35. Sy Montgomery, *Walking with the Great Apes* (Boston: Houghton Mifflin, 1991), p. 28.

36. John Muir, *The Story of My Boyhood and Youth* (San Francisco: Sierra Club Books, 1988).

37. Wilson, *Biophilia*, pp. 86–92.

38. Rachel Carson, *The Sense of Wonder* (New York: Harper & Row, 1987), p. 45.

39. James A. Nash, *Loving Nature: Ecological Integrity and Christian Responsibility* (Nashville: Abingdon Press, 1991), pp. 139–161.

40. Erich Fromm, *The Art of Loving* (New York: Harper & Row, 1989), p. 100.

41. Wendell Berry, *The Gift of Good Land* (San Francisco: North Point Press, 1981), p. 281.

42. Wilson, *Biophilia*, p. 145.

43. Aldo Leopold, *A Sand County Almanac* (New York: Ballantine, 1966), p. 246.

44. Simone Weil, *The Need for Roots* (New York: Harper Colophon, 1971), p. 43.

45. Jacquetta Hawkes, *A Land* (New York: Random House, 1951), p. 202.

46. Gary Snyder, *Turtle Island* (New York: New Directions, 1974), p. 101.

47. Lewis Mumford, *Values for Survival* (New York: Harcourt Brace, 1946), pp. 150–154.

48. Lewis Mumford, *The Culture of Cities* (New York: Harcourt Brace Jovanovich, 1970), p. 385.

49. Ibid., p. 386.

50. David W. Orr, "Education and the Ecological Design Arts," *Conservation Biology* 6(2) (June 1992):162–164.

51. Barry Lopez, "Renegotiating the Contracts," in Thomas J. Lyon (ed.), *This Incomperable Lande: A Book of American Nature Writing* (Boston: Houghton Mifflin, 1989), p. 383.

52. Gary Snyder, *The Practice of the Wild* (San Francisco: North Point Press, 1990), pp. 3–24.

53. Lewis Mumford, *The Conduct of Life* (New York: Harcourt Brace Jovanovich, 1970), p. 286.

54. See Paul Shepard and Barry Sanders, *The Sacred Paw* (New York: Viking Penguin, 1992), and Chapter 9 in this volume.

55. Peter Vitousek et al., "Human Appropriation of the Products of Photosynthesis," *Bioscience* 36(6) (June 1986):368–373.

56. Václav Havel, *Living in Truth* (London: Faber & Faber, 1989), p. 153.

57. Václav Havel, *Summer Meditations* (New York: Knopf, 1992), p. 6.

58. Erich Fromm, *The Sane Society* (New York: Fawcett Books, 1955).

59. Theodore Roszak, *The Voice of the Earth* (New York: Simon & Schuster, 1992), p. 320.

CHAPTER 15

Biophilia:

Unanswered Questions

Michael E. Soulé

T HIS CHAPTER IS a commentary on the discussions that led to this volume on biophilia—strictly the love of living nature, but more accurately the whole range of innately channeled human responses to living nature. Instead of attempting to represent the viewpoints recorded in the preceding chapters, here I wish to raise some issues that I find important or perplexing.

As in so many fields and so many debates, E. O. Wilson has done society a great service by drawing attention to a fascinating, if controversial, problem. In his book *Biophilia: The Human Bond with Other Species* (1984), Wilson penetrated to the fundamental question: To what degree and in what forms has evolution produced genetically based responses in humans, positive and negative, to biological and other environmental phenomena? Edward Wilson and Steven Kellert, by producing this volume, have achieved a notable advance in the analysis of this question.

For most of the contributors to this volume, there is little doubt about the Cenozoic's genetic imprimatur in the sequence of nucleotides that affect human physical characteristics; they would all agree that functional attributes, such as the articulation of the shoulder or the geographic distribution of genes that protect against malaria, are the genetic footprints of natural selection.

But can the same be said about human behavior? Most of this essay and some of the other chapters in this volume are unabashedly Neo-Darwinian, evolutionary, or sociobiological; they posit a genetic basis for human affinities and reactions to nature. At the same time the authors appear to ignore much of contemporary social criticism. By social criticism I refer to those movements, mostly but not exclusively of the left, that seek explanations for human nature and human differences in arguments that are nondeterministic, nonbiological, concerned with culture, social equity, and justice, and often influenced by Marxist thought. These movements, including certain forms of feminism, political ecology, deconstructionism, and multiculturalism, usually claim that differences between the sexes (they are real according to the currently dominant "difference feminists") and between human groups are environmental—that is, nongenetic. Parenthetically, many on the ideological right detest genetic determinism, as well, especially when it is used to explain the existence of "sinful" behavior such as criminality and homosexuality. The coexistence of these two mutually near-invisible academic cultures in modern intellectual discourse—Neo-Darwinist and Neo-Marxist humanist for short—may testify to the powerlessness of our educational systems to root out ignorance, to the tenacity of ideology, or to the fear of a paradigm in ascendency by one in retreat.

I suggest that it is not prudent for these two parties—who appear like two teams of mutually oblivious miners tunneling through the same mountain—to labor on in ignorance of each other. Somewhere, sometime, they should meet to challenge the premises, data, and interpretations of the other's faith, pretending as it were that we are all scholars in the same discipline publishing in the same journals. The only reason for avoiding engagement is that ideology is more precious than knowledge. Though it is

TABLE 15.1. *A Tentative Classification of Biophilic Responses*

	Variation	
Category of Response	Gender Differences	Ethnic/Cultural Differences
Affinity for scenery/habitat of certain kinds	Probable	Possible
Affinity for domestic animals	Possible	Possible
Affinity for dangerous, wild animals	Probable	Probable
Aversion to snakes, spiders, etc. (phobias)	Not likely	Probable
Aversion to cliffs, high places	Possible	Not likely

Note: Neurotic biophilias are not included.

naive to expect either of these cultures to admit error, let alone defeat, one can hope that the best thinkers from both sides of the mountain will meet, talk, and learn from each other.

Complexity

The major premise of biophilia is the existence of genetically based physiological (probably neural) structures that respond selectively to various plants, animals, and habitats. Like social behavior, though, biophilia is manifold in kinds and degree. There appear to be many bioresponsive behavioral systems, not just one biophilia. Wilson (Chapter 1), in fact, refers to a biophilia complex. Table 15.1 indicates some of the responses that have been discussed. The table also suggests that biophilic responses may vary between sexes and "races" or geographic subpopulations. Thus even as we search for evidence of universals, we must not ignore heterogeneity, for the patterns of variation often contain clues about mechanism and cause.

Habitat Selection

One of the most frequently discussed biophilic responses is habitat selection (Chapters 3 and 4). Even bacteria forage along chemical gradients, so it is nearly inconceivable to an evolutionist that primates would be bereft of mental structures capable of recognizing "good" places. The learning rules

(Chapters 1 and 3) for distinguishing dangerous from benign topographies, and for discriminating between potentially nourishing habitats and less promising sites, should be simple and deeply embossed in the brain and the genes that produce it. Caves, promontories, fresh waters, herds of animals, flocks of birds—all may fire fusillades of nerve impulses, as suggested by the work of Heerwagen and Orians (Chapter 4).

What are landscape aesthetics, then, if not a mirror of the Pliocene and Pleistocene? Or as Wilson suggests (Chapter 1) "with aesthetics we return to the central issue of biophilia." But habitat biophilia can have more than just aesthetic and recreational implications; Ulrich (Chapter 3) reviews the psychological benefits of outdoor activity and inactivity. One of these benefits appears to be a sense of well-being. A sense of well-being is not, some would say, far from religious feelings—grace, connectedness with nature. Many people believe that religious feeling can be evoked by certain arrangements of trees and rocks or by certain landscapes. Thus biophilia may be difficult to tease apart from what some people call a relationship with "spirit" or God. Hard data are usually lacking, however, on the evocation of religious or spiritual experiences by places and landscapes. Here we risk stepping across the nebulous line between affinities for organisms per se (biophilia) and the love of place, of scenery, and of the earth itself.

Returning to habitat, defined as association of species and landscapes, it would be surprising if all humans had identical, genetically based habitat preferences. There are several reasons to expect such differences between isolated human groups. First, thousands of generations have passed since the African savanna was our universal home, and some human groups have occupied other habitats such as deserts, dense forests, or treeless polar regions for many millennia. An interval of many thousands of years would allow even very small selection pressures to change genetically based habitat preferences. Second, it is likely that gender differences exist for positive biophilia as proposed in Table 15.1, and evidence is found in the work of Orians and Heerwagen (Chapter 4) for gender-based differences in landscape preferences. In summary, then, research is needed on cross-cultural and cross-racial differences before concluding that certain landscapes are universally preferred. Cross-species comparisons, using the four species of great apes, might also be illuminating.

The Biophilia Hypothesis

One subject that has received insufficient attention is the relationship between biophilia, on the one hand, and the different senses that mediate the perception of organisms and habitat on the other. One would expect that the sense involved, whether touch, smell, taste, hearing, or vision, might make a difference—not only because different centers of the brain are associated with different senses but also because these centers are under different degrees and kinds of influence from the rational neocortex.

Olfaction (chemical sense) is one of the oldest senses. There may exist atavistic networks in the mind that facilitate negative or positive responses to certain odors (decay, skunks, fecal material, flowers, baby skin, cooking food, riparian vegetation, wood, smoke). One might expect some of these responses to show little variation among mammals. Similar principles might apply to the sense of touch. The search for universal biophilial responses should probably begin with these deep and ancient sensory wellsprings.

Hearing and vision are more recent developments in evolution and are more closely associated with language. Many avian and mammalian alarm calls and other auditory and visual signals of animals are instantly recognized, even employed, by humans. A snarl needs no translation across species boundaries. But many interesting questions await study. What are the physiological correlates of drumming sounds? Are natural sounds (of animals) more relaxing than sounds of machinery? Are the songs and calls of seabirds less relaxing than savanna or forest bird sounds? How does complexity per se of sound and vision affect us? For example, are signals of higher complexity or higher fractal dimensionality more appealing? Are nonlinear or chaotic phenomena more appealing than linear structures and sounds? If so, are such preferences universal, innate, and limited to humans?

For those of us committed to the protection of biotic diversity, one important question is whether diversity per se is rewarding (pleasant, appealing)—and if so, through which senses and to what degree? Given two landscapes differing only in the number of species or the number of habitat types, for example, do people generally prefer the more diverse? Humans might prefer intermediate levels of complexity (information) rather than the extremes of either uniformity or diversity. "Overdiverse" habitats such as the tropical rain forest canopy or coral reefs in western Oceania, for instance, contain so many species that it would require a lifetime to learn all the species in groups, such as flowering plants or butterflies in the former and damselfishes in the latter. Such a high degree of diversity might be overwhelming, and innately less appealing, to hunter-gatherers than lower levels of environmental information.

To what degree are temporal and spatial scales a factor in such potential affinities? Highly diverse systems might be more attractive if the diversity were spatially or seasonally partitioned so that the human observer did not encounter it all at once. Again, the sensory modality might be relevant to the question of optimum levels of biodiversity. Could it be that visual variety is attractive but textural or tactile variety is not? Or is variety always the spice of life? The answers to such questions may have some bearing on the fate of life on earth.

Salience

An issue related to complexity and affinity is *salience*. To what extent is the appeal of a biological entity a question of its salience—its "standing out"? Salience may turn out to be culture- and gender-specific as suggested earlier for landscapes. Nabhan and St. Antoine (Chapter 7) mentioned that paleo-taxonomies include the rarest of the rare and even extinct species. Human beings value rare objects and remember them for a lifetime. The specialness of rare organisms (see also Diamond in Chapter 8) may suggest a learning rule hitherto not posited.

Size

Salience is often proportional to size. Out of sight, out of mind applies to most bacteria, algae, and fungi (except the psychedelic and larger ones)—despite their biogeochemical dominance and importance as sources for intuition and knowledge about ecology and symbiogenesis (Sagan and Margulis in Chapter 11). The biological basis for the size/salience correlation is not difficult to fathom. Katcher and Wilkins (Chapter 5) have pointed out that salience is associated with danger, and Nelson (Chapter 6) has noted that salience in some aboriginal cultures is associated with the species' totemic, magical, and spiritual qualities. For obvious reasons, the salience of both human prey and predators is proportional to their body mass: big and dangerous animals appear most prominently in art and myth (but see Lawrence in Chapter 10). As we shall see, the same bias may apply to pets.

Complexity and Power

Complexity per se is salient, too, and Gadgil (Chapter 12) reminds us that we may be in danger of confusing biophilia with the fascination for complexity. Many humans are more fascinated with complex artifacts such as cellular telephones, for example, than with simple ones such as megaphones. One might hypothesize that complexity is correlated with power, an eternal preoccupation of the male of our species. The apparent appeal of mechanical artifacts—from weapons to communication devices—is related to their use for the control of other people and resources.

And there is a corollary: The most preferred animals are the big, strong, fast, compliant, and intelligent species that give us the most control over the environment and other humans. This might explain why horses retain their popularity in industrialized societies. It would also explain the peculiar human delight with large dogs, even though few humans rely on them any longer for hauling, hunting, seeing, or even security. No doubt culture plays a major role in this category of salience, although such nongenetic influences do not rule out the possibility of a genetic basis for differential preferences.

The Genetics of Biophilia

An intrinsic, genetic predisposition to react to biological phenomena is the core premise of biophilia. As implied in the opening section of this essay, any mention of genetics always stirs controversy because ideologues of both left and right are often invested in environmental explanations of human behavior. Conservationists and animal rights enthusiasts, however, hope that biophilia is innate because this would add weight to the argument that nature and organisms are essential to personal well-being and growth (self-realization). An inborn need for nature, in other words, justifies conservation as a social and biological imperative and might even justify the keeping of companion animals.

A Predisposition to Learn

Wilson (Chapter 1) defines biophilia as a complex of weak learning rules. Ulrich (Chapter 3) refers to such differential receptiveness or predisposition to learning about biological entities as learning biases or biologically prepared learning. He adds that survival in prehistory (deep history) depended on knowledge of natural history, the brain having evolved long before farming and pastoral practices.

Genetic Mechanisms

Mention of genetics always begs the question of mechanism. One common approach to analyzing the genetic basis of complex traits (those affected by many genes) is the quantification of heritability. *Heritability* is a technical term that can mean the proportion of the overall (phenotypic) variation that is inherited in a simple (additive) way from the parents. There is always variation among individuals for behavioral and physiological traits. Some of this variation (such as differences in linguistic ability, tanning response to radiation, and visual acuity) is genetic, and the differences among people and particularly the degree of similarity of relatives are powerful tools for studying genetic mechanisms.

What are typical heritabilities for biophilic traits, and what do they mean? In his review of the literature, Ulrich (Chapter 3) informs us that, based on twin studies, the heritability for ophidophobia (fear of snakes) is 30 percent and that for agoraphobia (fear of being in open or public places) is 40 percent. These are typical heritability values for quantitative (polygenic) traits, including anatomical traits such as body size and shape. It should be noted, however, that heritability is often *inversely* correlated with fitness. The human trait with the highest estimated heritability (95 percent) is the number and pattern of fingerprint ridges, for example, a seemingly trivial characteristic from the standpoint of survival and reproduction. Yet so-called fitness characteristics often have heritabilities below 30 percent. The number of heads in humans has a heritability of close to zero, there being little or no manifest genetic variation for head number because any departure from the norm (one) is essentially lethal.

Ulrich (Chapter 3), reporting the work of Öhman and his colleagues, presents results suggesting that the conditioned physiological defense responses to certain dangerous (high-risk) stimuli are not quickly extinguished (forgotten), even when the stimuli that evoke the conditioned response are subliminal. The responses to snakes and spiders appear not to extinguish at all. Yet phobias that are conditioned by evolutionarily new dangers—a loaded revolver pointed at the subject or a frayed electric wire—though they are learned just as fast as the responses to dangerous animals, extinguish quickly. This difference in physiological extinction rates suggests that our species still retains the genetic imprint of selection on our pretechnological ancestors and that newer technological threats have yet to be incorporated into our DNA.

Variability Between Populations

Although different human groups have had vastly different natural and cultural environments in the last 100,000 to 200,000 years, a parsimonious premise would be that isolated groups do not differ in genetically based and selectively tuned biophilic responses. It might follow that genetically based predispositions for learning certain things, such as phobias, have not been eroded by random mutation or counterselection accompanying the relatively short interval since the beginning of agriculture and the lifting of prelapsarian selection pressures. This is a testable hypothesis because some peoples have been living in agricultural/pastoral and industrial societies for up to 10,000 years, while others have lived in predominantly hunting-gathering economies until this century.

The alternative hypothesis, simply put, is the existence of ethnic or "racial" variation in biophilic responses and the notion that this variation has a genetic basis. Wilson (Chapter 1) and Diamond (Chapter 8) both argue for the expectation of geographic (genetic) variation in biophilic behavior. Indeed, it would be most surprising if the hypothetical alleles affecting such responses were held by selection at the same frequencies in all populations, especially given that some human groups have occupied open savannas and others have kept to forests for millennia. The responses to snakes may provide support, if anecdotal, for this idea: fear of snakes appears to be powerful in European cultures, yet Diamond reports that in Papua New Guinea snakes are not especially feared; Gadgil reminds us that snakes are often revered in India. Thus there may be different kinds or degrees of ophidophobia.

Variability Within Populations

The existence of between-population variability in biophilic responses presupposes the existence of raw material for natural selection—namely, genetic differences among individuals within populations. Most phenotypes, when plotted as a frequency distribution, approximate a normal, bell-shaped curve, suggesting but not proving underlying additive genetic variation. We would, therefore, expect additive genetic variation for many biophilic responses, including the aesthetic ones such as habitat preference. Ulrich (pers. comm.) says there is a correlation between the kinds of posters preferred by people and the degree of danger to which they expose themselves, and others have hypothesized about a genetic basis for risk-taking personality traits. If this is true for the adventure factor in personality, it is likely to be true for the aesthetic-biophilic response as well, some people loving nature and others being less enamored. Indeed, Nelson (Chapter 6) reminds us that ecological sinners exist in hunter-gatherer groups, a point often made by Kent Redford of the University of Florida.

Sociobiology—the Quaternary Hypothesis

The question is not whether biophilic responses still reside in our DNA and therefore in our minds, but the degree to which primitive responses and behavior have been erased by a few millennia of agriculture and technology. Putting this question slightly differently: does a creature that was genetically molded in the Pliocene and Pleistocene depend for its mental health on Pleistocene-like experiences, Pleistocene-like social interactions, and Pleistocene-like companion creatures for solace?

This may be going too far, but Paul Shepard (Chapter 9) claims that mental health depends on acknowledging the big, scary, mythic creatures residing deep in the inarticulate limbic system. Shepard says that human beings need contact with real brutes to symbolize the inner ones: to suppress these inner brutes may be to suppress conflicts between the still unreconciled parts of the brain, an organ that evolved too quickly to be fully integrated. Shepard and the poet/writer Robert Bly are not the only thinkers who sense the need for the reconciliation of the inner brute with the outer planner.

Shepard dismisses the idea that inbred "goofy" dogs and cats can ever fulfill a need for psychic reconciliation. While not disagreeing, Katcher and Wilkins (Chapter 5) suggest that pets have another function that is equally important—stress reduction. The purring, licking, soft, and cuddly object is the ideal surrogate for the female breast, for the cozy fire in the hearth, and especially for a lost mate or child. Pets and other activators of the relaxation response, "goofy" though they may be, help many endure pain and loneliness, help many recover from stress and frustration, and reduce the cost of medical care.

The relaxation response raises a definitional issue. Biophilia is hardest to define at its edges. This is because there is a tendency to include in biophilia certain phenomena that are only remotely or marginally biological, including affinities for mountainous landscapes and water. As pointed out by Katcher and Wilkins in Chapter 5, there is little doubt that certain natural/organic stimuli have strongly therapeutic and prophylactic effects, and the exploitation of these effects is beneficial to society and individual health. Perhaps it is quibbling to worry about whether biophilia in the strict sense is part of the explanation for such responses. More problematic is the concern of many conservationists (such as Shepard in Chapter 9) that electronic substitutes of nature (such as virtual reality) will soon replace the need to experience real animals and real nature.

So if pets, fish tanks, and vistas of parks, waves, and scenes of antelope safely grazing in savannas—and, yes, even pictorial or electronic surrogates of these—reduce our systolic pressure and inhibit adrenals from secreting adrenaline, maybe we should not complain about their artificiality. But neither should we confuse a tranquilizing homeostatic mechanism with other, functionally distinct, components of the biophilic complex.

Several contributors indicate that language may have originated because of the pressure to communicate about the growing lists of useful medicinal herbs, prey animals, food plants, not to mention predators, and beneficial and harmful insects and fungi. Assuming that the mental structures of language may have coevolved with our need to classify the biota and pass on important biological information to kin and affiliates, it should not be surprising, as Shepard points out, that the first words taught to a baby are anatomical terms and the next are the names of animals. Hence both the ontogeny and phylogeny of speech and thought depend on animal morphology and taxonomy. It is no wonder that the first books for infants are full of quacks, woofs, baas, moos, and cock-a-doodle-doos. One wonders if there are parallels in aboriginal households. Stephen Kellert (Introduction) gives this subject erudite attention.

The Beautiful and the Ugly

Biophilia means the love of life. Classically, the ultimate object of love is the beautiful. Many have mentioned the compulsion of astronauts and cosmonauts for spending as much time as possible staring at the earth through the ship's portholes. Holmes Rolston (Chapter 13) suggests that we are genetically adapted to loving the earth, and he helps us to understand our correct relation to the planet by exposing the falsity of the selfish/altruistic dualism in our relation to the beautiful.

The irony, though, is that this love of nature, like many emotions, is not absolute. The question is not whether humans love the earth and its biota, but whether we love them enough. As David Orr observes, "something tugs at us, but the tug is weak." Scott McVay (Prelude) frames the perennial challenge: how to learn to be in right relation to the earth. (See also Orr in Chapter 14.) Even a cursory familiarity with environmental history would suggest that agricultural and industrial cultures cannot coexist harmoniously with nature. The challenges of daily life, the distractions of interreligious, interclass, intertribal, and interracial conflicts, desires for power and status, plain ordinary indolence, and the human tendency to respond only to crises while ignoring the gradual erosion of social and environmental quality (Ornstein and Ehrlich 1989)—all these and other sins are obstacles that prevent biophilia from making a sufficient difference.

Are there data on how much people actually value nature? Statistics on charitable giving in the United States are no cause for optimism. Americans appear to rank the recipients of their surplus dollars in the following order: religious institutions, health and other public benefit organizations, cultural institutions, and, finally, the environment (Soulé 1991). Churches receive two-thirds of the charitable donations; environmental charities (advocacy, nonprofits, and others) receive less than 2 percent. Apparently people support causes and institutions in proportion to the degree that they perceive such causes and institutions supporting them, both on earth and in heaven.

Charitable contributions, of course, may not be the best metric for estimating how people rank their concerns. There are sociological surveys and voting data, as well. Kellert's surveys (Introduction) inform us that most people would make some kind of sacrifice to protect nature. The question, though, is how big a sacrifice? Just as the rare act of heroism does not prove that humans are always selfless, the rare, individual act of environmental commitment does not demonstrate that society has the will to end the current wave of extinctions.

About 10 million Americans support various animal rights and animal welfare groups, but the popularity of animal rights and welfare should not be mistaken for environmentalism. Many animal rights enthusiasts oppose conservation efforts when these efforts require removing or controlling mammals, including feral cats, foxes, goats, pigs, and other introduced species. Instead of defending the rights of native plants, birds, fishes, reptiles, amphibians, and insects, many animal rights activists are concerned with the health of individual mammals—even where their presence does great harm. Some conservationists see this movement as biophilia gone awry.

I regretfully conclude, therefore, that life-affiliating and life-protecting instincts are real, but they have few benefits for wild nature. For most of us humans, entangled as we are in daily struggles of material and emotional survival, the threats to real nature, wild nature, are real but not salient. Our little acts of biophilia—buying bird seed, caring for our pets, nurturing our gardens—sustain us emotionally. But what we need now are big, selfless, and costly acts of biophilia to protect nature.

Believers in the transformative power of biophilia might respond that awareness of the *idea* of biophilia, by itself, can change people's behavior—

just as awareness of the ideas of civil disobedience and nonviolence eventually changed politics in a few Western nations. Perhaps biophilia will eventually lead to an extension of rights to other species (Nash 1985). But if the history of social movements is a guide, progress will be slow, and biophilia is not likely to create a revolution in public policy. The social inertia produced by contemporary demographic, political, and economic problems, especially in the tropics, will require a century or more to overcome (Meadows et al. 1992).

There are many signs, nevertheless, of a growing role for biophilia. Movements such as bioregionalism and other forms of biocentrism that can change the way people live (Grumbine 1992) are gaining adherents, and both the Biodiversity Convention signed by most nations in Rio de Janeiro and the likely survival of the Endangered Species Act augur well for ethical progress in humanity's relationship to nature.

A New Religion?

If biophilia is destined to become a powerful force for conservation, then it must become a religion-like movement. Only a new religion of nature, similar but even more powerful than the animal rights movement, can create the political momentum required to overcome the greed that gives rise to discord and strife and the anthropocentrism that underlies the intentional abuse of nature. Rolston's views (Chapter 13) appear to be concordant.

The social womb for such a "biophilism" could be bioregional communities that recapture tribal-hunter-gatherer-pagan wisdom, integrating it with relevant science, appropriate technology, family planning, and sustainable land use practices. Such communities already exist in the foothills of the Sierra Nevada and elsewhere. Until the far-off day when most humans reach this stage in cultural evolution, conservationists will have to go about business as usual.

REFERENCES

Grumbine, R. E. 1992. *Ghost Bears: Exploring the Biodiversity Crisis.* Washington: Island Press.

Meadows, D. H., D. L. Meadows, and J. Randers. 1992. *Beyond the Limits: Confronting Global Collapse, Envisioning a Sustainable Future*. Post Mills, Vt.: Chelsea Green.

Nash, R. 1985. "Rounding Out the American Revolution: Ethical Extension and the New Environmentalism." In M. I. Tobias (ed.), *Deep Ecology*. San Diego: Avant Books.

Ornstein, R. E., and P. R. Ehrlich. 1989. *New World, New Mind: Moving Toward Conscious Evolution*. New York: Doubleday.

Soulé, M. E. 1991. "Conservation: Tactics for a Constant Crisis." *Science* 253:744–750.

Wilson, E. O. 1984. *Biophilia: The Human Bond with Other Species*. Cambridge: Harvard University Press.

Coda

Stephen R. Kellert

THIS BOOK REPRESENTS the start rather than the conclusion of a journey of exploration. The various contributors have sought to examine elements of the question of how nature, particularly its living biota, has provided humans with an evolutionary basis for our species' physical, emotional, cognitive, and even spiritual development. Whatever accomplishments may have been achieved, this book is but one step toward understanding the complicated question of how human affect, intellect, language, culture, technology, and even ethics are molded by a basic human affinity for life and lifelike processes.

Our development as individuals and as a species constitutes more than just the struggle for physical survival. We aspire to something called fulfillment, as well, perhaps a striving after an ideal which every species, especially the more complex forms, appears to manifest as a kind of unrealized potential. The notion of biophilia suggests that this possibility for fulfillment and self-realization may be found in our relationship with the diversity of life around us.

We need other species not only for the promise of material and physical sustenance but, just as important, for the raw material they offer for our psychological and intellectual growth. We can survive the extinction and extirpation of many unique life-forms, just as we may endure polluted water, contaminated air, and toxic soil. Our persistence as a species can allow for far fewer nonhuman life-forms—"other nations, caught with ourselves in the net of life and time" (Beston 1990:394)—but will this impoverished condition permit us to prosper psychologically, spiritually, and materially as individuals and as a species?

The notion of biophilia sounds a profound note of skepticism regarding the human capacity to thrive in a biologically depauperate world that has countenanced and abetted in the massive destruction of life. The idea of biophilia asserts that the achievement of our fullest potential will depend on a matrix of complex and subtle emotional, intellectual, and physical interactions with a rich and diverse biota. Beston (1990:394) continues: "Whatever attitude to human existence [we] fashion for [ourselves], know that it is valid only if it be the shadow of an attitude to nature. . . . The ancient values of dignity, beauty, and poetry which sustain [us] are of nature's inspiration. . . . Do not dishonor the earth lest you dishonor the spirit of man."

The essays in this book represent a fledgling attempt to explore a more robust and complex theory of evolution than one based merely on the assertion of the physical struggle to survive. A twenty-first-century review of our effort may conclude that this was but a crude attempt to understand our psychological, aesthetic, spiritual, and physical dependence on nature. This group of scholars will retain a measure of pride, however, in their effort to establish a basis for the scientific exploration of the human affinity for life in its many myriad and mysterious forms.

REFERENCE

Beston, H. 1990. "The Outermost House: A Year of Life on the Great Beach of Cape Cod." In R. Finch and J. Elder (eds.), *The Norton Book of Nature Writing*. New York: W. W. Norton.

About the Contributors

JARED DIAMOND is a professor of physiology at the University of California Medical School working on laboratory studies on the evolutionary design of membrane transport mechanisms and undertaking field studies of New Guinea bird ecology and behavior. His popular science articles appear regularly in *Natural History*, *Discover*, and *Nature* magazines. His recent book, *The Third Chimpanzee* (HarperCollins, 1992), is an account of how language, art, and other apparently unique human characteristics arose from animal precursors.

MADHAV GADGIL studied biology in India before obtaining a Ph.D. from Harvard University. He is a professor at the Indian Institute of Science and conducts both field research and mathematical modeling in population biology, conservation biology, and human ecology. He is active in Indian nature conservation and ecodevelopment efforts. Dr. Gadgil is a fellow of the Indian National Science Academy and a foreign associate of the U.S. National Academy of Sciences. He has also served as a member of the Science Advisory Council to the prime minister of India.

JUDITH H. HEERWAGEN is a research assistant professor in psychosocial nursing and in the College of Architecture and Urban Planning at the University of Washington, Seattle. She is a psychologist whose work has focused on behavioral ecology and the application of evolutionary theory to human/environmental interactions. She has conducted research on environmental preferences and aesthetics and is currently studying the relationship between the environment and human well-being as it relates to gender and life transitions. She is presently a visiting scientist at the Battelle Human Affairs Research Center in Seattle, where she is investigating organizational ecology issues.

AARON KATCHER received his medical degree and psychiatric training at the University of Pennsylvania, where he is currently on the faculty in the Schools of Medicine, Dentistry, and Veterinary Medicine. His research interests have included the impact of emotion and dialogue on physiological states and the relationship between the social environment and disease. During the past twelve years, he has been investigating the influence of social relationships with animals and contact with nature on human behavior and health.

STEPHEN R. KELLERT is a professor at the Yale University School of Forestry and Environmental Studies. He has conducted extensive research and published widely on the subject of human values and perceptions relating to nature, particularly animals. He is a member of the board of directors of the Student Conservation Association, Defenders of Wildlife, and the Xerces Society. He has received awards from the Society for Conservation Biology, the International Foundation for Environmental Conservation, and the National Wildlife Federation.

ELIZABETH ATWOOD LAWRENCE, a veterinarian and anthropologist, received her V.M.D. degree from the University of Pennsylvania School of Veterinary Medicine and her Ph.D. degree in social anthropology from Brown University. She is professor of environmental studies at the Tufts University School of Veterinary Medicine, where she teaches and conducts research on human/animal relationships. Dr. Lawrence is the author of three books and numerous papers dealing with human/animal relationships. Her honors include the Elsie Clews Parsons Award of the American Ethnological Society, the James Moody Award of the Southern Anthropological Society, and the International Distinguished Scholar Award of the Association of Human-Animal Interaction Organizations.

LYNN MARGULIS is Distinguished University Professor, Department of Biology, at the University of Massachusetts. She had a Sherman Fairchild Fellowship at the California Institute of Technology and a Guggenheim Fellowship. She is a member of the U.S. National Academy of Sciences. She has contributed many original publications on cell biology and microbial evolution. She has also participated in the development of science teaching materials from elementary grades to graduate school. She has chaired the National Academy of Science's Space Science Board Committee on Planetary Biology and Chemical Evolution. She is also a member of the Commonwealth Book Fund advisory board and codirector of the Planetary Biology Internship Committee (NASA Life Sciences, Marine Biological Laboratory, Woods Hole).

SCOTT MCVAY has served as executive director of the Geraldine R. Dodge Foundation since 1976. Under his leadership, the Dodge Foundation has pioneered philanthropy for the welfare of animals, the arts, elementary and secondary education, and various public issues. He is the author of several papers on whales, dolphins, and porpoises and a graduate of Princeton University.

GARY PAUL NABHAN is research director and cofounder of Native Seeds/SEARCH, a Tucson-based conservation organization. He is a MacArthur Fellow and has been a Pew Scholar in Conservation and the Environment. His book *Gathering the Desert* won the Burroughs Medal for Nature Writing in 1986, and he is the author of five other books.

RICHARD NELSON is affiliate professor of anthropology at the University of Alaska, Fairbanks. He has published four books on relationships to nature among Alaskan native peoples. He was associate producer and writer for an award-winning public television series about Koyukon Indian life, with the same title as his book, *Make Prayers to the Raven*. His most recent book, *The Island Within* (Vintage Books), a personal exploration of nature and home, received the 1991 Burroughs Medal for Nature Writing. He is presently writing a book about deer and their relationships to people in modern America.

GORDON H. ORIANS is professor of zoology and environmental studies at the University of Washington, Seattle. His areas of research include the ecology of vertebrate social systems, plant/herbivore interactions, and the biology of rarity. His interest in science policy has been reflected in service on committees of the National Academy of Sciences and as past director of the Institute for Environmental Studies at the University of Washington. He has also conducted extensive studies of avian habitat selection and environmental use. This work stimulated his research on the value of evolutionary concepts for analyzing the evolutionary roots of human aesthetic responses to environmental variables.

DAVID W. ORR is professor and chair of environmental studies at Oberlin College and a cofounder of the environmental education Meadowcreek Project in Arkansas. He is the author of numerous publications; his most recent book is entitled *Ecological Literacy* (State University of New York Press, 1992). He is education editor for the journal *Conservation Biology*. He received a National Conservation Achievement Award in 1993 from the National Wildlife Federation, the Lyndhurst Prize in 1992, and an honorary doctorate from Arkansas College in 1990.

HOLMES ROLSTON III is University Distinguished Professor of Philosophy at Colorado State University. He has written six books, most recently *Philosophy Gone Wild* (Prometheus Books) and *Environmental Ethics* (Temple University Press). He is president of the International Society for Environmental Ethics.

DORION SAGAN is a writer. He has coauthored (with Lynn Margulis) *Origins of Sex, Garden of Microbial Delights, Microcosmos,* and *Mystery Dance: On the Evolution of Human Sexuality*. Mr. Sagan has written many articles and reviews in the fields of evolutionary biology and philosophy. He is author of the recently released paperback *Biospheres: Reproducing Planet Earth* (Bantam).

PAUL SHEPARD is the author of many books dealing with human evolution and ecology including, among others, *Nature and Madness, Thinking Animals,* and *The Sacred Paw*. He has been an active participant in the intellectual movement creating a cultural ecology of humans, and his essays and reviews have appeared widely. He is Avery Professor of Natural Philosophy at Pitzer College and the Claremont Graduate School.

MICHAEL E. SOULÉ is currently professor and chair of environmental studies at the University of California, Santa Cruz. His research interests include morphological and genetic variation in natural populations of animals, island biogeography, and conservation biology. He was the founder and first president of the Society for Conservation Biology and is a fellow of the American Association for the Advancement of Science.

SARA ST. ANTOINE received a master's degree from the Yale University School of Forestry and Environmental Studies. She is the author of several children's books and curriculum guides. Currently she is developing a story-based environmental education curriculum as part of an Echoing Green Public Service Fellowship.

ROGER S. ULRICH received his Ph.D. in behavioral geography and environmental psychology. Much of his research has investigated the influences of human experiences with natural and built environments on psychological well-being, physiological systems, behavior and effective functioning, and health-related indicators. He is professor of landscape architecture and urban planning at Texas A&M University and serves as associate dean for research of the College of Architecture.

GREGORY WILKINS received his doctorate from the University of Florida and has held a number of clinical, administrative, and research positions. Currently he is director of clinical training and research at the Brandywine Treatment Center. His primary research and training interests include disruptive behavior disorders, psychodiagnostics, psychometrics, and the use of pair theory and animal-facilitated therapy in residential treatment settings.

EDWARD O. WILSON is Frank B. Baird Jr. Professor of Science and curator of entomology at Harvard University, where he has been a member of the faculty since 1956. Wilson's research interests include population and behavioral biology, biogeography, and the study of social insects. Twice winner of the Pulitzer Prize in general nonfiction, he has won many scientific awards including the National Medal of Science, the Tyler Prize in Ecology, and the Crafoord Prize of the Royal Swedish Academy of Science. He is now actively involved in the conservation of biological diversity.

Index